HUMAN MUSCULOSKELETAL BIOMECHANICS

Edited by **Tarun Goswami**

Human Musculoskeletal Biomechanics
Edited by Tarun Goswami

Published by InTech

Janeza Trdine 9, 51000 Rijeka, Croatia

Publishing Process Manager Iva Simcic

Technical Editor Teodora Smiljanic

Cover Designer InTech Design Team

Additional hard copies can be obtained from orders@intechopen.com

Human Musculoskeletal Biomechanics, Edited by Tarun Goswami
p. cm.
ISBN 978-953-307-638-6

For EU product safety concerns:
IN TECH d.o.o., Prolaz Marije Krucifikse Kozulić 3, 51000 Rijeka, Croatia,
info@intechopen.com or visit our website at intechopen.com.

Meet the editor

Dr. Tarun Goswami graduated with a Doctor of Science degree in Mechanical Engineering from Helsinki University of Technology. He worked in the aircraft industry before transitioning to academia. He currently holds a joint appointment of Associate Professor in Biomedical, Industrial and Human Factors Engineering and Orthopaedic Surgery, Sports Medicine and Rehabilitation, at Wright State University, Dayton, Ohio. He directs Orthopaedic Biomechanics Laboratory, Spinal Research Group and Device Development Center. Students from undergraduate, graduate, medical and residency programs work with him on a range of biomechanical topics. His research is supported by Air Force and Army and other orthopaedic related companies. He has published more than 75 papers in archival journals and numerous abstracts and presentations in conferences. For some of his works on total ankle replacement, total toe replacement, spinal disc replacement, fracture fixation devices, pacemaker lead designs and other topics patents are pending. He teaches such courses as Introduction to Biomedical Engineering, Biomechanics and Bioenergetics, Biomaterials, and Experimental Orthopaedic Engineering.

Contents

Preface

The field of biomechanics has been evolving from the ancient Greeks times. Recent publications and research in biomechanics sky rocketed as the field of traditional biomechanics is creating new opportunities in diagnostics, therapy, rehabilitation, motion preservation, kinesiology, total joint replacement, biomechanics of living systems at small scale, and other areas.

Biomechanics now encompasses a range of fields. The book on Human Musculoskeletal Biomechanics is a broad topic and may provide the platform for newer text and editions as the research evolves and new results obtained. In the current form, the book covers four areas: 1) motion preservation, which will be useful in designing functional braces for different skeletal areas used in therapy and/or rehabilitation, 2) musculoskeletal biomechanics, which includes soft and hard tissue and their behavior under the actions of forces, motion, strain and modeling them analytically and experimentally, 3) nano-behavior, is another area which is developing where mechanical properties of living systems are determined that will be useful in developing treatment methods and understanding the small living systems such as viruses, and 4) vascular biomechanics, a new area that will also develop in the future with surgery.

Therefore, the book presents information on the four sections, in a concise format. Based on these sections, new courses may be developed at graduate level or some of the concepts used to teach undergraduate students in biomedical engineering. Since the book will be available on open access, its use will be free to students, and to introduce this topic as a new course, if desired. The four sections presented in this book will continue to challenge both the researchers and students in the future and therefore, creation of new knowledge.

Dr. Tarun Goswami
D.Sc. Equity Advisor - College of Engineering and Computer Science;
Founding Director - Device Development Center;
Director - Damage Tolerance and Probabilisitic Life Prediction of Materials Center;
Focus Area Chair - Ph.D. in Engineering - Medical and Biological Systems;
Associate Professor of Biomedical, Industrial, and Human Factors Engineering,
Wright State University;
Associate Professor, Department of Orthopaedic Surgery, Sports Medicine & Rehabilitation
USA

Part 1

Motion Preservation

1

A Task-Level Biomechanical Framework for Motion Analysis and Control Synthesis

Vincent De Sapio and Richard Chen
Sandia National Laboratories *
USA

1. Introduction

The behavioral richness exhibited in natural human motion results from the complex interplay of biomechanical and neurological factors. The biomechanical factors involve the kinematics and dynamics of the musculoskeletal system while the neurological factors involve the sensorimotor integration performed by the central nervous system (CNS). An adequate understanding of these factors is a prerequisite to understanding the overall effect on human motion as well as providing a means for synthesizing human motion.

The fields of neuroscience, biomechanics, robotics, and computer graphics provide motivation, as well as tools, for understanding human motion. In neuroscience, fundamental scientific understanding drives the motivation to understand human motion, whereas, in biomechanics, clinical applications often form the driving motivation. These clinical applications involve the use of movement analysis and simulation tools to help direct patient rehabilitation as well as predict the effects of surgery on movement.

In addition to the clinical desire to analyze movement there has been an emerging desire in recent years to synthetically generate human-like motion in both simulated and physical settings. In computer graphics this desire is directed toward autonomously generating realistic motion for virtual actors. The intent is to direct these virtual actors using high-level goal directed commands for which low-level motion control is automatically generated.

Motivated by similar desires, the robotics community seeks a high-level control framework for robotic systems. With the recent advent of complex humanoid robots this challenge has grown more demanding. Consistent with their anthropomorphic design, humanoid robots are intended to operate in a human-like manner within man-made environments and to promote interaction with their biological counterparts. To achieve this, common control strategies have involved generating joint space trajectories or learning specific motions, but these approaches require extensive motion planning computations and do not generalize well to related tasks.

1.1 Human motion control

The basic constituents of the human motor system include the biomechanical plant and the CNS. A high-level block diagram, sufficient for our present purposes, is depicted in Fig. 1. Based on some specified task the CNS performs motor planning which culminates in low-level control issued as a motor command to the biomechanical plant. This motor planning

*Sandia National Laboratories is a multi-program laboratory managed and operated by Sandia Corporation, a wholly owned subsidiary of Lockheed Martin Corporation, for the U.S. Department of Energy's National Nuclear Security Administration under contract DE-AC04-94AL85000.

and control occurs based on the integration of sensory information from proprioceptors distributed throughout the musculoskeletal system. Some knowledge of the biomechanical plant is also assumed to be encoded in the CNS.

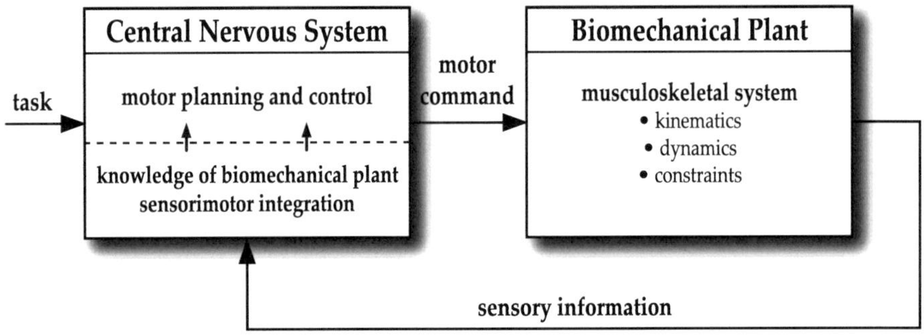

Fig. 1. Motor control involves the task-driven action of the central nervous system (CNS) on the biomechanical plant. Given proprioceptive information the CNS performs motor planning which results in the issuance of motor commands.

While the biomechanical plant can be decomposed and understood in reasonable detail the processes of the CNS are understood more vaguely. As a consequence, while Fig. 1 provides a conceptual framework, with regard to the CNS it lacks enough precision to be useful as a functional model. For this reason it is appropriate to consider some more basic analogs. To this end we will consider the most basic analog which is still useful, that of a joint space model, followed by task/posture analogs.

1.2 Joint space motion control

Joint space control is the earliest and still most common form of feedback control in robotic systems. In this scheme a task is specified in some natural coordinate system associated with the robot and environment. Based on a knowledge of the robot kinematics, the robot controller performs inverse kinematic computations to arrive at a posture or trajectory in terms of joint angles. The joint command is issued to the servo motors which execute the motion.

While this method of controlling a robot is effective it requires the computation of inverse kinematics. Additionally, it does not make use of any knowledge of the robot's dynamics. The method of computed torque is an enhancement of the basic joint space control approach in which the controller does make use of the robot's dynamics. However, the control is still encoded in joint space rather than a more natural task space description. As an analog to the human motor system this joint space encoding may constitute a deficiency since a number of studies, Buneo et al. (2002); Sabes (2000); Scholz & Schöner (1999); Shenoy et al. (2003), suggest a task-oriented spatial encoding of planning and control. Rather than using inverse kinematic transformations this task-oriented encoding is accomplished through visumotor transformations from retinal coordinates to hand- or body-centered spatial coordinates.

1.3 Task/posture motion control

Motivated by evidence for a task-oriented spatial encoding of motion by the CNS we now consider a task/posture control model. This is depicted in Fig. 2 and represents a generalization beyond a strict joint space motor control model. In this case the control is encoded in the same native space in which the task is expressed. This obviates the need for inverse kinematics to *convert* the task description into a joint space description. The dynamics of the robot plant are expressed in task space with a complementary description of the posture (see Fig. 3). The controller exploits this decomposed structure to yield separate task and posture control terms. As such, the posture term can be chosen to minimize some criterion, consistent with the execution of the task. The motor command can then be issued in the appropriate actuator space (e.g., motor torques).

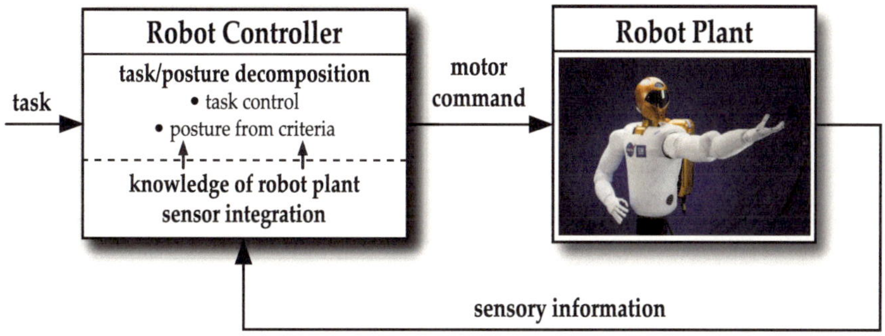

Fig. 2. Task space motion control model where the dynamics are decomposed into complementary task and posture spaces. The posture control can be chosen using an auxiliary criterion which can be optimized consistent with the execution of the task (image courtesy of NASA).

As alluded to earlier this task/posture model represents a more general abstraction than the joint space model, and may be more suitable for purposes of modeling and understanding human motor control. The notions of task and posture are directly applicable to human motion control and, as we shall see, can be specifically interpreted in terms of physiological criteria.

1.4 Task/posture approach for biomechanical systems

Up to this point we have considered robotic control models as a means of addressing human motor control. In a more general sense the challenge of synthesizing low-level human motion control from high-level commands can be addressed by integrating approaches from the

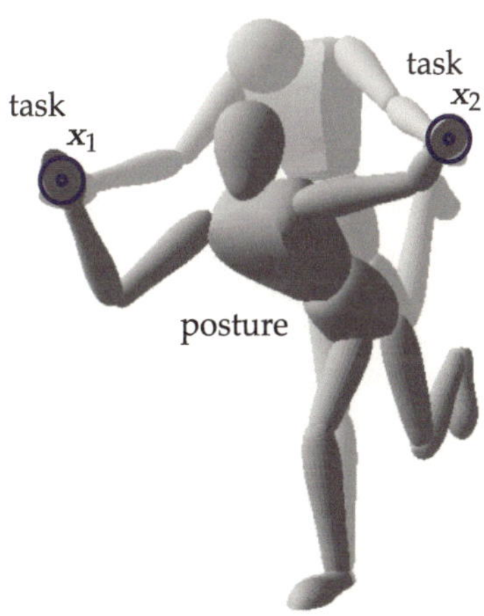

Fig. 3. A task description with complementary task consistent postures. Redundancy with respect to task introduces task dynamics as well as posture dynamics.

biomechanics and robotics communities. The biomechanics community has investigated the phenomenon of neuromuscular dynamics and control through the use of computational muscle models. This characterization allows for the description of muscle strength limitations, activation delays, and overall muscle contraction dynamics. Properly accounting for these characteristics is critical to authentically simulating human motion. In a complementary manner, the robotics community has investigated the task-level feedback control of robots using the operational space approach. This approach recasts the dynamics of the robotic system into a relevant task space description. This provides a natural mechanism for specifying high-level motion commands that can be executed using feedback control.

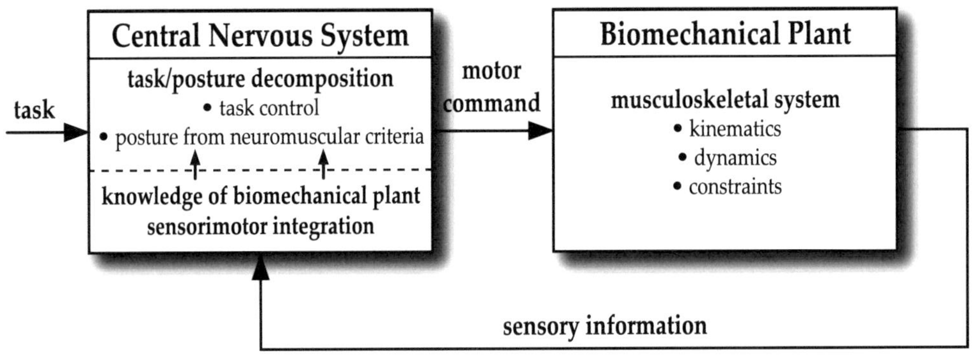

Fig. 4. Task/posture motion control model for biomechanical systems. In addition to task control, neuromuscular criteria are used to control the posture by minimizing neuromuscular cost, consistent with the execution of the task.

Fig. 4 depicts a task/posture model of motor control which integrates robotic and biomechanical approaches. In this model the CNS is seen to affect motor control using task/posture decomposition. While the control is task-driven the task consistent postures are driven by neuromuscular criteria. In other words, while the CNS issues motor commands to achieve some task it is assumed that this is being done in a way that minimizes some neuromuscular cost (subject to the task requirements). While the precise nature of what, if anything, is being minimized by the CNS is difficult to directly infer, computational muscle models can be used to evaluate particular hypothetical effort criteria. Predicted postural behavior associated with minimizing these criteria can then be compared with actual postures from subject trials to validate the applicability of the criteria.

Through the combined utilization of task-level constrained motion strategies and computational muscle models this chapter addresses motion control with application to human motion synthesis. A coherent framework is presented for the management of motion tasks, physical constraints, and neuromuscular criteria. The subsequent sections will address the constituent elements of this framework and will be divided into (i) task-based modeling and analysis and (ii) posture-based modeling and analysis.

2. Task-based modeling and analysis of biomechanical systems

In this section we present a task-based formulation for application to biomechanical systems. In the overall framework this addresses the highlighted element of Fig. 5. The focus is on task control in the presence of constraints.

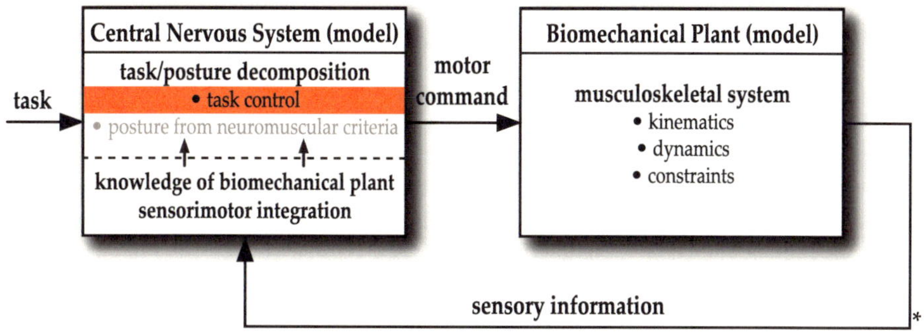

Fig. 5. Task/posture motion control model for biomechanical systems highlighting task control in the presence of constraints. In this chapter the human shoulder complex will be investigated using this approach.

2.1 Configuration space dynamics

The equations of motion for a multibody system that is unconstrained with respect to configuration space can be expressed by the Euler-Lagrange equations,

$$\frac{d}{dt}\frac{\partial \mathcal{L}}{\partial \dot{\boldsymbol{q}}} - \frac{\partial \mathcal{L}}{\partial \boldsymbol{q}} = \boldsymbol{\tau}, \tag{1}$$

where $\mathcal{L} = \mathcal{L}(\boldsymbol{q}, \dot{\boldsymbol{q}})$ is the Lagrangian of the system, $\boldsymbol{q} \in \mathbb{R}^n$ is the vector of generalized coordinates, and $\boldsymbol{\tau} \in \mathbb{R}^n$ is the vector of generalized actuator forces (torques). In standard matrix form the equations of motion can be expressed as,

$$\boldsymbol{M}(\boldsymbol{q})\,\ddot{\boldsymbol{q}} + \boldsymbol{b}(\boldsymbol{q}, \dot{\boldsymbol{q}}) + \boldsymbol{f}(\boldsymbol{q}, \dot{\boldsymbol{q}}) = \boldsymbol{B}(\boldsymbol{q}, \dot{\boldsymbol{q}})^T \boldsymbol{u}, \tag{2}$$

where $\boldsymbol{u} \in \mathbb{R}^k$ is the vector of control inputs, $\boldsymbol{B}(\boldsymbol{q}, \dot{\boldsymbol{q}})^T \in \mathbb{R}^{n\times k}$ is the matrix mapping control inputs to generalized actuator forces, $\boldsymbol{M}(\boldsymbol{q}) \in \mathbb{R}^{n\times n}$ is the configuration space mass matrix, $\boldsymbol{b}(\boldsymbol{q}, \dot{\boldsymbol{q}}) \in \mathbb{R}^n$ is the vector of centrifugal and Coriolis terms, and $\boldsymbol{f}(\boldsymbol{q}, \dot{\boldsymbol{q}}) \in \mathbb{R}^n$ is the vector of generalized applied forces. We will often use a modified and more specialized form of (2) common in robotics,

$$\boldsymbol{M}(\boldsymbol{q})\,\ddot{\boldsymbol{q}} + \boldsymbol{b}(\boldsymbol{q}, \dot{\boldsymbol{q}}) + \boldsymbol{g}(\boldsymbol{q}) = \boldsymbol{\tau}, \tag{3}$$

where $\boldsymbol{g}(\boldsymbol{q}) \in \mathbb{R}^n$ is the vector of gravity terms. The form of (3) assumes that the generalized actuator forces can be directly interpreted as control inputs; that is, $\boldsymbol{\tau} = \boldsymbol{B}^T\boldsymbol{u} = \boldsymbol{u}$, i.e., $\boldsymbol{B}^T = \mathbf{1}$. Additionally, the generalized applied forces are assumed to be restricted to gravity terms; that is, $\boldsymbol{f}(\boldsymbol{q}, \dot{\boldsymbol{q}}) = \boldsymbol{g}(\boldsymbol{q})$.

We now introduce a set of m_C holonomic and scleronomic constraint equations, $\phi(\boldsymbol{q}) = \mathbf{0} \in \mathbb{R}^{m_C}$, that are satisfied on a $p = n - m_C$ dimensional manifold, Q^p, in configuration space, $Q = \mathbb{R}^n$. The gradient of ϕ yields the constraint matrix,

$$\boldsymbol{\Phi}(\boldsymbol{q}) = \frac{\partial \phi}{\partial \boldsymbol{q}} \in \mathbb{R}^{m_C \times n}. \tag{4}$$

Adjoining the constraints to (3) by introducing a set of constraint forces yields the dynamic equation in the familiar multiplier form,

$$\boldsymbol{M}\ddot{\boldsymbol{q}} + \boldsymbol{b} + \boldsymbol{g} - \boldsymbol{\Phi}^T\boldsymbol{\lambda} = \boldsymbol{\tau}, \tag{5}$$

subject to,

$$\boldsymbol{\Phi}\ddot{\boldsymbol{q}} + \dot{\boldsymbol{\Phi}}\dot{\boldsymbol{q}} = \mathbf{0} \qquad (\boldsymbol{\phi} = \mathbf{0}, \quad \boldsymbol{\Phi}\dot{\boldsymbol{q}} = \mathbf{0}), \tag{6}$$

where $\boldsymbol{\lambda}$ is a vector of unknown Lagrange multipliers.

2.2 Task space dynamics and control

In the previous section we considered configuration space descriptions of the dynamics of constrained multibody systems. Our objective is to reformulate these descriptions in the context of task space, Khatib (1987); Khatib (1995). This will provide the foundation for constrained task-level control to be discussed in the next section.

Given a branching chain system and defining a set of m task, or operational space, coordinates, $\mathbf{x} \in \mathbb{R}^m$ we define the task Jacobian as,

$$\boldsymbol{J}(\boldsymbol{q}) = \frac{\partial \mathbf{x}}{\partial \boldsymbol{q}} \in \mathbb{R}^{m \times n}. \tag{7}$$

The generalized force (or control torque) in (3) can then composed as $\boldsymbol{J}^T\boldsymbol{f}$, where $\boldsymbol{f} \in \mathbb{R}^m$ is the task, or operational space, force. In the redundant case an additional term needs to complement the task term in order to realize any arbitrary generalized force. We will refer to this term as the null space term and it can be composed as $\boldsymbol{N}^T\boldsymbol{\tau}_0$, where $\boldsymbol{\tau}_0$ is an arbitrary generalized force and $\boldsymbol{N}(\boldsymbol{q})^T \in \mathbb{R}^{n \times n}$ is the null space projection matrix. We then have the following set of unconstrained task, or operational space, equations of motion, Khatib (1987),

$$\boldsymbol{\Lambda}(\boldsymbol{q})\,\ddot{\mathbf{x}} + \boldsymbol{\mu}(\boldsymbol{q}, \dot{\boldsymbol{q}}) + \boldsymbol{p}(\boldsymbol{q}) = \boldsymbol{f}, \tag{8}$$

where $\boldsymbol{\Lambda}(\boldsymbol{q}) \in \mathbb{R}^{m \times m}$ is the operational space mass matrix, $\boldsymbol{\mu}(\boldsymbol{q}, \dot{\boldsymbol{q}}) \in \mathbb{R}^m$ is the operational space centrifugal and Coriolis force vector, and $\boldsymbol{p}(\boldsymbol{q}) \in \mathbb{R}^m$ is the operational space gravity vector. These terms are given by,

$$\boldsymbol{\Lambda}(\boldsymbol{q}) = (\boldsymbol{J}\boldsymbol{M}^{-1}\boldsymbol{J}^T)^{-1}, \tag{9}$$

$$\boldsymbol{\mu}(\boldsymbol{q}, \dot{\boldsymbol{q}}) = \boldsymbol{\Lambda}\boldsymbol{J}\boldsymbol{M}^{-1}\boldsymbol{b} - \boldsymbol{\Lambda}\dot{\boldsymbol{J}}\dot{\boldsymbol{q}}, \tag{10}$$

$$\boldsymbol{p}(\boldsymbol{q}) = \boldsymbol{\Lambda}\boldsymbol{J}\boldsymbol{M}^{-1}\boldsymbol{g}, \tag{11}$$

$$\boldsymbol{N}(\boldsymbol{q})^T = \mathbf{1} - \boldsymbol{J}^T\boldsymbol{\Lambda}\boldsymbol{J}\boldsymbol{M}^{-1}. \tag{12}$$

Thus, the overall dynamics of our multibody system, given by,

$$\boldsymbol{M}\ddot{\boldsymbol{q}} + \boldsymbol{b} + \boldsymbol{g} = \boldsymbol{J}^T\boldsymbol{f} + \boldsymbol{N}^T\boldsymbol{\tau}_0 = \boldsymbol{\tau}, \tag{13}$$

can be mapped into task space,

$$\boldsymbol{M}\ddot{\boldsymbol{q}} + \boldsymbol{b} + \boldsymbol{g} = \boldsymbol{\tau} \quad \overset{\bar{\boldsymbol{J}}^T}{\rightarrow} \quad \boldsymbol{f} = \boldsymbol{\Lambda}\ddot{\mathbf{x}} + \boldsymbol{\mu} + \boldsymbol{p}, \tag{14}$$

where $\bar{\boldsymbol{J}}$ is the dynamically consistent inverse of the task Jacobian,

$$\bar{\boldsymbol{J}} = \boldsymbol{M}^{-1}\boldsymbol{J}^T\boldsymbol{\Lambda}. \tag{15}$$

In a complementary manner the overall dynamics can be mapped into the task consistent null space (or self-motion space) using $\boldsymbol{N}^T$.

The overall dynamics can be expressed as,

$$\boldsymbol{J}^T(\boldsymbol{\Lambda}\ddot{\mathbf{x}} + \boldsymbol{\mu} + \boldsymbol{p}) + \boldsymbol{N}^T\boldsymbol{\tau}_o = \boldsymbol{\tau}. \tag{16}$$

A controller employing (8) would be assumed to have imperfect knowledge of the system. Therefore, (8) should reflect estimates for the inertial and gravitational terms. Additionally, a control law needs to be incorporated. To this end we replace $\ddot{\boldsymbol{x}}$ in (8) with the input of the decoupled system, Khatib (1995) $\boldsymbol{f}^\star$, to yield the dynamic compensation equation,

$$\boldsymbol{f} = \widehat{\boldsymbol{\Lambda}}\boldsymbol{f}^\star + \widehat{\boldsymbol{\mu}} + \widehat{\mathbf{p}}, \tag{17}$$

where the $\widehat{\cdot}$ represents estimates of the dynamic properties. Thus our control torque is,

$$\boldsymbol{\tau} = \boldsymbol{J}^T(\widehat{\boldsymbol{\Lambda}}\boldsymbol{f}^\star + \widehat{\boldsymbol{\mu}} + \widehat{\mathbf{p}}) + \widehat{\boldsymbol{N}}^T\boldsymbol{\tau}_o. \tag{18}$$

Any suitable control law can be chosen to serve as input of the decoupled system. In particular, we can choose a linear proportional-derivative (PD) control law of the form,

$$\boldsymbol{f}^\star = \boldsymbol{K}_p(\mathbf{x}_d - \mathbf{x}) + \boldsymbol{K}_v(\dot{\mathbf{x}}_d - \dot{\mathbf{x}}) + \ddot{\mathbf{x}}_d, \tag{19}$$

where $\mathbf{x}_d$ are reference values for the task coordinates and $\boldsymbol{K}_p$ and $\boldsymbol{K}_v$ are gain matrices. As we introduced a set of m_C holonomic and scleronomic constraint equations to the configuration space dynamics we can do the same for the task space dynamics. Mapping (5) and (6) into an appropriate task/constraint space yields, De Sapio et al. (2006),

$$\boldsymbol{\Lambda}\ddot{\mathbf{x}} + \boldsymbol{\mu} + \boldsymbol{p} - \bar{\boldsymbol{J}}^T\boldsymbol{\Phi}^T(\boldsymbol{\alpha} + \boldsymbol{\rho}) = \bar{\boldsymbol{J}}^T\boldsymbol{\Theta}^T\boldsymbol{\tau}. \tag{20}$$

The term $\boldsymbol{\alpha}(\boldsymbol{q}, \dot{\boldsymbol{q}}) \in \mathbb{R}^{m_C}$ is the vector of centrifugal and Coriolis forces projected at the constraint, and $\boldsymbol{\rho}(\boldsymbol{q}) \in \mathbb{R}^{m_C}$ is the vector of gravity forces projected at the constraint. These terms are given by,

$$\boldsymbol{\alpha}(\boldsymbol{q}, \dot{\boldsymbol{q}}) = \boldsymbol{H}\boldsymbol{\Phi}\boldsymbol{M}^{-1}\boldsymbol{b} - \boldsymbol{H}\dot{\boldsymbol{\Phi}}\dot{\boldsymbol{q}}, \tag{21}$$

$$\boldsymbol{\rho}(\boldsymbol{q}) = \boldsymbol{H}\boldsymbol{\Phi}\boldsymbol{M}^{-1}\boldsymbol{g}, \tag{22}$$

where $\boldsymbol{H}(\boldsymbol{q}) \in \mathbb{R}^{m_C \times m_C}$ is the constraint space mass matrix which reflects the system inertia projected at the constraint,

$$\boldsymbol{H}(\boldsymbol{q}) = (\boldsymbol{\Phi}\boldsymbol{M}^{-1}\boldsymbol{\Phi}^T)^{-1}. \tag{23}$$

The constraint null space projection matrix, $\boldsymbol{\Theta}(\boldsymbol{q})^T \in \mathbb{R}^{n \times n}$, is given by,

$$\boldsymbol{\Theta}(\boldsymbol{q})^T = \mathbf{1} - \boldsymbol{\Phi}^T\bar{\boldsymbol{\Phi}}^T, \tag{24}$$

where, $\bar{\boldsymbol{\Phi}}$, is the dynamically consistent inverse of $\boldsymbol{\Phi}$,

$$\bar{\boldsymbol{\Phi}} = \boldsymbol{M}^{-1}\boldsymbol{\Phi}^T\boldsymbol{H}. \tag{25}$$

The control equation can be expressed as,

$$\bar{\boldsymbol{J}}^T\widehat{\boldsymbol{\Theta}}^T\boldsymbol{\tau} = \widehat{\boldsymbol{\Lambda}}\boldsymbol{f}^\star + \widehat{\boldsymbol{\mu}} + \widehat{\mathbf{p}} - \bar{\boldsymbol{J}}^T\boldsymbol{\Phi}^T(\widehat{\boldsymbol{\alpha}} + \widehat{\boldsymbol{\rho}}), \tag{26}$$

where the linear control law of (19) can be used.

It is noted that (20) does not expose the constraint forces (Lagrange multipliers). An alternate form of the constrained task space dynamics is, De Sapio & Park (2010); De Sapio (2011),

$$\boldsymbol{\Theta}^T \boldsymbol{J}^T(\boldsymbol{\Lambda}_c\ddot{\mathbf{x}} + \boldsymbol{\mu}_c + \boldsymbol{p}_c) + \boldsymbol{\Phi}^T(\boldsymbol{\alpha} + \boldsymbol{\rho}) + \boldsymbol{N}_c^T \boldsymbol{\tau}_o - \boldsymbol{\Phi}^T \boldsymbol{\lambda} = \boldsymbol{\tau}. \tag{27}$$

The term $\boldsymbol{\Lambda}_c(\boldsymbol{q}) \in \mathbb{R}^{m\times m}$ is the task/constraint space mass matrix, $\boldsymbol{\mu}_c(\boldsymbol{q},\dot{\boldsymbol{q}}) \in \mathbb{R}^m$ is the task/constraint space centrifugal and Coriolis force vector, $\boldsymbol{p}_c(\boldsymbol{q}) \in \mathbb{R}^m$ is the task/constraint space gravity vector, and $\boldsymbol{N}_c(\boldsymbol{q})^T \in \mathbb{R}^{n\times n}$ is the task/constraint null space projection matrix. These terms are given by,

$$\boldsymbol{\Lambda}_c(\boldsymbol{q}) = (\boldsymbol{J}\boldsymbol{M}^{-1}\boldsymbol{\Theta}^T\boldsymbol{J}^T)^{-1}, \tag{28}$$

$$\boldsymbol{\mu}_c(\boldsymbol{q},\dot{\boldsymbol{q}}) = \boldsymbol{\Lambda}_c\boldsymbol{J}\boldsymbol{M}^{-1}\boldsymbol{\Theta}^T\boldsymbol{b} - \boldsymbol{\Lambda}_c(\dot{\boldsymbol{J}} - \boldsymbol{J}\boldsymbol{M}^{-1}\boldsymbol{\Phi}^T\boldsymbol{H}\dot{\boldsymbol{\Phi}})\dot{\boldsymbol{q}}, \tag{29}$$

$$\boldsymbol{p}_c(\boldsymbol{q}) = \boldsymbol{\Lambda}_c\boldsymbol{J}\boldsymbol{M}^{-1}\boldsymbol{\Theta}^T\boldsymbol{g}, \tag{30}$$

$$\boldsymbol{N}_c(\boldsymbol{q})^T = \boldsymbol{\Theta}^T(\mathbf{1} - \boldsymbol{J}^T\boldsymbol{\Lambda}_c\boldsymbol{J}\boldsymbol{\Theta}\boldsymbol{M}^{-1}). \tag{31}$$

Equation (27) expresses the control torque as a function of the task accelerations, $\ddot{\mathbf{x}}$, the kinematic and dynamic properties, and the constraint forces, $\boldsymbol{\lambda}$. The control equation can be expressed as,

$$\boldsymbol{\tau} + \boldsymbol{\Phi}^T\boldsymbol{\lambda} = \widehat{\boldsymbol{\Theta}}^T \boldsymbol{J}^T(\widehat{\boldsymbol{\Lambda}}_c\boldsymbol{f}^\star + \widehat{\boldsymbol{\mu}}_c + \widehat{\boldsymbol{p}}_c) + \boldsymbol{\Phi}^T(\widehat{\boldsymbol{\alpha}} + \widehat{\boldsymbol{\rho}}) + \widehat{\boldsymbol{N}}_c^T\boldsymbol{\tau}_o, \tag{32}$$

where the linear control law of (19) can be used. These equations need to be complemented by a passivity condition on any unactuated joints,

$$\boldsymbol{S}_p\boldsymbol{\tau} = \mathbf{0}, \tag{33}$$

where $\boldsymbol{S}_p \in \mathbb{R}^{(n-k)\times n}$ is a selection matrix that identifies the passive (unactuated) joints. We can express (32) as,

$$\boldsymbol{\tau} + \boldsymbol{\Phi}^T\boldsymbol{S}_u^T\boldsymbol{\lambda}_u = \widehat{\boldsymbol{\Theta}}^T \boldsymbol{J}^T(\widehat{\boldsymbol{\Lambda}}_c\boldsymbol{f}^\star + \widehat{\boldsymbol{\mu}}_c + \widehat{\boldsymbol{p}}_c) + \boldsymbol{\Phi}^T(\widehat{\boldsymbol{\alpha}} + \widehat{\boldsymbol{\rho}} - \boldsymbol{S}_c^T\boldsymbol{\lambda}_{c_d}) + \widehat{\boldsymbol{N}}_c^T\boldsymbol{\tau}_o, \tag{34}$$

where $\boldsymbol{S}_c \in \mathbb{R}^{(k-p)\times m_C}$ is a selection matrix used to select the controlled constraint forces and $\boldsymbol{S}_u \in \mathbb{R}^{(n-k)\times m_C}$ is a selection matrix used to select the uncontrolled constraint forces. The terms $\boldsymbol{\lambda}_{c_d}$ and $\boldsymbol{\lambda}_u$ are the vectors of controlled and uncontrolled constraint forces, respectively, selected out of the full vector of constraint forces. The term $\boldsymbol{\lambda}_{c_d}$ is specified as part of the control reference, along with $\mathbf{x}_d$, $\dot{\mathbf{x}}_d$, and $\ddot{\mathbf{x}}_d$. We can solve for the control torque,

$$\boldsymbol{\tau} = (\mathbf{1} - \boldsymbol{\Phi}^T\boldsymbol{S}_u^T(\boldsymbol{S}_p\boldsymbol{\Phi}^T\boldsymbol{S}_u^T)^{-1}\boldsymbol{S}_p)\boldsymbol{h}(\boldsymbol{q},\dot{\boldsymbol{q}}), \tag{35}$$

where,

$$\boldsymbol{h}(\boldsymbol{q},\dot{\boldsymbol{q}}) \triangleq \widehat{\boldsymbol{\Theta}}^T \boldsymbol{J}^T(\widehat{\boldsymbol{\Lambda}}_c\boldsymbol{f}^\star + \widehat{\boldsymbol{\mu}}_c + \widehat{\boldsymbol{p}}_c) + \boldsymbol{\Phi}^T(\widehat{\boldsymbol{\alpha}} + \widehat{\boldsymbol{\rho}} - \boldsymbol{S}_c^T\boldsymbol{\lambda}_{c_d}) + \widehat{\boldsymbol{N}}_c^T\boldsymbol{\tau}_o. \tag{36}$$

2.3 Task-level control applied to biomechanical Systems

In this section we apply the task-level control formulation, presented in the previous section, to a biomechanical subsystem. This formulation possesses particular efficacy in addressing

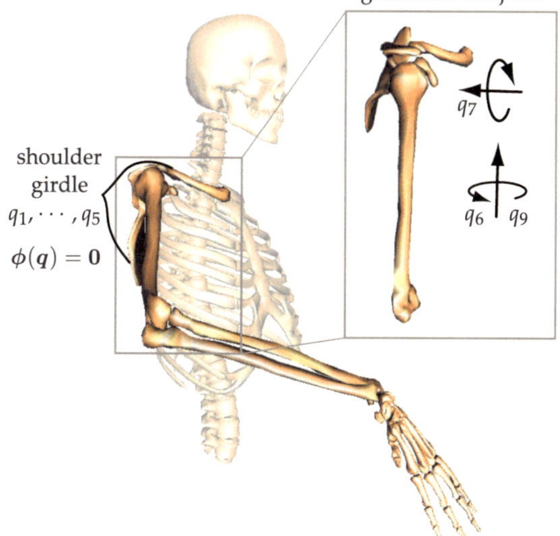

Fig. 6. Reparameterization of the model of Holzbaur et al. Five holonomic constraints couple the movement of the shoulder girdle with the glenohumeral rotations.

complicated systems that involve holonomic constraints. We will choose the human shoulder complex as an illustrative example of this.

Perhaps the most kinematically complicated subsystem in the human skeletal system is the shoulder complex. While the purpose of the shoulder complex is to produce spherical articulation of the humerus, the resultant motion does not exclusively involve motion of the glenohumeral joint. The shoulder girdle, which is comprised of the clavicle and scapula, connects the glenohumeral joint to the torso and produces some of the motion associated with the overall movement of the humerus. While this motion is small compared to the glenohumeral motion its impact on overall arm function is significant, Klopčar & Lenarčič (2001); Lenarčič et al. (2000). This impact is not only associated with the influence of the shoulder girdle on the skeletal kinematics of the shoulder complex, but also its influence on the routing and performance of muscles spanning the shoulder. As a consequence, shoulder kinematics is tightly coupled to the behavior of muscles spanning the shoulder. In turn, the action of these muscles (moments induced about the joints) influences the overall musculoskeletal dynamics of the shoulder. For the aforementioned reasons, when modeling the human shoulder it is important to model the kinematically coupled interactions between the shoulder girdle and the glenohumeral joint.

We can apply a constrained task-level approach to the control of a holonomically constrained shoulder model. This is based on work of De Sapio et al. (2006). The constrained task-level formulation has been updated to the one presented in the previous section. We reparameterized the model of Holzbaur et al. (2005) to include a total of $n = 13$ generalized coordinates (9 for the shoulder complex and 4 for the elbow and wrist) to describe the unconstrained configuration of the arm. As shown in Fig. 6, the coordinates q_6, q_7, and q_9 correspond to the independent coordinates for the shoulder complex used in Holzbaur et al. (2005); elevation plane, elevation angle, and shoulder rotation, respectively.

Five holonomic constraints need to be imposed to properly constrain the motion of the shoulder girdle. With an additional constraint at the glenohumeral joint we have a total of $m_C = 6$ constraints. This yields $p = n - m_C = 7$ degrees of kinematic freedom (3 for the shoulder complex and 4 for the elbow and wrist). These constraint equations, $\phi(\boldsymbol{q}) = \mathbf{0}$, are given by.

$$\phi(\boldsymbol{q}) = \begin{pmatrix} q_1 - b_1 q_6 - c_1 q_7 \\ q_2 - b_2 q_6 - c_2 q_7 \\ q_3 - b_3 q_6 - c_3 q_7 \\ q_4 - b_4 q_6 - c_4 q_7 \\ q_5 - b_5 q_6 - c_5 q_7 \\ q_8 + q_6 \end{pmatrix} = \mathbf{0}, \tag{37}$$

where the constraint constants, $\boldsymbol{b}$ and $\boldsymbol{c}$, associated with the dependency on humerus elevation plane and elevation angle were obtained from the regression analysis of de Groot and Brand de Groot & Brand (2001).

2.4 Simulated control implementation

Defining a humeral orientation, or pointing, task we have,

$$\mathbf{x}(\boldsymbol{q}) = \begin{pmatrix} q_6 \; q_7 \; q_9 \end{pmatrix}^T. \tag{38}$$

We will not control any of the constraint forces so our control equations consist of,

$$\boldsymbol{\tau} + \boldsymbol{\Phi}^T \boldsymbol{\lambda} = \widehat{\boldsymbol{\Theta}}^T \boldsymbol{J}^T (\widehat{\boldsymbol{\Lambda}}_c \boldsymbol{f}^\star + \widehat{\boldsymbol{\mu}}_c + \widehat{\mathbf{p}}_c) + \boldsymbol{\Phi}^T (\widehat{\boldsymbol{\alpha}} + \widehat{\boldsymbol{\rho}}) + \widehat{\boldsymbol{N}}_c^T \boldsymbol{\tau}_o, \tag{39}$$

$$\boldsymbol{f}^\star = \boldsymbol{K}_p(\mathbf{x}_d - \mathbf{x}) + \boldsymbol{K}_v(\dot{\mathbf{x}}_d - \dot{\mathbf{x}}) + \ddot{\mathbf{x}}_d, \tag{40}$$

$$\boldsymbol{S}_p \boldsymbol{\tau} = \mathbf{0}, \tag{41}$$

where $\boldsymbol{S}_p$ accounts for the unactuated (passive) joints, $q_1, \cdots, q_5$, and q_8,

$$\boldsymbol{S}_p = \begin{pmatrix} 1\,0\,0\,0\,0\,0\,0\,0\,0 \\ 0\,1\,0\,0\,0\,0\,0\,0\,0 \\ 0\,0\,1\,0\,0\,0\,0\,0\,0 \\ 0\,0\,0\,1\,0\,0\,0\,0\,0 \\ 0\,0\,0\,0\,1\,0\,0\,0\,0 \\ 0\,0\,0\,0\,0\,0\,0\,1\,0 \end{pmatrix}. \tag{42}$$

Fig. 7 displays simulation plots for the shoulder complex under a goal position command. The controller was applied to both the constrained shoulder model and a simple model with only glenohumeral articulation (motion of the scapula and clavicle not coupled to glenohumeral motion). The glenohumeral joint control torques associated with the constrained and simple shoulder models, performing identical humeral pointing tasks, differ over their respective time histories. This is particularly true for shoulder elevation angle and elevation plane.

2.5 Muscle-based actuation

In the previous section the simulation of the shoulder complex was actuated with joint torque actuators. In reality biomechanical systems are actuated by a set of musculotendon actuators. Hill-type lumped parameter models for muscle-tendon pairs yield equations of state which

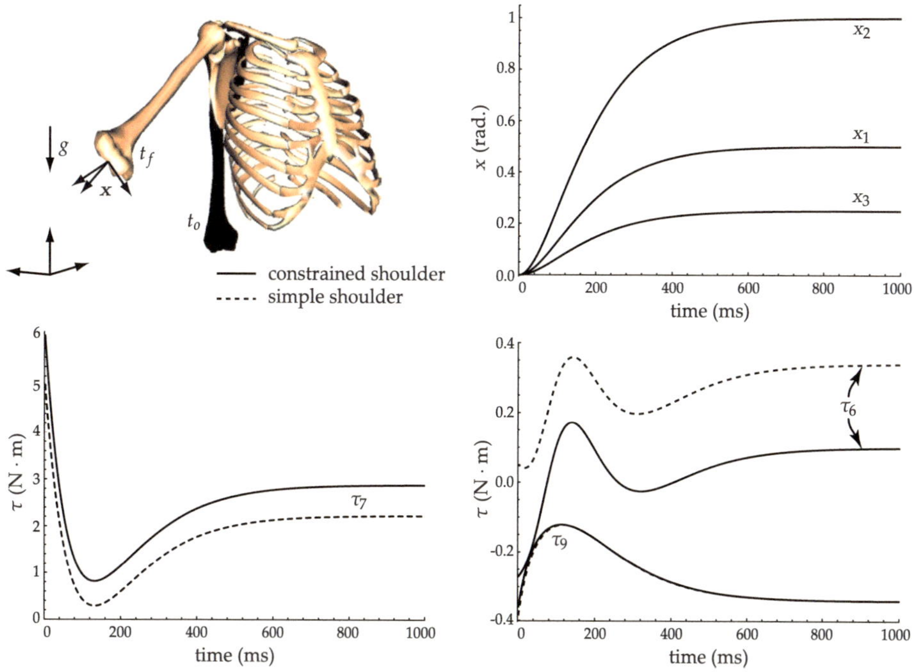

Fig. 7. (Top) Time response of humeral pointing during execution of a goal command for constrained and simple shoulder models. Appropriate dynamic compensation accounts for the control task, $\boldsymbol{x}$, and the shoulder girdle constraints, ϕ. The control gains are $k_p = 100$ and $k_v = 20$. (Bottom) Glenohumeral joint control torques as predicted by the constrained and simple shoulder models. The inclusion of shoulder girdle constraints influences the resulting torques, particularly for shoulder elevation plane, q_6, and elevation angle, q_7.

describe musculotendon behavior, Zajac (1993). Given a set of r musculotendon actuators we can express the vector of musculotendon forces as $\boldsymbol{f} = \boldsymbol{f}(\boldsymbol{l}, \dot{\boldsymbol{l}}, \boldsymbol{a}) \in \mathbb{R}^r$, where $\boldsymbol{l} \in \mathbb{R}^r$ are the muscle lengths whose behavior is described by a state equation and $\boldsymbol{a} \in \mathbb{R}^r$ are the muscle activations, which reflect the level of motor unit recruitment for a given muscle. Activation is a normalized quantity, that is $a_i \in [0,1]$. By using either a stiff tendon model or a steady state evaluation of the musculotendon forces we can express $\boldsymbol{f} = \boldsymbol{f}(\boldsymbol{q}, \dot{\boldsymbol{q}}, \boldsymbol{a}) = \boldsymbol{F}(\boldsymbol{q}, \dot{\boldsymbol{q}})\boldsymbol{a}$, where $\boldsymbol{F}(\boldsymbol{q}, \dot{\boldsymbol{q}}) \in \mathbb{R}^{r \times r}$ is a diagonal matrix mapping muscle activation, $\boldsymbol{a}$, to muscle force, $\boldsymbol{f}$. The joint moments induced by these musculotendon forces are,

$$\boldsymbol{\tau} = -\boldsymbol{L}(\boldsymbol{q})^T \boldsymbol{f} = -\boldsymbol{L}(\boldsymbol{q})^T \boldsymbol{F}(\boldsymbol{q}, \dot{\boldsymbol{q}})\boldsymbol{a} = \boldsymbol{B}(\boldsymbol{q}, \dot{\boldsymbol{q}})^T \boldsymbol{a}, \tag{43}$$

where $\boldsymbol{L}(\boldsymbol{q}) = \partial \boldsymbol{l} / \partial \boldsymbol{q} \in \mathbb{R}^{r \times n}$ is the musculotendon path Jacobian and $\boldsymbol{B}(\boldsymbol{q}, \dot{\boldsymbol{q}})^T \in \mathbb{R}^{n \times r}$ maps muscle activation, $\boldsymbol{a}$, to joint torque, $\boldsymbol{\tau}$. Equation (5) can thus be expressed in terms of muscle actuation,

$$\boldsymbol{M}\ddot{\boldsymbol{q}} + \boldsymbol{b} + \boldsymbol{g} - \boldsymbol{\Phi}^T \boldsymbol{\lambda} = \boldsymbol{B}^T \boldsymbol{a}. \tag{44}$$

We can then express the control equation as (26),

$$\bar{\boldsymbol{J}}^T \widehat{\boldsymbol{\Theta}}^T \boldsymbol{B}^T \boldsymbol{a} = \widehat{\boldsymbol{\Lambda}} \boldsymbol{f}^\star + \widehat{\boldsymbol{\mu}} + \widehat{\boldsymbol{p}} - \bar{\boldsymbol{J}}^T \boldsymbol{\Phi}^T (\widehat{\boldsymbol{\alpha}} + \widehat{\boldsymbol{\rho}}). \tag{45}$$

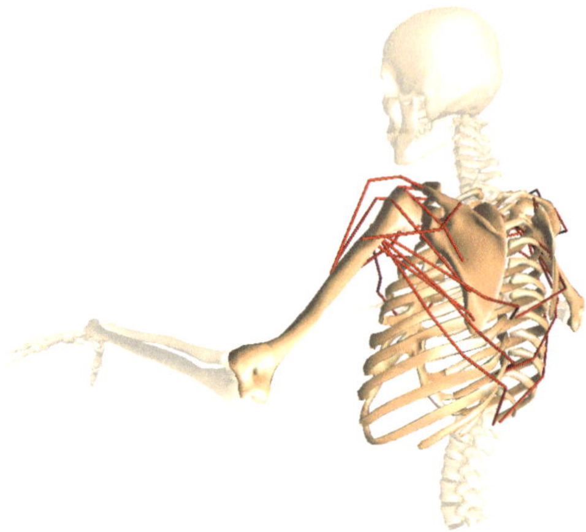

Fig. 8. Muscle paths spanning the shoulder complex. Muscle moment arms are determined from the muscle path data Holzbaur et al. (2005). The motion of the shoulder girdle influences the moment arms about the glenohumeral joint.

Due to both kinematic redundancy and actuator redundancy there will typically be many solutions for $\boldsymbol{a}$. Using a static optimization procedure, Thelen et al. (2003), this can be resolved by finding the solution which minimizes $\|\boldsymbol{a}\|^2$ given $a_i \in [0,1]$. This corresponds to minimizing the instantaneous muscle effort. The use of $\|\boldsymbol{a}\|^2$ and similar cost measures have been suggested in a number of sources, Anderson & Pandy (2001); Crowninshield & Brand (1981).

In Section 2.4 we observed that the constrained shoulder model, which involves kinematic coupling between the humerus, scapula and clavicle, differs from the simple shoulder model with regard to the control torques that are required to achieve a desired motion control task. The constrained model also differs from the simple model in the degree to which the system of muscles are able to generate control forces to achieve a desired motion control task. This is due to the influence of the constrained motion between the humerus, scapula and clavicle on the muscle forces and muscle moment arms about the glenohumeral joint (see Fig. 8).

An example of this is shown in Fig. 9. Predicted muscle moment arms, muscle forces, and moment generating capacities for the deltoid muscles are compared for the simple and constrained shoulder models. The muscle path and force-length data were taken from the study of Holzbaur et al. (2005). In the constrained shoulder model the motions of the scapula and clavicle are highly coupled to humerus elevation angle (q_7 coordinate), whereas, in the simple shoulder model the motion of the scapula and clavicle are not coupled to glenohumeral motion. The paths of the deltoid muscles are affected by the constrained motion of the humerus, scapula, and clavicle. This results in significant differences in moment arms predicted by the two models, with the constrained model often generating moment arms of substantially larger magnitude than the simple model.

Additionally, the predicted isometric muscle forces (computed at full activation) generated by the two models differ. The resulting moment generating capacities of the constrained model are often substantially larger in magnitude than the simple model. This implies that the simple model, which excludes the constrained shoulder girdle motion, typically underestimates the moment generating capacities of muscles that span the shoulder, since Holzbaur et al. (2005) demonstrated correlation between predicted and experimental moment generating capacities

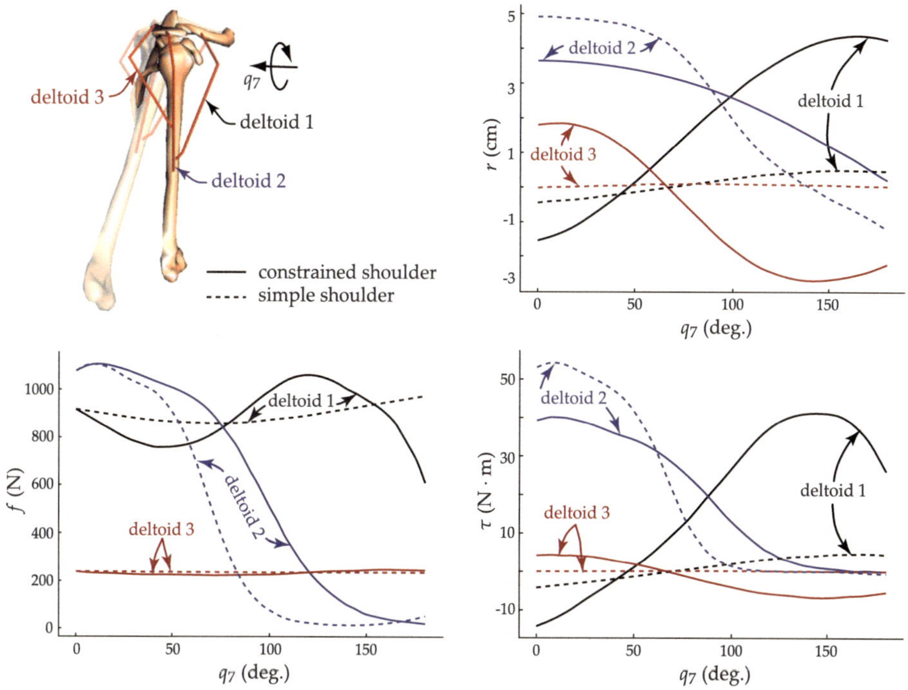

Fig. 9. (Top) Muscle moment arms for the deltoid muscles, as predicted by the constrained and simple shoulder models. The constrained model typically generates moment arms of substantially larger magnitude than those of the simple model. (Bottom) Muscle forces and moment generating capacities for the deltoid muscles. The resulting moment generating capacities associated with the constrained model are typically larger in magnitude than those associated with the simple model.

for the constrained model. This is critical in various applications involving the study and synthesis of human movement, Khatib et al. (2004).

3. Posture-based modeling and analysis of biomechanical systems

In this section we present a muscle effort criterion for the prediction of upper limb postures. In the overall framework this addresses the highlighted element of Fig. 10. The focus is on developing a neuromuscular criterion and a methodology for synthesizing posture in the presence of that criterion.

A particularly relevant class of human movements involves targeted reaching. Given a specific target the prediction of kinematically redundant upper limb motion is a problem of choosing one of a multitude of control solutions, all of which yield kinematically feasible configurations. It has been observed that humans resolve this redundancy problem in a relatively consistent manner, Kang et al. (2005); Lacquaniti & Soechting (1982). For this reason general mathematical models have proven to be valuable tools for motor control prediction across human subjects.

Approaches for predicting human arm movement have been categorized into posture-based and trajectory-based (or transport-based) models, Hermens & Gielen (2004); Vetter et al. (2002). Posture-based models are predicated upon the assumption of Donders' law. Specifically, Donders' law postulates that final arm configuration is dependent only on

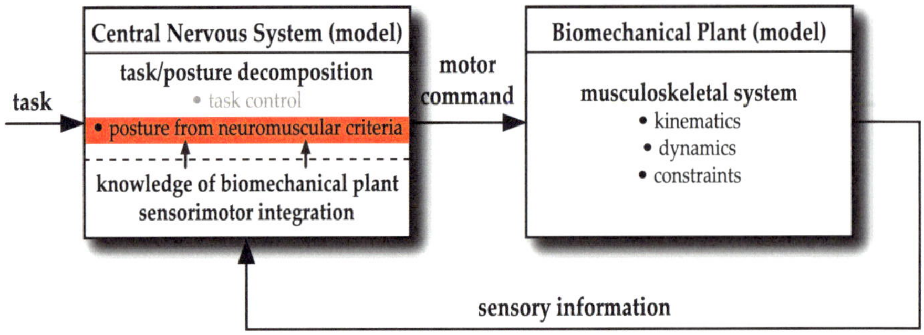

Fig. 10. Task/posture motion control model for biomechanical systems highlighting posture control from neuromuscular criteria.

final hand position and is independent of initial (or past) arm configurations. Thus, the fundamental characteristic of posture-based models is path independence in predicting equilibrium arm postures. In these models the postulated behavior of the central nervous system (CNS) can be said to execute movements based strictly on control variables (e.g., hand position). Conversely, trajectory-based models, which include the minimum work model, Soechting et al. (1995), and the minimum torque-change model, Uno et al. (1989), are characterized by dependence of final arm configuration on the final hand position, the starting configuration, and the choice of a specific optimal path parameterized over time (i.e., past arm configurations).

Many of the models for predicting human arm movement, including the minimum work model and the minimum torque-change model, do not involve any direct inclusion of muscular properties such as routing kinematics and strength properties. Even models described as employing biomechanical variables, Kang et al. (2005), typically employ only variables derivable purely from skeletal kinematics and not musculoskeletal physiology. It is felt that the utilization of a model-based characterization of muscle systems, which accounts for muscle kinematic and strength properties, is critical to authentically simulating human motion since all human motion is predicated upon physiological capabilities.

3.1 Biomechanical effort minimization

We begin with a general consideration of biomechanical effort measures. An instantaneous effort measure can be used in a trajectory-based model of movement by seeking a trajectory, consistent with task constraints, that minimizes the integral of that measure over the time interval of motion. Alternatively, the instantaneous effort measure can be used in a posture-based model by seeking a static posture, consistent with the target constraint, which minimizes the *static* form of the measure.

Proceeding from Section 2.5 we express the joint torques in terms of muscle activations,

$$\boldsymbol{\tau} = -\boldsymbol{L}(\boldsymbol{q})^T \boldsymbol{f} = -\boldsymbol{L}(\boldsymbol{q})^T \boldsymbol{F}(\boldsymbol{q}, \dot{\boldsymbol{q}})\boldsymbol{a} = \boldsymbol{B}(\boldsymbol{q}, \dot{\boldsymbol{q}})^T \boldsymbol{a}. \tag{46}$$

Due to the fact that there are typically more muscles spanning a set of joints than the number of generalized coordinates used to describe those joints this equation will have an infinite set of solutions for $\boldsymbol{a}$. Choosing the solution, $\boldsymbol{a}_o$, which has the smallest magnitude (least norm) yields,

$$\boldsymbol{a}_o = \boldsymbol{B}^{T+} \boldsymbol{\tau} = \boldsymbol{B}(\boldsymbol{B}^T \boldsymbol{B})^{-1} \boldsymbol{\tau}, \tag{47}$$

where $\boldsymbol{B}^{T+}$ is the pseudoinverse of $\boldsymbol{B}^{T}$. Our instantaneous muscle effort measure can then be expressed as,

$$U = \|\boldsymbol{a}_o\|^2 = \boldsymbol{\tau}^T(\boldsymbol{B}^T\boldsymbol{B})^{-1}\boldsymbol{\tau}. \tag{48}$$

Expressing this effort measure in constituent terms and dissecting the structure we have,

$$U = \boldsymbol{\tau}^T [\overbrace{\underbrace{\boldsymbol{L}^T}_{\text{kinematics}} \underbrace{(\boldsymbol{F}\boldsymbol{F}^T)}_{\text{kinetics}} \underbrace{\boldsymbol{L}}_{\text{kinematics}}}^{\text{muscular capacity}}]^{-1}\boldsymbol{\tau}. \tag{49}$$

This allows us to gain some physical insight into what is being measured. The terms inside the brackets represent a measure of the net capacity of the muscles. This is a combination of the force generating kinetics of the muscles as well as the mechanical advantage of the muscles, as determined by the muscle routing kinematics. The terms outside of the brackets represent the kinetic torque requirements of the task and posture.

It is noted that the solution of (46) expressed in (47) corresponds to a constrained minimization of $\|\boldsymbol{a}\|^2$, however, this solution does not enforce the constraint that muscle activation must be positive (muscles can only produce tensile forces). Imposing inequality constraints, $0 \leq a_i \leq 1$, on the activations requires a quadratic programming (QP) approach for performing the constrained minimization. In this case the solution of (46) which minimizes $\|\boldsymbol{a}\|^2$ and satisfies $0 \leq a_i \leq 1$ can be represented in shorthand as,

$$\boldsymbol{a}_o = \mathrm{QP}(\boldsymbol{B}^T, \boldsymbol{\tau}, \|\boldsymbol{a}\|^2, 0 \leq a_i \leq 1), \tag{50}$$

where QP($\square$) represents the output of a quadratic programming function (e.g., `quadprog()` in the Matlab optimization toolbox). Our muscle effort criterion is then $U = \|\boldsymbol{a}_o\|^2$, where $\boldsymbol{a}_o$ is given by (50). Despite the preferred use of quadratic programming for computational purposes, (49) provides valuable insights at a conceptual level.

3.2 Posture-based criteria

For posture-based analysis the static form of the instantaneous muscle effort measure can be constructed by noting that $\dot{\boldsymbol{q}} \to \boldsymbol{0}$, thus eliminating the dependency of U on $\dot{\boldsymbol{q}}$. This also implies that $\boldsymbol{\tau} \to \boldsymbol{g}$. Thus, the static form, $U(\boldsymbol{q})$, of (48) is,

$$U(\boldsymbol{q}) = \boldsymbol{g}(\boldsymbol{q})^T[\boldsymbol{B}(\boldsymbol{q})^T\boldsymbol{B}(\boldsymbol{q})^T]^{-1}\boldsymbol{g}(\boldsymbol{q}). \tag{51}$$

Alternatively, imposing the inequality constraints on the activations we have $U = \|\boldsymbol{a}_o\|^2$ where,

$$\boldsymbol{a}_o = \mathrm{QP}(\boldsymbol{B}(\boldsymbol{q})^T, \boldsymbol{g}(\boldsymbol{q}), \|\boldsymbol{a}\|^2, 0 \leq a_i \leq 1). \tag{52}$$

To find a task consistent static configuration which minimizes $U(\boldsymbol{q})$, we first define the self-motion manifold associated with a fixed task point, $\mathbf{x}_o$. This is given by $M(\mathbf{x}_o) = \{\boldsymbol{q} \,|\, \mathbf{x}(\boldsymbol{q}) = \mathbf{x}_o\}$ where $\mathbf{x}(\boldsymbol{q})$ is the operational point of the kinematic chain (e.g., the position of the hand). For each $\boldsymbol{q}$ on $M(\mathbf{x}_o)$ we can compute $U(\boldsymbol{q}) = \|\boldsymbol{a}_o\|^2$ by solving the quadratic programming problem of (52). The minimal effort task consistent configuration is then the configuration, $\boldsymbol{q}$, for which $U(\boldsymbol{q})$ is minimized on $M(\mathbf{x}_o)$. Figure 11 illustrates changes in the predicted posture associated with minimal muscle effort as weight at the hand is varied.

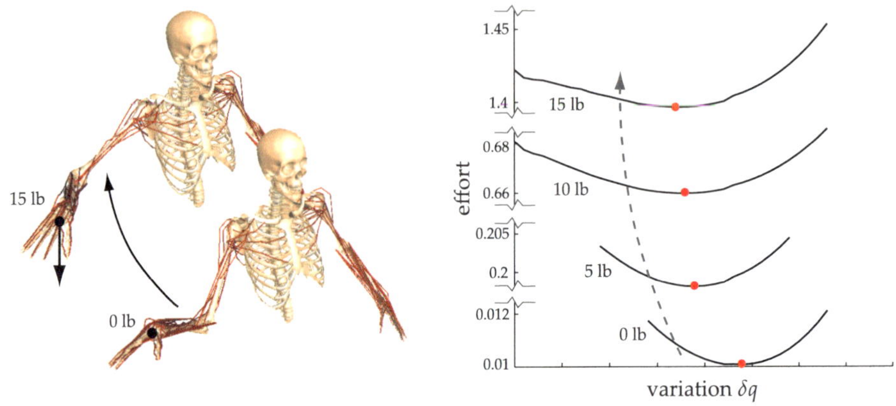

Fig. 11. Muscle effort variation and predicted minimal efforts associated with different weights in hand. The weight at the hand was projected into joint space and added to the gravity vector associated with the limb segments. The effect is that the predicted posture, associated with the minimum of the muscle effort curve, shifts as weight is added. Each point on each of the curves was computed by solving a quadratic programming problem.

3.3 Sphere methods for quadratic programming

Quadratic programming addresses the general minimization of a quadratic function subject to a combination of equality and inequality constraints. It can be formally stated as: Minimize the objective function, $z(\boldsymbol{x})$, with respect to $\boldsymbol{x}$, where,

$$z(\boldsymbol{x}) = \frac{1}{2}\boldsymbol{x}^T \boldsymbol{D}\boldsymbol{x} + \boldsymbol{d}^T\boldsymbol{x}, \tag{53}$$

subject to,

$$\boldsymbol{A}\boldsymbol{x} \geq \boldsymbol{b}, \tag{54}$$

$$\boldsymbol{C}\boldsymbol{x} = \boldsymbol{y}. \tag{55}$$

We assume that $\boldsymbol{D}$ is symmetric positive definite and that the polytope defined by $\boldsymbol{A}\boldsymbol{x} \geq \boldsymbol{b}$ is convex. In the case of muscle effort minimization we have the specific form,

$$z(\boldsymbol{a}) = \frac{1}{2}\boldsymbol{a}^T\boldsymbol{a}, \tag{56}$$

subject to,

$$\begin{pmatrix} \mathbf{1}_{r\times r} \\ -\mathbf{1}_{r\times r} \end{pmatrix} \boldsymbol{a} \geq \begin{pmatrix} \mathbf{0}_{r\times 1} \\ -\mathbf{1}_{r\times 1} \end{pmatrix} \tag{57}$$

$$\boldsymbol{B}(\boldsymbol{q})^T\boldsymbol{a} = \boldsymbol{g}(\boldsymbol{q}), \tag{58}$$

where $\mathbf{1}_{r\times r}$ is the $r \times r$ identity matrix, $\mathbf{0}_{r\times 1}$ is a column vector of *zeros*, and $\mathbf{1}_{r\times 1}$ is a column vector of *ones*. Clearly, the quadratic form (56) is positive definite and the polytope (57) is convex. For the procedure of muscle effort minimization this QP problem is repeatedly solved for different values of $\boldsymbol{q}$ on $M(\mathbf{x}_o)$, generating the function $U(\boldsymbol{q})$. A line search over $M(\mathbf{x}_o)$ then yields $\boldsymbol{q}_o$ where $U(\boldsymbol{q}_o)$ represents the minimum of U on the self-motion manifold.

Since this QP problem needs to be solved repeatedly we would like an efficient method for solving it. There are a number of interior point method (IPM) solvers that addresses QP problems. We have implemented one based on the sphere method approach. This approach was initially developed for linear programming (LP) problems, Murty (2006); Murty (2010b), but has been extended for QP problems, Murty (2010a). Our implementation of the sphere method approach for QP will be described here and is based on the approach of Murty et al.
We begin with the general problem of minimizing (53) subject to (54) and (55). It is noted that the equality constraints, $\boldsymbol{Cx} = \boldsymbol{y}$, can be represented as the inequality constraints.

$$\boldsymbol{Cx} > \boldsymbol{y} - \boldsymbol{\epsilon}, \tag{59}$$

$$\boldsymbol{Cx} < \boldsymbol{y} + \boldsymbol{\epsilon}. \tag{60}$$

where ϵ is a vector of small positive tolerances. Consequently, we consider all constraints, both equality and inequality, as being represented by $\boldsymbol{Ax} \geq \boldsymbol{b}$. These constraints describe a polytope K. A simple check can be made to determine if the unconstrained minimum of the objective function is interior to the polytope. If this is the case then the solution to the QP problem is trivial. Assuming that this is not the case we proceed by noting that the facetal hyperplanes defined by, $\boldsymbol{Ax} = \boldsymbol{b}$, can be represented as,

$$\boldsymbol{v}_i^T \boldsymbol{x} = b_i \qquad \text{for } i = 1, \cdots, m, \tag{61}$$

where $\{\boldsymbol{v}_1, \cdots, \boldsymbol{v}_m\}$ are the inward normals of the facetal planes and,

$$\boldsymbol{A} = \begin{pmatrix} \boldsymbol{v}_1^T \\ \vdots \\ \boldsymbol{v}_m^T \end{pmatrix}. \tag{62}$$

We normalize (61) by dividing both sides by $\|\boldsymbol{v}_i\|$. Thus,

$$\hat{\boldsymbol{v}}_i = \frac{\boldsymbol{v}_i}{\|\boldsymbol{v}_i\|}, \qquad \hat{b}_i = \frac{b_i}{\|\boldsymbol{v}_i\|}, \qquad \hat{\boldsymbol{A}} = \begin{pmatrix} \hat{\boldsymbol{v}}_1^T \\ \vdots \\ \hat{\boldsymbol{v}}_m^T \end{pmatrix}. \tag{63}$$

Following these normalizations we perform centering steps from some initial point, $\boldsymbol{x}_i$, interior to the polytope. Two types of centering steps are performed. One is termed a line search from facetal normals (LSFN), the other is termed a line search from computed profitable directions (LSCPD). First, the touching set, $T(\boldsymbol{x})$, at the current point, $\boldsymbol{x}$ (initially $\boldsymbol{x}_i$) is computed. This is the set of facetal hyperplanes which are touched by the largest hypersphere that can be inscribed in the polytope, centered at the current point, $\boldsymbol{x}$.
For the LSFN step the facetal unit normals, $\{\hat{\boldsymbol{v}}_1, \cdots, \hat{\boldsymbol{v}}_m\}$, are iterated through until one is found, $\hat{\boldsymbol{y}}$, such that,

$$\hat{\boldsymbol{v}}_i^T \hat{\boldsymbol{y}} > 0 \qquad \text{for all } i \in T(\boldsymbol{x}), \tag{64}$$

and such that it reduces the objective function, that is,

$$-\left[\nabla z(\boldsymbol{x})\right]^T \hat{\boldsymbol{y}} > 0, \tag{65}$$

where $\nabla z(\boldsymbol{x}) = \boldsymbol{Dx} + \boldsymbol{d}$. Given a profitable direction, $\hat{\boldsymbol{y}}$, that meets these criteria a line search is performed to move along this profitable direction until a point is reached for which

the inscribed sphere at that point is a maximum. A backtracking line search has been implemented for this. The line search is terminated at any point where (65) is not satisfied (no longer descending). This LSFN step is repeated as long as profitable directions meeting the criteria are found.
For the LSCPD step the linear system,

$$\hat{\boldsymbol{v}}_i^T \boldsymbol{y}_1 = 1 \quad \text{and} \quad -[\nabla z(\boldsymbol{x})]^T \boldsymbol{y}_1 = 0 \qquad \text{for all } i \in T(\boldsymbol{x}), \tag{66}$$

is solved for a direction $\boldsymbol{y}_1$ and the linear system,

$$\hat{\boldsymbol{v}}_i^T \boldsymbol{y}_2 = 0 \quad \text{and} \quad -[\nabla z(\boldsymbol{x})]^T \boldsymbol{y}_2 = 1 \qquad \text{for all } i \in T(\boldsymbol{x}), \tag{67}$$

is solved for a direction $\boldsymbol{y}_2$. Backtracking line searches are performed sequentially in both of these unit directions, $\hat{\boldsymbol{y}}_1$ and $\hat{\boldsymbol{y}}_2$, until a point is reached for which the inscribed sphere at that point is a maximum. Again, the line search is terminated at any point where (65) is not satisfied. This LSCPD step is repeated until the incremental reduction in the objective function falls below some tolerance. The final output of the centering steps will be labeled $\boldsymbol{x}_r$.
Following the centering steps, descent steps are performed. For a given iteration, a single descent step is chosen based on the best performance of a set different descent steps, in reducing the objective function. All of these descent steps terminate at the boundary of the polytope. Given a unit descent direction $\hat{\boldsymbol{y}}$ the distance along this direction to the polytope boundary is given by,

$$\delta = \min \left(\frac{\hat{\boldsymbol{v}}_i^T \boldsymbol{x}_r - \hat{b}_i}{\hat{\boldsymbol{v}}_i^T \hat{\boldsymbol{y}}} \right) \quad \text{over } i, \quad \text{such that,} \quad \hat{\boldsymbol{v}}_i^T \hat{\boldsymbol{y}} < 0. \tag{68}$$

These candidate descent directions are as follows:

- D1: Choose $\boldsymbol{y} = -\nabla z(\boldsymbol{x}_r)$. Move from $\boldsymbol{x}_r$ along $\hat{\boldsymbol{y}}$ to the boundary of K.
- D2: Choose $\boldsymbol{y}$ to be the direction defined by the displacement vector between the previous two centering locations, $\boldsymbol{y} = \boldsymbol{x}_r - \boldsymbol{x}_{r-1}$. Move from $\boldsymbol{x}_r$ along $\hat{\boldsymbol{y}}$ to the boundary of K.
- D3: Define directions associated with projecting $-\nabla z(\boldsymbol{x}_r)$ on each of the facetal hyperplanes in the touching set. These directions are given by,

$$\boldsymbol{y}_i = -(\mathbf{1} - \hat{\boldsymbol{v}}_i \hat{\boldsymbol{v}}_i^T) \nabla z(\boldsymbol{x}_r) \quad \forall i \in T(\boldsymbol{x}_r). \tag{69}$$

  Move from $\boldsymbol{x}_r$ along $\hat{\boldsymbol{y}}_i$, $\forall i \in T(\boldsymbol{x}_r)$, to the boundary of K. Of these $|T(\boldsymbol{x}_r)|$ descents retain the one that results in the greatest reduction in the objective function.
- D4: Choose $\boldsymbol{y}$ to be the average of the directions from D3. Move from $\boldsymbol{x}_r$ along $\hat{\boldsymbol{y}}$ to the boundary of K. The average of the directions from D3 is given by,

$$\boldsymbol{y} = \sum_{i \in T(\boldsymbol{x}_r)} \frac{-(\mathbf{1} - \hat{\boldsymbol{v}}_i \hat{\boldsymbol{v}}_i^T) \nabla z(\boldsymbol{x}_r)}{|T(\boldsymbol{x}_r)|}. \tag{70}$$

- D5: Compute the touching point, $\boldsymbol{x}_r^i$ associated with $\boldsymbol{x}_r$. This is the point on each facetal hyperplane in the touching set where the maximum inscribed hypersphere, centered at $\boldsymbol{x}_r$, touches. These points are given by,

$$\boldsymbol{x}_r^i = \boldsymbol{x}_r + \hat{\boldsymbol{v}}_i (b_i - \hat{\boldsymbol{v}}_i^T \boldsymbol{x}_r) \quad \forall i \in T(\boldsymbol{x}_r). \tag{71}$$

The near touching point is defined as a point on the line segment between $\boldsymbol{x}_r$ and $\boldsymbol{x}_r^i$.

$$\tilde{\boldsymbol{x}}_r^i = \epsilon \boldsymbol{x}_r + (1-\epsilon)\boldsymbol{x}_r^i \quad \forall i \in T(\boldsymbol{x}_r), \tag{72}$$

where *epsilon* is a small tolerance (e.g., ≈ 0.1). Projecting $-\nabla z(\tilde{\boldsymbol{x}}_r^i)$ on each of the facetal hyperplanes in the touching set yields,

$$\boldsymbol{y}_i = -(\mathbf{1} - \hat{\boldsymbol{v}}_i \hat{\boldsymbol{v}}_i^T)\nabla z(\tilde{\boldsymbol{x}}_r^i) \quad \forall i \in T(\boldsymbol{x}_r). \tag{73}$$

Move from $\tilde{\boldsymbol{x}}_r^i$ along $\hat{\boldsymbol{y}}_i$, $\forall i \in T(\boldsymbol{x}_r)$, to the boundary of K. Of these $|T(\boldsymbol{x}_r)|$ descents retain the one that results in the greatest reduction in the objective function.

The output of D1 through D5 that results in the greatest reduction in the objective function is used to yield the new point $\boldsymbol{x}$. The centering and descent steps are repeated until some solution tolerance is met. In subsequent iterations the feasible set K shrinks based on the objective tangent hyperplane passing thorough $\boldsymbol{x}$. That is, the constraints are appended to include the objective tangent hyperplane passing through the current $\boldsymbol{x}$,

$$\hat{\boldsymbol{A}} = \begin{pmatrix} -[\nabla z(\boldsymbol{x})]^T \\ \hat{\boldsymbol{v}}_1^T \\ \vdots \\ \hat{\boldsymbol{v}}_m^T \end{pmatrix} \quad \text{and} \quad \hat{\boldsymbol{b}} = \begin{pmatrix} -[\nabla z(\boldsymbol{x})]^T \boldsymbol{x} \\ \hat{b}_1 \\ \vdots \\ \hat{b}_m \end{pmatrix}. \tag{74}$$

Fig. 12 illustrates some of the general steps for centering and descent in this algorithm. The algorithm has been implemented in Matlab and in C++ on problems involving thousands of variables and constraints. It performs favorably in terms of accuracy and speed as compared with Matlab's `quadprog()` IPM routine. Quantitative benchmarking is planned for the future.

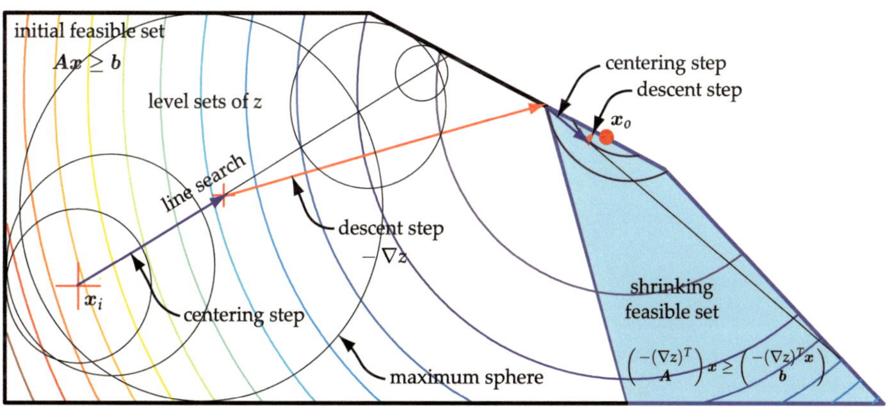

Fig. 12. An illustration of the centering and descent steps associated with the sphere method implemented for QP problems.

3.4 Least action of cost criteria

We now pose the problem of minimizing a cost criterion subject to a motion control task. This is detailed in De Sapio et al. (2008). We can perform this for an instantaneous potential-based criterion, $U(\boldsymbol{q})$, by using a gradient descent method in conjunction with the task/posture decomposition of (13). Given our overall control torque,

$$\boldsymbol{\tau} = \boldsymbol{J}^T \boldsymbol{f} + \boldsymbol{N}^T \boldsymbol{\tau}_p, \tag{75}$$

the posture term, $\boldsymbol{\tau}_p$, can be chosen to correspond to the gradient descent, $-\partial U / \partial \boldsymbol{q}$, of our cost criterion. In this case the equations of motion are,

$$\boldsymbol{M}\ddot{\boldsymbol{q}} + \boldsymbol{b} + \boldsymbol{g} = \boldsymbol{J}^T \boldsymbol{f} - \boldsymbol{N}^T \frac{\partial U}{\partial \boldsymbol{q}}, \tag{76}$$

subject to the task $\ddot{\mathbf{x}}(\boldsymbol{q}) = \ddot{\mathbf{x}}_d(t)$. We complement (76) with the task space control law given by (17) and (19).
Gradient descent seeks to reduce an instantaneous criterion rather than extremize a criterion over an integration interval. To address this latter case we define the action integral associated with a cost criterion, as in De Sapio et al. (2008),

$$I \triangleq \int_{t_o}^{t_f} U(\boldsymbol{q}, \dot{\boldsymbol{q}})\, dt. \tag{77}$$

If no task trajectory constraints are specified we have,

$$\begin{gathered} \delta I = 0, \\ \forall\, \delta \mid \delta \boldsymbol{q}(t_o) = \delta \boldsymbol{q}(t_f) = \mathbf{0}. \end{gathered} \tag{78}$$

Equations (77) and (78) result in the Euler-Lagrange equations,

$$\frac{d}{dt} \frac{\partial U}{\partial \dot{\boldsymbol{q}}} - \frac{\partial U}{\partial \boldsymbol{q}} = \mathbf{0}. \tag{79}$$

Imposing rheonomic task trajectory constraints, $\mathbf{x}(\boldsymbol{q}) = \mathbf{x}_d(t)$, implies,

$$\begin{gathered} \delta J = 0, \\ \forall\, \delta \mid \delta \boldsymbol{q}(t_o) = \delta \boldsymbol{q}(t_f) = \mathbf{0}, \quad \text{and } \boldsymbol{J}\, \delta \boldsymbol{q} = \mathbf{0}, \end{gathered} \tag{80}$$

which yields the system,

$$\frac{d}{dt} \frac{\partial U}{\partial \dot{\boldsymbol{q}}} - \frac{\partial U}{\partial \boldsymbol{q}} = \boldsymbol{J}^T \boldsymbol{\lambda}, \tag{81}$$

or,

$$\boldsymbol{M}_U \ddot{\boldsymbol{q}} + \boldsymbol{b}_U + \boldsymbol{g}_U = \boldsymbol{J}^T \boldsymbol{\lambda}, \tag{82}$$

subject to $\ddot{\mathbf{x}}(\boldsymbol{q}) = \ddot{\mathbf{x}}_d(t)$. Projecting (82) into task space yields the operational space equations for this system,

$$\boldsymbol{\Lambda}_U(\boldsymbol{q}, \dot{\boldsymbol{q}})\, \ddot{\mathbf{x}} + \boldsymbol{\mu}_U(\boldsymbol{q}, \dot{\boldsymbol{q}}) + \boldsymbol{p}_U(\boldsymbol{q}) = \boldsymbol{\lambda}, \tag{83}$$

where $\boldsymbol{\Lambda}_U$, $\boldsymbol{\mu}_U$, and $\boldsymbol{p}_U$ are analogous to $\boldsymbol{\Lambda}$, $\boldsymbol{\mu}$, and $\boldsymbol{p}$, but with $\boldsymbol{M}$, $\boldsymbol{b}$, and $\boldsymbol{g}$ replaced by $\boldsymbol{M}_U$, $\boldsymbol{b}_U$, and $\boldsymbol{g}_U$. Applying constraint stabilization, the trajectory constraints can be expressed as,

$$\ddot{\mathbf{x}} = \boldsymbol{\lambda}^\star = \ddot{\mathbf{x}}_d(t) + \beta[\dot{\mathbf{x}}_d(t) - \dot{\mathbf{x}}] + \alpha[\mathbf{x}_d(t) - \mathbf{x}], \tag{84}$$

and the constraint stabilized system is,

$$\boldsymbol{\lambda} = \boldsymbol{\Lambda}_U \boldsymbol{\lambda}^{\star} + \boldsymbol{\mu}_U + \boldsymbol{p}_U. \tag{85}$$

Two examples from De Sapio et al. (2008) can be used to illustrate the approaches described. First we consider a simplified $n = 3$ degree-of-freedom model of the human arm actuated by $r = 14$ muscles. The system is kinematically redundant with respect to the $m = 2$ degree-of-freedom task of positioning the hand. The muscle attachment and force-length data were taken from the study of Holzbaur et al. (2005). We wish to control the hand to move to a target location, $\mathbf{x}_f$, while minimizing an instantaneous muscle effort criterion defined as,

$$U(\boldsymbol{q}) \triangleq \boldsymbol{g}^T(\boldsymbol{B}^T\boldsymbol{B})^{-1}\boldsymbol{g}, \tag{86}$$

where $\boldsymbol{B}(\boldsymbol{q}) = -\boldsymbol{L}(\boldsymbol{q})^T\boldsymbol{F}(\boldsymbol{q})$ and the muscle forces are modeled as $\boldsymbol{f}(\boldsymbol{q},\boldsymbol{a}) = \boldsymbol{F}(\boldsymbol{q})\boldsymbol{a}$, where,

$$\boldsymbol{F} = \operatorname{diag}\left(f_{o_i} e^{-5\left(\frac{l_i(\boldsymbol{q})}{l_{o_i}}-1\right)^2} \right). \tag{87}$$

The term, f_{o_i}, represents the maximum isometric force for the ith muscle and l_{o_i} represents the optimal fiber length for the ith muscle. No task trajectory, $\mathbf{x}_d(t)$, will be specified, just the final target location, $\mathbf{x}_f$.

We have the following control equations,

$$\boldsymbol{f}^{\star} = k_p(\mathbf{x}_f - \mathbf{x}) - k_v\dot{\mathbf{x}}, \tag{88}$$

$$\boldsymbol{\tau} = \boldsymbol{J}^T\boldsymbol{f}^{\star} + \widehat{\boldsymbol{g}} - \widehat{\boldsymbol{N}}^T(k_e\frac{\partial U}{\partial \boldsymbol{q}} + k_d\dot{\boldsymbol{q}}). \tag{89}$$

In this case no model of the dynamic properties is included in (**??**). Also, the terms $\ddot{\mathbf{x}}_d(t)$ and $\dot{\mathbf{x}}_d(t)$ have been omitted in (88) and $\mathbf{x}_d(t)$ has been replaced by the final target location, $\mathbf{x}_f$, since the goal is to move to a target location without specifying a trajectory. To the posture space portion of (89) we have added a dissipative term, $k_d\dot{\boldsymbol{q}}$, and a gain, k_e, on the gradient descent term. Finally, the gravity vector, $\boldsymbol{g}$, is perfectly compensated for in the overall control. Fig. 13 displays time histories of joint motion, hand motion, and muscle effort for a simulation run. We can see that the controller achieves the final target objective while the null space control simultaneously seeks to reduce the instantaneous muscle effort (consistent with the task requirement). It is recalled that no compensation for the dynamics (except for gravity) was included in (89). Thus, there is no feedback linearization present in the control. Normally, perfect feedback linearization without explicit trajectory tracking would produce straight line motion to the goal. In the absence of feedback linearization non-straight line motion results.

We now seek a trajectory which moves the hand to a target location (see Fig. 14), while extremizing muscle action,

$$I \triangleq \int_{t_o}^{t_f} U(\boldsymbol{q},\dot{\boldsymbol{q}})\, dt. \tag{90}$$

In this case we will define the instantaneous muscle effort criterion as,

$$U(\boldsymbol{q},\dot{\boldsymbol{q}}) = \sum_{i=1}^{r}\left(\frac{l_i - l_{o_i}}{l_{o_i}}\right)^2 + \sum_{i=1}^{r}\left(\frac{\dot{l}_i}{v_{o_i}}\right)^2 + \dot{q}_3^2, \tag{91}$$

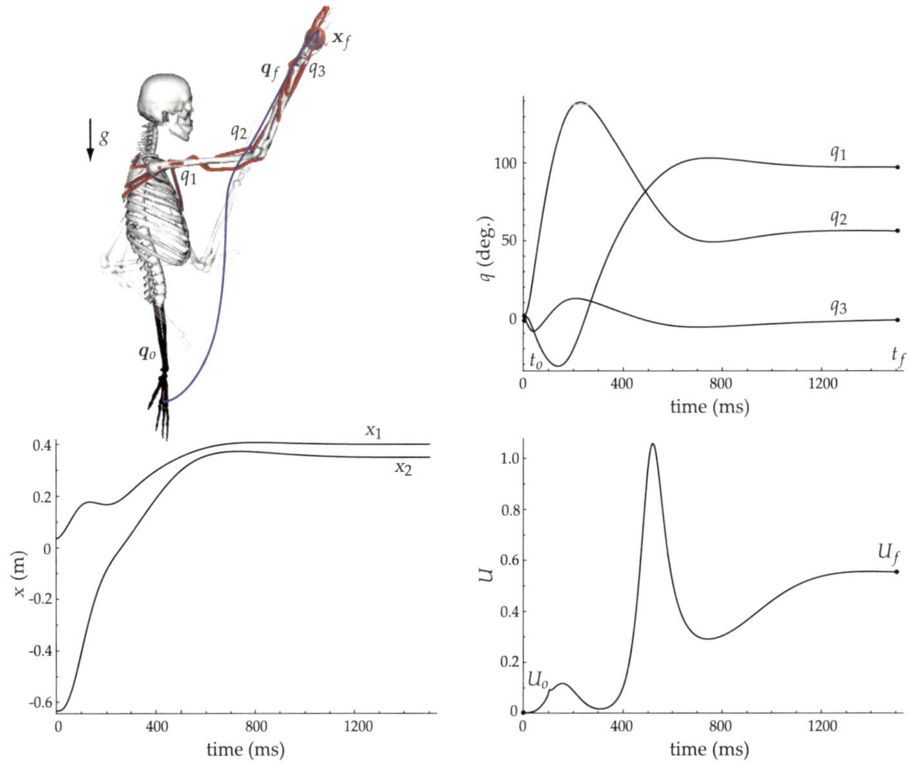

Fig. 13. A redundant muscle-actuated model of the human arm. Initial and final configurations, $\boldsymbol{q}(t_o)$ and $\boldsymbol{q}(t_f)$, associated with gradient descent movement to a target, $\mathbf{x}_f$, are shown. (Top) Time history of the arm motion to the target. Motion corresponds to gradient descent of the muscle effort, subject to the task requirement. (Bottom) Time history of hand trajectory and muscle effort criterion, $U(\boldsymbol{q}) = \boldsymbol{g}^T(\boldsymbol{B}^T\boldsymbol{B})^{-1}\boldsymbol{g}$, associated with gradient descent for human arm model. The null space control seeks to reduce the muscle effort but is also constrained by the task requirement.

where l_{o_i} represents the optimal fiber length for the ith muscle and v_{o_i} represents the maximum contraction velocity for the ith muscle.
Under task constraints the system which extremizes the muscle action is given by,

$$\boldsymbol{\lambda}^\star = \alpha(\mathbf{x}_f - \mathbf{x}) - \beta\dot{\mathbf{x}}, \tag{92}$$

$$\boldsymbol{\lambda} = \boldsymbol{\Lambda}_U\boldsymbol{\lambda}^\star + \boldsymbol{\mu}_U + \boldsymbol{p}_U. \tag{93}$$

and (82). The solution yields the muscle action extremizing path between configurations $\boldsymbol{q}(t_o)$ and $\boldsymbol{q}(t_f)$, given the hand target constraint. Fig. 14 displays time histories of joint motion, hand motion, and muscle effort for a simulation run. The straight line motion of the hand results from the feedback linearization employed.

4. Task/posture control for neural prosthetics

If we return to our initial description of the human motor system depicted in Fig. 1 we can add an outer loop associated with the high-level task reasoning and planning functions of

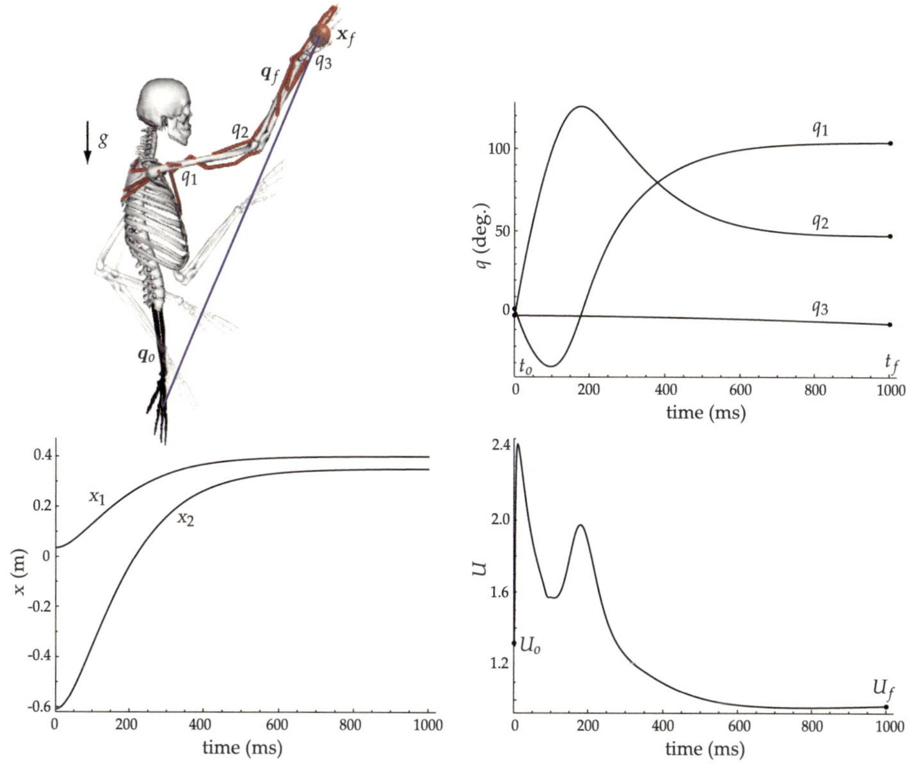

Fig. 14. (Top) Time history of the arm motion between configurations $\boldsymbol{q}(t_o)$ and $\boldsymbol{q}(t_f)$. Motion corresponds to extremization of the muscle effort action integral. (Bottom) Time history of hand trajectory for human arm model and time history of muscle effort criterion associated with extremizing the action integral of (91).

the brain. This is depicted in Fig. 15. In this abstraction motion control is divided into a task generative phase and a motor execution phase. The abstraction depicted in Fig. 15 has relevance not only to the basic understanding of the biomechanics and control of movement but also to the design of engineered systems that augment physiological systems.

Neural prosthetics and brain-computer interfaces have emerged as compelling technologies for the inference of cognitive motor intent using neuroimaging techniques. These techniques can be invasive, as in the case of a brain implant, or non-invasive, as in the case of electroencephalography (EEG). In either case the goal of these techniques is to restore or augment a degree of motor functionality to an individual. This is accomplished through the prediction of motor intent, based on inference from neuroimaging data, and subsequent realization of that intent through a robotic prosthesis. This inference involves decoding the neural encoding manifested in the neuroimaging data. As referenced earlier, current research suggests a task-oriented spatial encoding of motor intent. Based on this premise exciting work has been done to control robotic devices by decoding motor intent.

Current breakthroughs in motor-based brain-computer interfaces can be furthered by the implementation of more sophisticated control theoretic algorithms. Using existing invasive or non-invasive neuroimaging techniques it is believed that the performance of computer controlled robotic devices can be enhanced using a task/posture control framework where, in addition to the inference of task-oriented objectives, postural control objectives can also

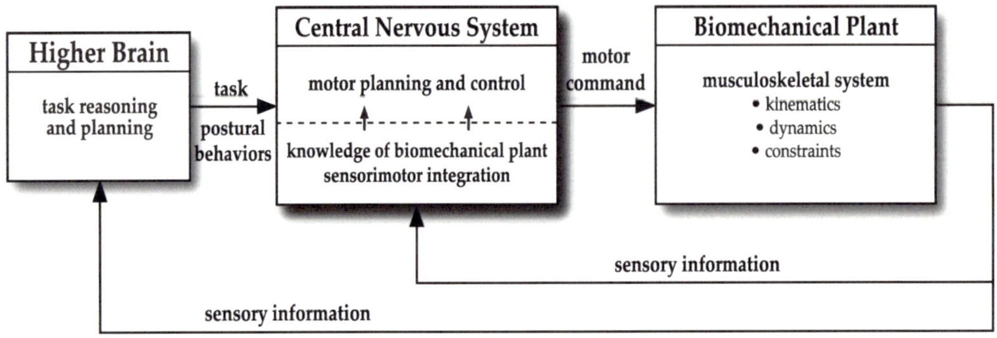

Fig. 15. An outer loop represents the high-level task reasoning and planning functions of the brain. This feeds into the lower-level motor control functions involving the task-driven action of the central nervous system (CNS) on the biomechanical plant.

be inferred from the neuroimgaing data and used as the control reference for the robotic prosthesis. Some of the approaches presented in the previous sections are relevant to the realization of such a neural-based task/posture control framework, as depicted in Fig. 16.

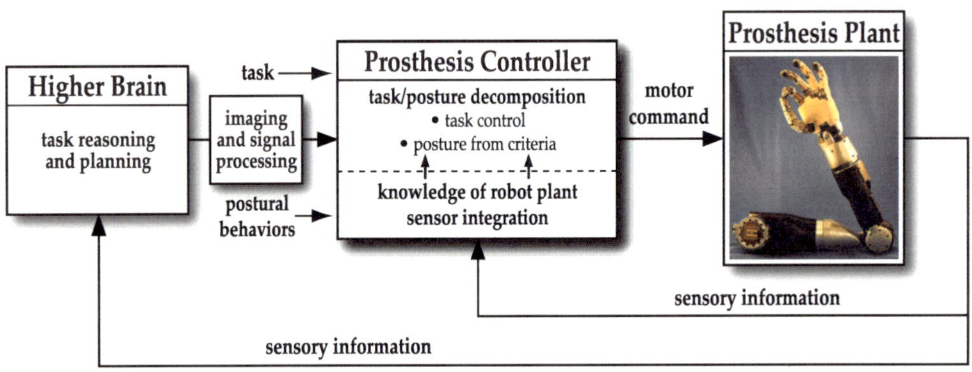

Fig. 16. Task and postural motion intent is inferred from the brain using neuroimaging technologies. The prosthesis controller realizes this intent using a task/posture decomposition. Ultimately, the motor commands are used to control a robotic prosthesis (robot prosthesis image courtesy of DARPA).

Such a framework would involve two principal components: (1) the application of existing signal processing and machine learning methods to the inference of both task-level motor intent as well as postural intent/behavior from neuroimaging data, and (2) control system design and implementation to realize the inferred motor intent on a robotic prosthesis. To complement the neuroimaging data both neuromuscular data in the form of electromyography (EMG) measurements, as well as computational neuromuscular models can be employed in such a framework. This would allow inference and synthesis of control laws based on neuromuscular criteria such as the minimization of neuromuscular effort, etc.

5. Conclusion

A framework has been presented for the analysis and synthesis of human motion through the management of motion tasks, physical constraints, and neuromuscular criteria. The constituents of this framework include a task-level control methodology for constrained systems as well as a muscle effort criterion for the prediction of postures. The constrained task-level control methodology presented exploits the symmetry between task-level control and constrained dynamics. This approach can be applied to the motion control of systems with persistent holonomic constraints as well as to the motion control of systems which undergo intermittent contact with the environment, as in locomotive biomechanical and robotic systems which make intermittent ground contact.

With regard to posture synthesis a posture-based muscle effort criterion for predicting upper limb motion has been implemented. This criterion characterizes effort expenditure in terms of musculoskeletal parameters, rather than just skeletal parameters as with many previous criteria. As with any posture-based model this one is based upon the assumption of Donders' Law. In other words, the final arm configuration is assumed to be independent of initial or prior arm configurations and is only dependent on hand position (the control variable) and the instantaneous physiological criterion. Good correlation between natural reaching postures and those predicted by the proposed posture-based muscle effort criterion have been shown De Sapio, Warren & Khatib (2006); Khatib et al. (2009). Additionally, an analytical procedure has been outlined for the analysis of trajectory-based effort minimization using gradient descent and least action methods. We have also outlined how our task/posture approach might be employed in neural prosthetics and brain-computer interfaces.

6. References

Anderson, F. C. & Pandy, M. G. (2001). Static and dynamic optimization solutions for gait are practically equivalent, *Journal of Biomechanics* 34(2): 153–161.

Buneo, C. A., Jarvis, M. R., Batista, A. P. & Andersen, R. A. (2002). Direct visuomotor transformations for reaching, *Nature* 416: 632–636.

Crowninshield, R. & Brand, R. (1981). A physiologically based criterion of muscle force prediction in locomotion, *Journal of Biomechanics* 14: 793–801.

de Groot, J. H. & Brand, R. (2001). A three-dimensional regression model of the shoulder rhythm, *IEEE Transactions on Biomedical Engineering* 16: 735–743.

De Sapio, V. (2011). Task-level control of motion and constraint forces in holonomically constrained robotic systems, *Proceedings of the 18th World Congress of the International Federation of Automatic Control*. To appear.

De Sapio, V., Holzbaur, K. & Khatib, O. (2006). The control of kinematically constrained shoulder complexes: Physiological and humanoid examples, *Proceedings of the 2006 IEEE International Conference on Robotics and Automation*, Vol. 1, IEEE, pp. 2952–2959.

De Sapio, V., Khatib, O. & Delp, S. (2006). Task-level approaches for the control of constrained multibody systems, *Multibody System Dynamics* 16(1): 73–102.

De Sapio, V., Khatib, O. & Delp, S. (2008). Least action principles and their application to constrained and task-level problems in robotics and biomechanics, *Multibody System Dynamics* 19(3): 303–322.

De Sapio, V. & Park, J. (2010). Multitask constrained motion control using a mass-weighted orthogonal decomposition, *Journal of Applied Mechanics* 77(4): 041004 (10 pp.).

De Sapio, V., Warren, J. & Khatib, O. (2006). Predicting reaching postures using a kinematically constrained shoulder model, *in* J. Lenarčič & B. Roth (eds), *On Advances in Robot Kinematics*, first edn, Springer, pp. 209–218.

Hermens, F. & Gielen, S. (2004). Posture-based or trajectory-based movement planning: a comparison of direct and indirect pointing movements, *Experimental Brain Research* 159(3): 340–348.

Holzbaur, K. R. S., Murray, W. M. & Delp, S. L. (2005). A model of the upper extremity for simulating musculoskeletal surgery and analyzing neuromuscular control, *Annals of Biomedical Engineering* 33(6): 829–840.

Kang, T., He, J. & Tillery, S. I. H. (2005). Determining natural arm configuration along a reaching trajectory, *Experimental Brain Research* 167(3): 352–361.

Khatib, O. (1987). A unified approach to motion and force control of robot manipulators: The operational space formulation., *International Journal of Robotics Research* 3(1): 43–53.

Khatib, O. (1995). Inertial properties in robotic manipulation: An object level framework, *International Journal of Robotics Research* 14(1): 19–36.

Khatib, O., Demircan, E., De Sapio, V., Sentis, L., Besier, T. & Delp, S. (2009). Robotics-based synthesis of human motion, *Journal of Physiology - Paris* 103(3–5): 211–219.

Khatib, O., Warren, J., De Sapio, V., & Sentis, L. (2004). Human-like motion from physiologically-based potential energies, *in* J. Lenarčič & C. Galletti (eds), *On Advances in Robot Kinematics*, first edn, Kluwer, pp. 149–163.

Klopčar, N. & Lenarčič, J. (2001). Biomechanical considerations on the design of a humanoid shoulder girdle, *Proceedings of the 2001 IEEE/ASME International Conference on Advanced Intelligent Mechatronics*, Vol. 1, IEEE, pp. 255–259.

Lacquaniti, F. & Soechting, J. F. (1982). Coordination of arm and wrist motion during a reaching task, *Journal of Neuroscience* 2(4): 399–408.

Lenarčič, J., Stanišić, M. M. & Parenti-Castelli, V. (2000). Kinematic design of a humanoid robotic shoulder complex, *Proceedings of the 2000 IEEE International Conference on Robotics and Automation*, Vol. 1, IEEE, pp. 27–32.

Murty, K. G. (2006). A new practically efficient interior point method for lp, *Algorithmic Operations Research* 1: 3–19.

Murty, K. G. (2010a). Quadratic programming models, *Optimization for Decision Making*, Springer Verlag, pp. 445–476.

Murty, K. G. (2010b). Sphere methods for lp, *Algorithmic Operations Research* 5: 21–33.

Sabes, P. N. (2000). The planning and control of reaching movements, *Current Opinion in Neurobiology* 10: 740–746.

Scholz, J. P. & Schöner, G. (1999). The uncontrolled manifold concept: identifying control variables for a functional task, *Experimental Brain Research* 126: 289–306.

Shenoy, K. V., Meeker, D., Cao, S., Kureshi, S. A., Pesaran, B., Buneo, C. A., Batista, A. P., Mitra, P. P., Burdick, J. W. & Andersen, R. A. (2003). Neural prosthetic control signals from plan activity, *NeuroReport* 14: 591–596.

Soechting, J. F., Buneo, C. A., Herrmann, U. & Flanders, M. (1995). Moving effortlessly in three dimensions: Does donders law apply to arm movement?, *Journal of Neuroscience* 15(9): 6271–6280.

Thelen, D. G., Anderson, F. C. & Delp, S. L. (2003). Generating dynamic simulations of movement using computed muscle control, *Journal of Biomechanics* 36: 321–328.

Uno, Y., Kawato, M. & Suzuki, R. (1989). Formation and control of optimal trajectory in human multijoint arm movement, *Biological Cybernetics* 61: 89–101.

Vetter, P., Flash, T. & Wolpert, D. M. (2002). Planning movements in a simple redundant task, *Current Biology* 12(6): 488–491.

Zajac, F. E. (1993). Muscle coordination of movement: a perspective, *Journal of Biomechanics* 26: 109–124.

2

European Braces for Conservative Scoliosis Treatment

Theodoros B. Grivas
Orthopaedic and Spinal Surgeon, Director of Orthopaedics and Trauma Department, "Tzanio" General Hospital of Piraeus, Piraeus
Greece

1. Introduction

Several published articles suggest that an untreated progressive idiopathic scoliosis (IS) curve may present a poor prognosis into adulthood including back pain, pulmonary compromise, cor pulmonale, psychosocial effects, and even death [Rowe 1998, Danielsson et al 2006, Danielsson et al. 2007, Weinstein et al. 1981, Weinstein and Ponsetty 1983, Weinstein et al 2003]. Bracing, even though it hasn't gained complete acceptance, has been the basis of non-operative treatment for IS for nearly 60 years, [Negrini et al. 2009, 2010a,b, Schiller et al. 2010].

The majority of publications in the peer review literature refer to braces used in North America, [Schiller et al 2010], and there is a lack of systematic examination of the braces commonly used in Europe. The aim of this report, based on peer review publications on the issue, is to concisely describe the European braces which are widely used, focusing on their history, design rationale, indications, biomechanics, outcomes and comparison between them. Cheneau Brace, the two Cheneau derivative braces, namely the Rigo System Cheneau and the ScoliOlogiC® "Chêneau light", the Lyonnaise Brace, the Dynamic Derotating Brace (DDB) the TriaC brace, the Sforzesco brace and the Progressive Action Short Brace PASB will be described.

2. Biomechanics of brace action used for conservative treatment in spinal deformity

The brace as a mean of spinal deformity conservative treatment should be based on the following general principles:

1. Prevention of asymmetric compressive forces related to passive posture
2. Reduction of the secondary muscle imbalance
3. Prevention of the lordosing reactive forces (passive posture, repeated forward bending movements)
4. Prevention of asymmetric torsional forces from gait
5. Production of dynamic detorsional forces involving breathing mechanics. [Rigo & Grivas 2010]

Understanding the biomechanics of brace action is most important. The brace applies external corrective forces to the trunk with the aim to halt the curve progression or to correct

it during growth, [The Scoliosis Research Society Brace Manual, Rigo et al. 2006, Grivas et al. 2003, 2010, Negrini et al.2010a] or to avoid further progression of an already established pathological curve in adulthood.
To achieve these goals, rigid supports or elastic bands can be used [Coillard et al. 2003, Wong et al 2008] and braces can be custom-made or prefabricated [Weiss et al 2008, Sankar et al 2007, Wong et 2005a, 2005b].
The spinal correction is accomplished by the application of mechanical forces with the intention to reduce the pathological compression on given parts of the vertebral column (usually the concave side), while increasing it on others, (usually the convex side). This will result in a more symmetrical and natural loading and will make possible proper spinal growth [Lupparelli et al. 2002, Castro 2003, Weiss & Hawes 2004]. It will also prevent progressive degeneration of the spine [Lupparelli et al. 2002, Stokes et al 2006, Stokes 2008].
Although this is an old concept, the theory has been reinforced over time and for IS was recently summarized in the "vicious cycle" hypothesis [Stokes et al 2006], where it is proposed that lateral spinal curvature produces asymmetrical loading of the skeletally immature spine through movement and neuromuscular control, which in turn causes asymmetrical growth and hence progressive wedging deformity. In this respect, the role of the intervertebral discs in the progression of IS and in its possible correction using bracing has also recently been considered [Grivas et al 2006 Grivas et al 2008a]. Conversely, bracing could establish a useful "virtuous cycle", and as a result could lead to gradual reduction of the asymmetry present in scoliosis [Rigo et al 2006, Rigo et al 2008]. In accordance with these theories, a novel concept describing a comprehensive model of IS progression, based on the patho-biomechanics of the deforming "three joint complex" was also recently presented [Grivas et al 2009].
An alternative hypothesis suggests that the use of braces leads to neuro-motor reorganization caused by the changes in external and proprioceptive inputs and movement resulting from the constraint of bracing [Coillard et al 2002, Odermattet al 2003, Negrini et al 2006, Smania et al 2008]. According to this hypothesis, braces are considered the drivers of movement while they increase external and internal bodily sensations. This permanently changes motor behaviours, even when the brace is removed, and can have a long-term effect on bone formation. This hypothesis can be easily applied also at all pathologies and ages; can be considered correct in terms of trunk behaviour and neuro-muscular organization, while its possible effect on growing bone needs further investigation. Two other interesting and significant concepts to explain the actions of the brace have been discussed. One suggests that the brace provides mechanical support to the body (passive component), while the other suggests that the patient pulls his/her body away from pressure sites (active component) to correct the curve. Such divergent theories illustrate the complexity of this problem, but the most important point of brace treatment is to provide the three dimensional correction of the spinal deformity, and methodologies must be developed with this in mind [Negrini & Grivas 2010, Bagnall et 2009].

3. Treatment management principles and outcome description

The analysis of the treatment management principles and outcome description is beyond the scope of this chapter, which describes the European braces in use. However it was considered that it would be very useful to cite them, at least epigrammatically and give to the reader the existing useful references.

Key elements of the recommendations are efficacy and compliance. The latter stem from the planned treatment, as well as from the responses of the patient and family. It is also highly related to behaviour of the treating team.
The recommendations concerning the standards of management of idiopathic scoliosis with bracing, with the aim to increase efficacy and compliance to treatment are extensively described in a recent SOSORT Consensus paper. It is recommended to professionals engaged in patient care to follow the guidelines of this Consensus in their clinical practice. The SOSORT criteria should also be used along with the published criteria for bracing proposed by SRS, [Negrini et al 2009b,e, Thompson and Richards 2008, Richards et al. 2005].
Several other major issues in brace management apart compliance are currently also in discussion, namely, the pressure being applied, the treatment time and the bending radiographs. These topics are also discussed in the recently published editorial on "Scoliosis" Journal Brace Technology Thematic Series by Negrini & Grivas 2010.

4. European braces for conservative scoliosis treatment

4.1 Cheneau brace

Dr Jacques Chêneau built the brace during the 60's. In 1972 the first patient's results were obtained and officially presented in 1979 at Bratislava. Initially the brace was named Cheneau-Toulouse-Munster Brace as well. Now it is accepted and used worldwide. Useful information on the brace and its philosophy can be found in http://cheneau.info. It is a rigid brace providing three-dimensional correction, **Figure 1.** The mechanisms of Chêneau Brace correction are a) passive mechanisms, namely 1) convex to concave tissue transfer, achieved by multiple three-point system acting in 3D, with the aim of curve hypercorrection, 2) elongation and unloading by the "cherry stone" effect, 3) Derotation of the thorax, 4) bending and b) active mechanisms, namely 1) vertebral growth acting as a corrective factor, 2) asymmetrically guided respiratory movements of the rib-cage, 3) repositioning of the spatial arrangement of the trunk muscles to provide their physiological action and 4) anti-gravitational effect, [Kotwicki & Cheneau 2008 a,b]. This brace opens anteriorly. After some modifications made by Dr Jacques Chêneau, since 1996 the brace is divided in 54 zones and provides large free spaces opposite to pressure sites. The hump should be pressed on 1/3 of the surface of apex. The corresponding dodging site involves 4/5 of the surface of the concave side of curve. Each of the remaining two pressure parts of the three-point system presses on 1/5 of the surface of the concave side. They are the apexes of the neighboring curves. Dodging opposite the latter sites allows movements and straightening of the curve in an active way. It is not permitted to hinder any of the three dodging areas, that is, the middle 4/5 of concave side and the 1/3 over and under the apex. Regarding the outcomes of brace application, the Cheneau-Toulouse-Munster brace has been found to decrease the coronal shift forward, the coronal tilt, the axial rotation, and to increase the sagittal shift forward and the sagittal vertebral tilt (3-D correction), [Périé et al. 2001], obtaining an average primary correction 41% (thoracic, lumbar, double) (n = 52 patients) and a long term correction 14.2% thoracic, 9.2% lumbar double curves: 5.5% in thoracic & 5.6% in lumbar, [Hopf & Heine 1985]. In a recent report, at the end of treatment there was an improvement of Cobb angle correction of about 23% and after 5 years there was a stabilization of about 15 % (p value < 0.05). Therefore, based on this study, it could be stated that conservative treatment with Cheneau brace not only stops progression, but it also reverses the scoliotic curve [Cinnella et al. 2009]. The effectiveness of Cheneau brace in the management of IS was also recently analysed in a prospective observational study, [Zaborowska-Sapeta et al 2011]. It reports the results of treatment according to SOSORT and SRS recommendations on 79 patients (58 girls and 21 boys) with progressive IS, treated with Cheneau brace and physiotherapy, with initial Cobb angle between 20 and 45 degrees, no previous brace treatment, Risser 4 or more at the final evaluation and minimum one year follow-up after weaning the brace. Achieving 50 degrees of Cobb angle was considered surgical recommendation. At follow-up 20 patients (25.3%) improved, 18 patients (22.8%) were stable, 31 patients (39.2%) progressed below 50 degrees and 10 patients (12.7%) progressed beyond 50 degrees (2 of these 10 patients progressed beyond 60 degrees). Progression concerned the younger and less skeletally mature patients, [Zaborowska-Sapeta et al 2011].

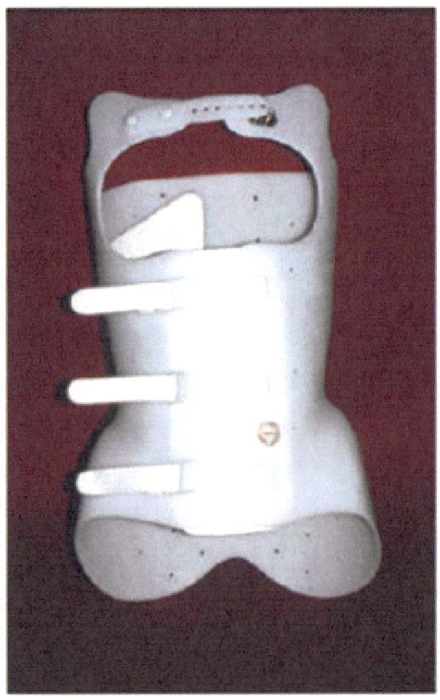

Fig. 1. The Cheneau brace

4.2 Cheneau brace derivatives (Rigo system Cheneau brace, ScoliOlogiC® "Chêneau light" and the Gensingen brace™)

4.2.1 Rigo system cheneau brace

This brace was developed by Dr Manuel Rigo during the early 90s in Instituto Èlena Salvá in Barcelona, Spain. The German - Spanish collaboration for brace production and information on manufacturing can be obtained at: http://www.ortholutions.de/start_english.php. The RSCB, **Figure 2,** is based on the Chêneau Brace, and it is able to produce the required combined forces to correct scoliosis in 3D. The blueprint of the brace is based on the idiopathic scoliosis curve classification correlated with brace treatment introduced by Dr Rigo [Rigo et al 2010a]. The classification includes radiological as well as clinical criteria. The radiological criteria are utilized to differentiate five basic types of curves including: (I) imbalanced thoracic (or three curves pattern), (II) true double (or four curve pattern), (III) balanced thoracic and false double (non 3 non 4), (IV) single lumbar and (V) single thoracolumbar. In addition to the radiological criteria, the Rigo Classification incorporates the curve pattern according to SRS terminology, the balance/imbalance at the transitional point, and L4-5 counter-tilting. The principles of correction of the five basic types of curves are also described by Dr. Rigo, [Rigo et al 2010a]. Biomechanically the RSC brace offers regional derotation. The rib cage and spine are de-rotated. The brace derotates the thoracic section against the lumbar section, with a counter-rotation pad at the upper thoracic region [Rigo and Weiss 2008]. The brace also produces physiological sagittal profile. Initial reports on outcomes using this brace indicated a 31.1% primary Cobb angle correction and 22.2% primary torsion angle correction. At a follow up of 16.8 months 54% of curves were stable, 27% improved and 19% progressed, [Rigo et al 2002]. In patients with long thoracic curves treated with a recently described RSC brace design (three-curve-scoliosis brace with pelvis open) there was 76.7 % in-brace Cobb angle correction and 55.9% in-brace axial rotation correction [Rigo & Gallo 2009]. The latter pattern is easy to correct according to the principles and it can not be compared to "Chêneau light" cohort, which in addition contains double curve patterns which correct least [Weiss et al 2007].

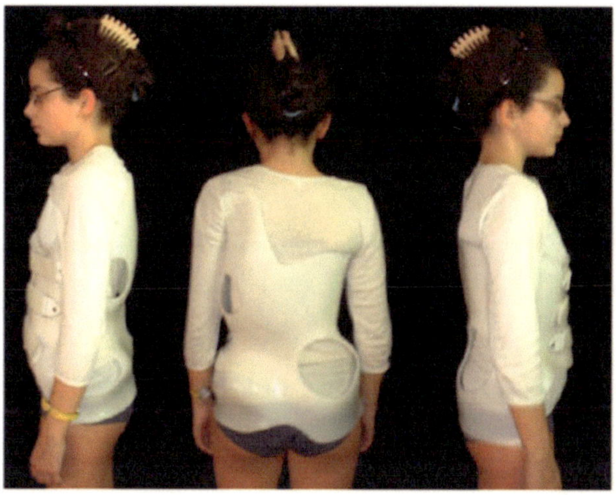

Fig. 2. Rigo System Cheneau Brace

4.2.2 ScoliOlogiC® "Chêneau light"

The brace, **Figure 3**, was invented by Dr. Hans-Rudolf Weiss. The application for the patent was presented in April 2005 and the first braces were built in May 2005. Useful information on the brace can be obtained in http://www.koob-scolitech.com/scoliologic.php and [Weiss et al 2007 and Weiss & Werkmann 2010].

The ScoliOlogiC® off the shelf bracing system enables the CPO to construct a light brace for scoliosis correction from a variety of pattern specific shells to be connected to an anterior and a posterior upright. This brace, when finally adjusted is called Chêneau light™ brace. The advantage of this new bracing system is that the brace is available immediately, is easily adjustable and that it can also be easily modified. This avoids construction periods of sometimes more than 6 weeks, where the curve may drastically increase during periods of fast growth. The disadvantage of this bracing system is that there is a wide variability of possibilities to arrange the different shells during adjustment, [Weiss & Werkmann 2010].

Weiss et al, 2007, reported 51% correction of Cobb angle (Cobb angle in the whole group of patients was reduced by an average of 16,4 degrees), 62 % correction for lumbar & thoracolumbar curve pattern, 36 % correction for thoracic scoliosis and 50 % correction for double major curve pattern. The correction effect correlated negatively with age (r = -0,24; p = 0,014), negatively with the Risser stage (-0,29; p = 0,0096) and negatively with Cobb angle before treatment (r = -0,43; p < 0,0001) [Weiss et al 2007].

The reduction of material in the Chêneau light® brace seems to increase patient's comfort and reduces the stress patients may suffer from whilst in the brace. 80% of the adolescent population of scoliosis patients can be braced with the Chêneau light™ brace. In certain patterns of curvature and in the younger population with an age of less than 11 years, other approaches have to be used, such as plaster based bracing or the application of CAD/CAM based orthoses [Weiss & Werkmann 2010].

There are blueprints to build a RSC® or a Chêneau light® brace according to the conservative treatment of AIS classification by Dr M. Rigo and Dr. HR Weiss [Rigo & Weiss 2008].

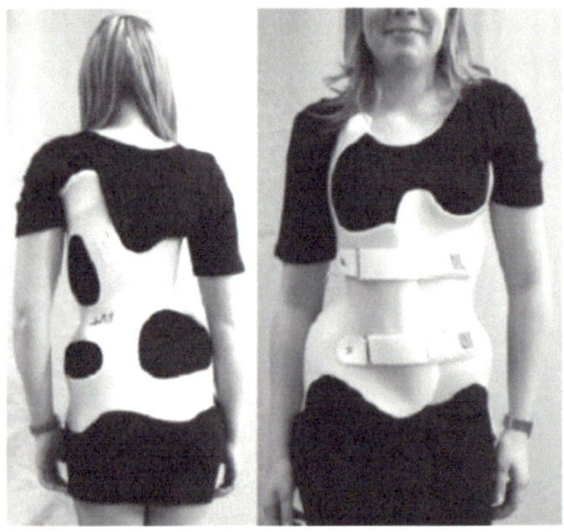

Fig. 3. ScoliOlogiC® "Chêneau light"

4.2.3 The Gensingen brace™

From the experience obtained through the Chêneau light® brace a new CAD/CAM brace has been designed and applied by Dr. Weiss since 2009 which is called the Gensingen brace®. It is extensively described, [Weiss 2010a,b]

This is a new asymmetric Chêneau style CAD/CAM derivate. It has been designed to overcome some problems the designer experienced with other Chêneau CAD/CAM systems over his medical praxis during the recent years, Figure 4.

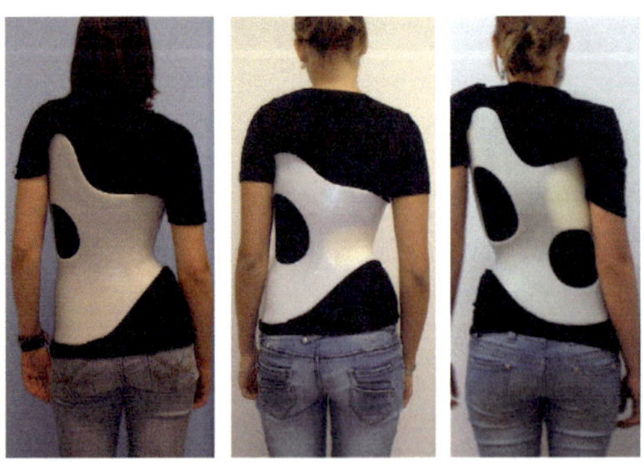

Fig. 4. Different Gensingen braces™ on the patient's body, from Weiss 2010b].

The Gensingen brace™ is adjusted according to the same principles of correction as the Chêneau light™ brace, therefore similar results in both brace types are expected, [Weiss 2010b].
As it is reported, the majority of patients treated by Dr Weiss choose the Chêneau light™ brace. However the Gensingen brace™ is used in curvature patterns a Chêneau light™ brace is not available for, or for curvatures exceeding 50°. Therefore, a direct comparison of the results achieved with both brace types will not be possible in the near future. The so far documented results are based on case reports showing sufficient in-brace corrections in certain curve patterns and in bigger curves as well, [Weiss 2010b].
According to the patients' reports the Gensingen brace™ is comfortable to wear, when adjusted properly, [Weiss 2010b].

4.3 Lyonnaise brace

It is an adjustable rigid brace, without any collar, Figure 5. The Lyon Brace was created by Pierre Stagnara in 1947. Allègre and Lecante modified it to its present form using aluminium bars and plexidur (a high rigidity material) in 1958.
The brace features several characteristics in order to allow for the child's growth of up to seven centimeters and increase in weight of seven kilograms. It is active because of the rigidity of the PMM (polymetacrylate of methyl) structure. The child's body shape is stimulated and the active axial auto correction decreases the pressures of the valve on the trunk. It is decompressive due to the effect of extension between the two pelvic and scapular girdles which decreases the pressure on the intervertebral disc and allows a better effectiveness of the pushes in the other planes. It is symmetrical and additionally to the aesthetic aspect, the brace is easier to build. It is stable and its stability of both the shoulder and pelvic girdles facilitates the intermediate 3D corrections. It is transparent and usually, it is not necessary to use "pads"; so the pressure of the shells on the skin can be directly controlled, [de Mauroy et al. 2011].

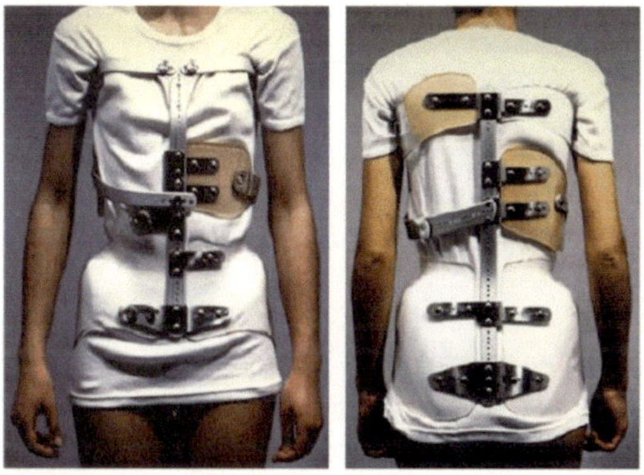

Fig. 5. The Lyonnaise Brace

The bars of the brace are made of radio see-through duralumin, the faceplate and joint of high steel and the thermo malleable plastic is made of polymetacrylate of methyl. The treatment using Lyonnaise Brace is based on two main principles of treatment. An initial plaster cast to stretch the deep ligaments before the application of Lyon brace and the subsequent application of the adjustable brace. The blueprint is designed according to

Lenke's idiopathic scoliosis classification and there are 14 design types, figure 6. The indications for this brace are scoliotics 11-15 years old. It is not applied earlier to prevent tubular deformation of the thorax. The reported results detail an effectivity index (results of 1338 scoliosis treated in France and in Italy based on SRS - SOSORT treatment criteria 2 years after the weaning of the brace) 0,97 for lumbar curve, 0,88 for thoraco-lumbar curve and 0,80 for thoracic curve. The Cobb angle correction is reported for thoracic (n=285 cases) correction 12%, double major (n=351 cases) 10% and 25% respectively, thoraco-lumbar (n=279 cases) 24%, lumbar (n= 450 cases) 36%. Results are also obtained on cosmesis (hump in mm). The rib hump is better corrected than the Cobb angle, which is reduced by 1/3 at the thoracic level and by more than 50% at the lumbar level. The esthetical aspect is always better than the radiographs. In 1338 treated scoliotics, 67.19 % improved, 27.80 % were stable and 5.00 % deteriorated, [de Mauroy et al 2008a,b].

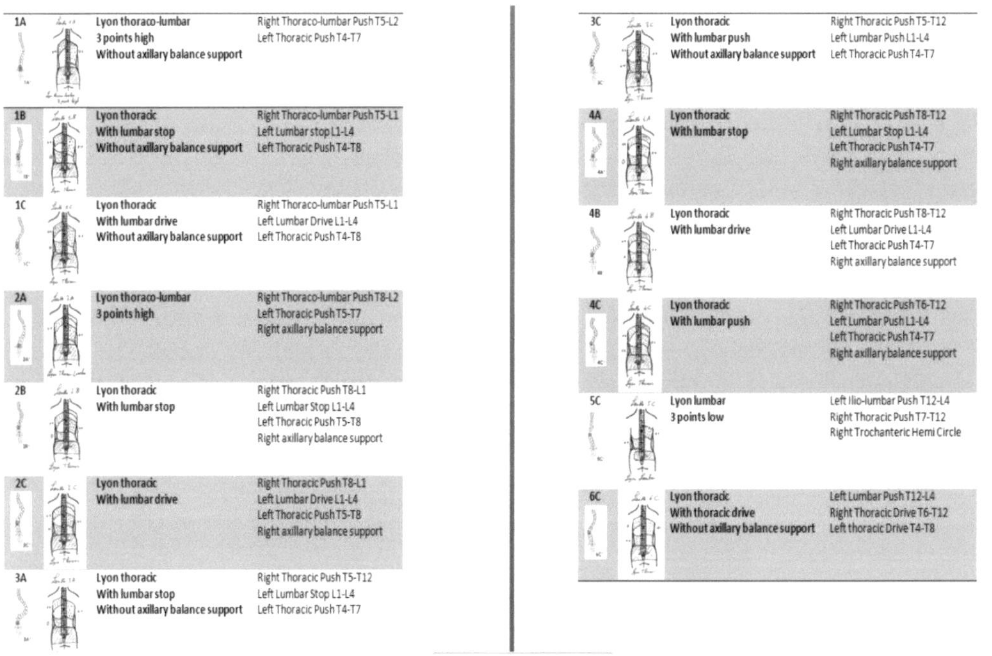

Fig. 6. Blueprints corresponding to 14 types of the Lenke's classification for a right thoracic and left lumbar scoliosis

4.4 Dynamic Derotating Brace (DDB)

The dynamic derotation brace (DDB) was designed in Greece in 1982, as a modification of the Boston brace. It is a custom-made, underarm spinal orthosis featuring aluminium blades set to produce derotating and anti-rotating effects on the thorax and trunk of patients with scoliosis. It is indicated for the non-operative correction of most curves, barring the very high thoracic ones, (when the apex vertebra is T5 or above) [Grivas et al. 2010].

This brace was developed by the late Dr D. Antoniou and Dr J Valavanis at Athens. The first official announcement of Dynamic Derotating Brace (DDB) took place at the 21st common meeting of SRS and BSS, 1986. It is made of polypropylene with a soft foam polyethylene lining, Figure 7a,b. This brace opens posteriorly, [Antoniou et al 1986].

The key feature of the DDB is the addition of the aluminium-made derotating blades posteriorly. These function as a force couple, which is added to the side forces exerted by the brace itself. Corrective forces are also directed through pads. One or more of previously proposed pathomechanical models of scoliosis may underline the corrective function of the DDB: it may act directly on the apical intervertebral disc, effecting correction through the Heuter-Volkman principle; the blades may produce an anti-rotatory element against the deforming "spiral composite muscle trunk rotator"; or it may alter the neuro-motor response by constantly providing new somatosensory input to the patient.

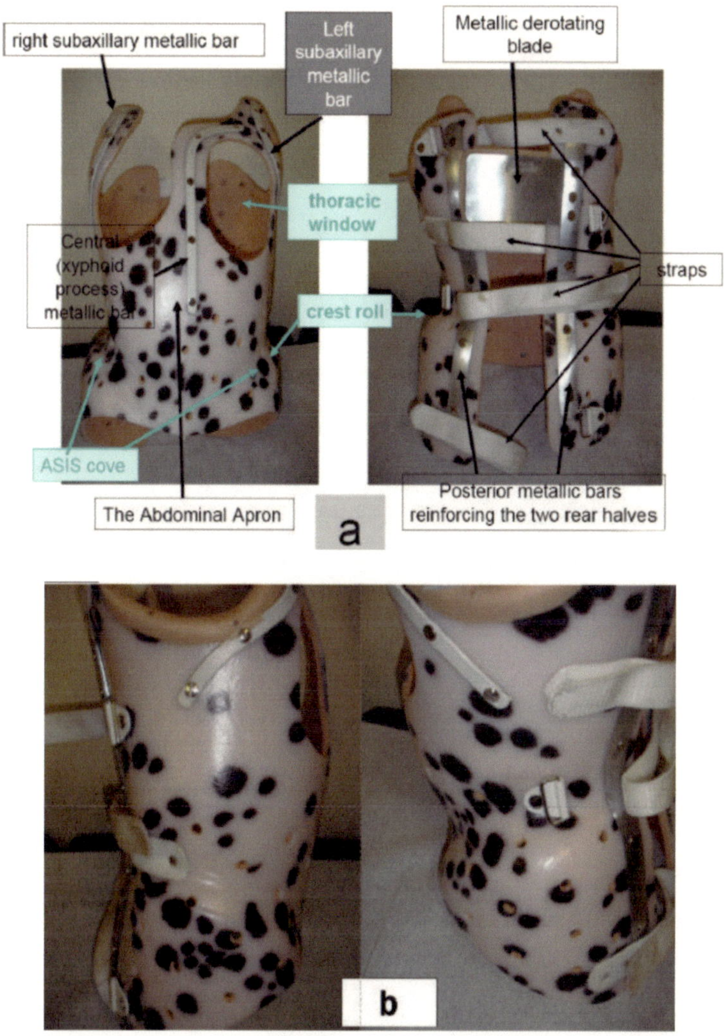

Fig. 7. a,b. The dynamic derotation brace (DDB). (7a: anterior and posterior view), The DDB extends from underneath the axillae to the pelvis, 7b. The dynamic derotation brace (DDB), lateral view of a DDB.

More specifically, in this TLSO type brace, the anti-rotatory blades act as springs - anti-rotatory devices, maintaining constant correcting forces at the pressure areas of the brace and, at the same time, produce movements in opposite directions of the two side-halves of the brace. The de-rotating metal blades are attached to the rear side of the brace corresponding to the most protruding part of the thorax (hump) or the trunk of the patient. They become active when their free ends are located underneath the opposite side of the brace and the brace is tightened by its straps, [Valavanis et al 1995]. The forces applied by the de-rotating blades are added to the side forces exerted by the brace, and changing of the backward angle of the blades can modify them.

There are three main types of DDB designs. The thoracic/thoracolumbar module, whose main indications are thoracic or thoracolumbar curves. It encompasses one or two de-rotatory blades, attached opposite to the thoracic or thoracolumbar hump, figure 8 The lumbar module, used in lumbar curves, is constructed with one de-rotatory blade, located opposite to the lumbar loin hump, figure 9 [Grivas et al. 2010] The double curve module, figure 10, used in patients with double major curves, is supplied with two de-rotatory blades, placed over the thoracic hump and lumbar loin hump each. Each blade acts on the contralateral posterior half of the brace. A major difference of the lumbar curve pattern module with the thoracic/thoracolumbar curve pattern and double major curve pattern modules is the longer trochanteric and the reduced thoracic extension of the former, as seen in figure 6, compared with the later two modules (see figures 8,9 and 10). The positioning of the derotation blade also differs according to the curve pattern module as described. As noted previously, the pads are always placed against the apex of the hump. [Grivas et al. 2010].

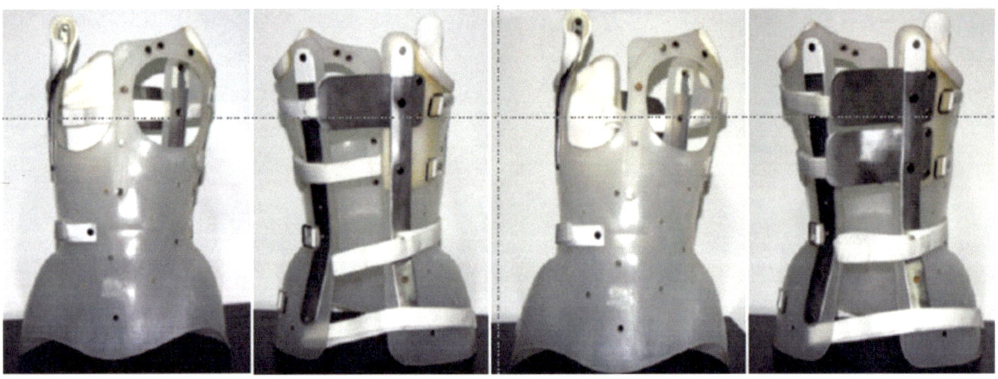

Fig. 8. The dynamic derotation brace (DDB), the thoracic/thoracolumbar module.

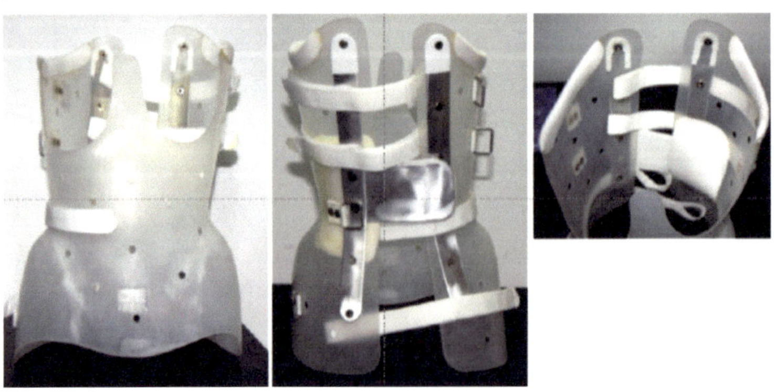

Fig. 9. The dynamic derotation brace (DDB), the Lumbar module.

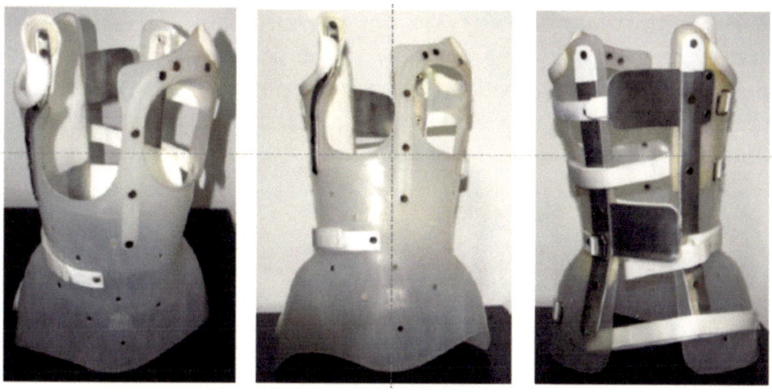

Fig. 10. Dynamic derotation brace (DDB), the double curve module

The conservative treatment of IS using the DDB has shown favorable results. The published outcomes reports detail an overall initial Cobb angle correction of 49.54% and at 2 years follow up a correction of 44.10%, [Andoniou et al. 1992 , Valavanis et al 1995]. It was also reported that the overall 35.70% of curves improved, 46,42% were stable and 7.83% worsened - increased, [Grivas et al 2003]. Thoracic curves appear more resistant to both angular and rotatory correction. As far as the cosmesis is concerned (Angle Trunk Inclination - ATI - hump), DDB improves the cosmetic appearance of the back of IS children with all but right thoracic curves, [Grivas et al 2008b]. Study on quality of life after conservative treatment of AIS using DDB with the Brace Questionnaire (BrQ), which is specific for conservative treatment, revealed an influence on school activity and social functioning, but not on general health perception, physical functioning, emotional functioning, vitality, bodily pain, self-esteem or aesthetics, [Vasiliadis & Grivas 2008, Vasiliadis et al 2006a, Vasiliadis et al 2006b].

The published outcome data on the DDB support the authors' belief that the incorporation of aluminium blades to other orthoses would likely improve their efficacy, [Grivas et al. 2010].

4.5 TriaC brace

This brace was developed by Dr Albert Gerrit Veldhuizen in Nederland. The name TriaC derives from the three C's of Comfort, Control, and Cosmesis. The TriaC orthosis has a flexible coupling module connecting a thoracic and a lumbar part, Figure 11. The TriaC brace exerts a transverse force system, consisting of an anterior progression force counteracted by a posterior force and torque, acts on the vertebrae of a scoliotic spine. In the frontal plane the force system in the TriaC brace is in accordance with the force system of the conventional braces. However, in the sagittal plane the force system only acts in the thoracic region. As a result, there is no pelvic tilt, and it provides flexibility without affecting the correction forces during body motion, [Veldhuizen 1985, Veldhuizen et al 2002]. The introducers suggest that the inclusion criteria are: IS with a Cobb-angle between 20 and 40 degrees, in skeletally immature scoliotics, with Risser 0-1 status, pre-menarche, post-menarche\1 year, in primary thoracic apex between the 7th and 11th thoracic vertebra and primary lumbar apex between the 2nd and 5th lumbar vertebra, in flexible spinal column as evidenced by at least 40% correction on bending films [Bulthuis et al 2008].

Some other studies suggest that the TriaC™-Brace represents an alternative exclusively for the correction of lumbar curves [Zeh et al 2008]. An initial 22% correction is reported for the primary curves within the brace and 35% for the secondary curves. The improvement remained after bracing and in a mean follow up of 1.6 years, as long as it was above a threshold of 20%. In 76% of the patients there was control or net correction of IS curves [Bulthuis et al 2008]. It is stated that the TriaC brace significantly alters the predicted natural history of AIS, [Bulthuis et al 2008].

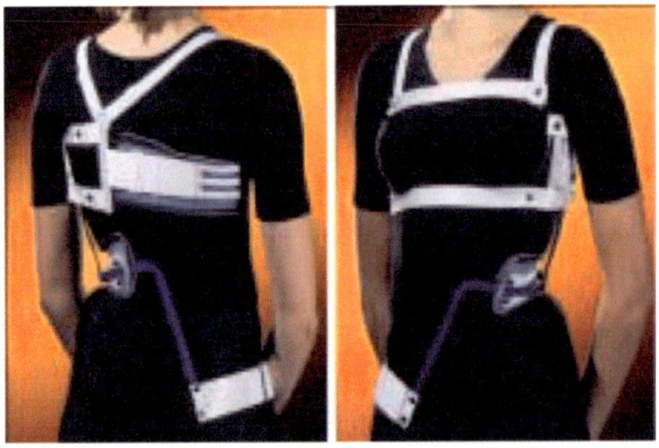

Fig. 11. The TriaC brace

4.6 Sforzesco brace

The Sforzesco brace was developed by Stefano Negrini together with the CPO Gianfranco Marchini in 2004, in Milan, Italy, based on the SPoRT concept (Symmetric, Patient- Oriented, Rigid, Three-Dimensional, Active). The Sforzesco brace combines characteristics of the Risser cast and the Lyon, Chêneau-Sibilla and Milwaukee braces, Figure 12.

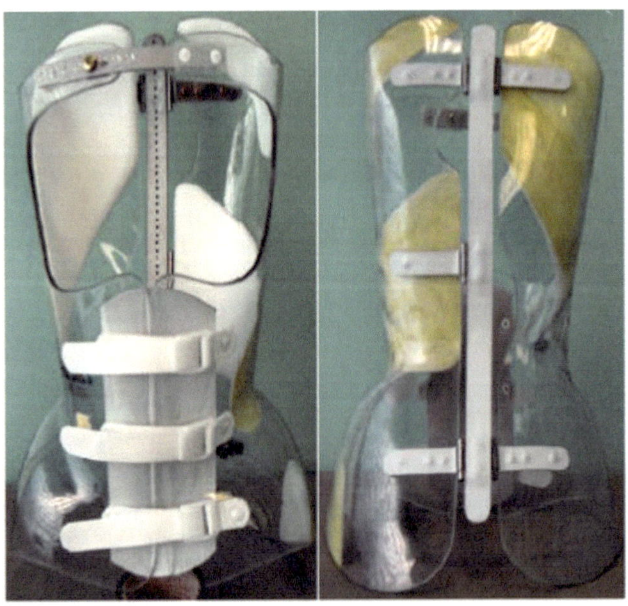

Fig. 12. The Sforzesco brace

Its main action is to push scoliosis from the pelvis up, so to deflex, derotate and restore the sagittal plane (three-dimensional action). For more information please visit http://isico.it/approach/default.htm. Results have been published superior to the Lyon brace [Negrini & Marchini 2007] and similar to the Risser cast with less side-effects [Negrini et al 2008a], making of the Sforzesco brace, according to authors, an instrument for worst cases [Negrini et al 2008a, Negrini et al 2009d]. It is based on the efficacy and acceptability correction principles. 1. Efficacy: a) the active brace: the patient is allowed (encouraged) to move freely, b) mechanical efficacy, achieved through pushes, escapes, stops and drivers (the last being a newly developed concept with this brace) c) versatility and adaptability; d) teamwork: MDs, CPOs, PTs patient & family, e) compliance. 2. Acceptability: a) body design and minimal visibility, b) maximal freedom in the Activities of Daily Life, c) assumption of responsibility and d) a cognitive-behavioural approach. The authors reported results on various outcomes (Cobb degrees and aesthetics) [Negrini et al 2009d, Negrini et al 2009b, Negrini et al 2008b, Negrini et al 2009e, Zaina et al 2009].

4.7 Progressive Action Short Brace (PASB)

The Progressive Action Short Brace (PASB) is used since 1976, for the treatment of thoraco-lumbar and lumbar idiopathic curves. It is a custom-made thoraco-lumbar-sacral orthosis (TLSO) brace of original design, devised by Dr. Lorenzo Aulisa, in Italy, Figure 13. The PASB is only indicated for the treatment of thoraco-lumbar and lumbar curves. The brace is informed by the principle that a constrained spine dynamics can achieve correction of a

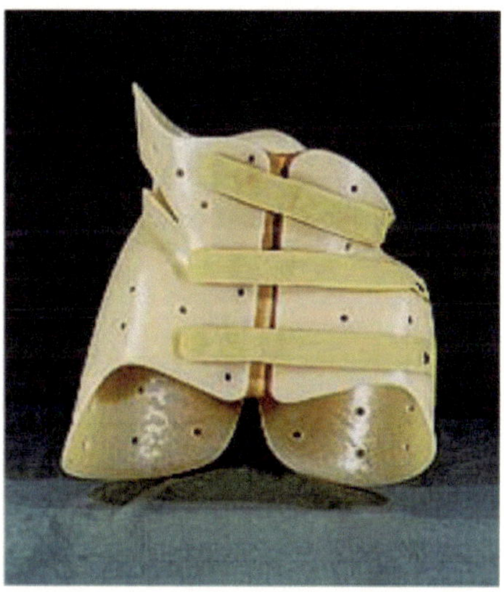

Fig. 13. The Progressive Action Short Brace (PASB)

curve, by inverting the abnormal load distribution during growth. The practical application of the biomechanical principles of the PASB is achieved through two operative phases. A plaster cast phase precedes the brace application. At this stage, external forces are exerted to correct the deformity that is elongation, lateral deflection and derotation. This procedure allows obtaining transversal sections represented by asymmetric ellipsis. The finishing touch of the cast establishes the real geometry of the plastic brace. One or sometimes two casts, in relation to the curve rigidity, are manufactured before switching to the custom-made polypropylene orthosis of the second phase of treatment, [Di Benedetto et al 1981, Aulisa et al 2009]. Aulisa et al, 2009 reported Cobb angle and Perdriole torsion angle readings of the treated thoraco-lumbar and lumbar curves. The pre treatment Cobb mean value was 29,30 degrees ± 5,16 SD and the initial apical rotation 12.70 degrees ± 6,14 SD. The immediate Cobb correction was 14,67 ± 7,65 SD and the apical rotation correction at follow up 8,95 degrees ± 5,82. Overall curve correction was noted at 94% of patients, curve stabilization in 6% of patients, [Aulisa et al 2009].

5. Conclusions

The treatment of adolescent idiopathic scoliosis (AIS) aims to stop the progression of the deformity and to improve the aesthetic appearance, trunk balance and quality of life [Negrini et al 2006]. Several centers in Europe offer full treatment, ranging from prevention (School screening), bracing with or without the use of exercises and surgery. The study and improvement of braces will ultimately improve the outcomes using the specific braces. As far as the conservative treatment with braces is concerned, there is a variety of outcomes reported in literature, [Rigo et al 2003, Maruyama et al 2003, Negrini et al 2008b, Weiss & Goodall 2008, Dolan & Weinstein 2007]. Poor results can be due to poor bracing and this could be verified through in-brace radiographs to assess the obtained correction. Poor results can also be due to improper management of the patient, a factor that can ultimately influence compliance. The latter has not been yet sufficiently stressed in literature despite its critical role in the efficacy of any treatment [Landauer et al. 2003, Negrini et al. 2009a]. Finally the documentation of all the critical aspects (history, design rationale, indications, biomechanics, outcomes and comparison between braces) of the European braces widely used will enable to draw attention to their pros and cons with the final aim not only to improve the braces, but also to offer a better conservative treatment for scoliosis.

6. Acknowledgements

We thank: Dr Jacques Chêneau and Dr Jean Claude de Mauroy for their consensus on the text, Dr Hans-Rudolf Weiss, Dr Stefano Nergini, and Prof. Lorenzo Aulisa and Dr Angelo Aulisa who read the text and made useful suggestions. We also express our appreciation for allowing the use of figures depicting their braces.

7. References

Andoniou D, Valavanis J et al: The effectiveness of our bracing system in the conservative treatment of idiopathic scoliosis. *J Bone Joint Surg*, 74-B. Suppl, I, (1992), 86.

Antoniou D, Valavanis J, Zachariou C, Smyrnis P. 1986 "Dynamic Derotation Brace (DDB). A new aspect for the conservative treatment of Idiopathic Scoliosis" (1986), *Presentation in 21st common meeting of SRS and BSS.*

Aulisa AG, Guzzanti V, Galli M, Perisano C, Falciglia F and Aulisa L. Treatment of thoraco-lumbar curves in adolescent females affected by idiopathic scoliosis with a progressive action short brace (PASB): assessment of results according to the SRS committee on bracing and nonoperative management standardization criteria *Scoliosis,* (2009), 4:21 doi: 10.1186/1748-7161-4-21.

Bagnall KM, Grivas TB, Alos N, Asher M, Aubin CE, Burwell RG, Dangerfield PH, Edouard T, Hill D, Lou E, et al: The International Research Society of Spinal Deformities (IRSSD) and its contribution to science. Scoliosis 2009, 4(1):28.

Bulthuis GJ , Veldhuizen AG, Z Nijenbanning G. Clinical effect of continuous corrective force delivery in the non-operative treatment of idiopathic scoliosis: a prospective cohort study of the triac-brace *Eur Spine J* 17 (2008), 231–239.

Castro FP Jr: Adolescent idiopathic scoliosis, bracing, and the Hueter-Volkmann principle. Spine J 2003, 3(3):180-185.

Cinnella P. Muratore M. Testa E. Bondente P.G. The Treatment of adolescent idiopathic scoliosis with Cheneau brace: long term outcome. *Oral Presentation at Lyon 2009 SOSORT Meeting.*

Coillard C, Leroux MA, Badeaux J, Rivard CH: SPINECOR: a new therapeutic approach for idiopathic scoliosis. Stud Health Technol Inform 2002, 88:215-217.

Coillard C, Leroux MA, Zabjek KF, Rivard CH: SpineCor–a non-rigid brace for the treatment of idiopathic scoliosis: post-treatment results. Eur Spine J 2003, 12(2):141-148.

Danielsson AJ, Romberg K, Nachemson AL. Spinal range of motion, muscle endurance, and back pain and function at least 20 years after fusion or brace treatment for adolescent idiopathic scoliosis: a case-control study. *Spine,* 31 (2006), 275–283.

Danielsson, A.J.; Hasserius, R.; Ohlin, A. & Nachemson A.L. (2007). A prospective study of brace treatment versus observation alone in adolescent idiopathic scoliosis: a follow-up mean of 16 years after maturity. *Spine,* (2007 Sep 15), Vol.32, No.20, pp. 2198-2207, ISSN 0362-2436

de Mauroy JC, Fender P, Tato B, Lusenti P, Ferracane G. Lyon brace. *Stud Health Technol Inform.* 135 (2008), 327-340.

de Mauroy JC, Lecante C, Barral F, Daureu D, Gualerzi S, Gagliano R. The Lyon brace. *Disabil Rehabil Assist Technol.* 3(3) (2008), 139-145.

de Mauroy JC, Lecante C, Barral F (2011) "Brace Technology" Thematic Series –The Lyon approach to the conservative treatment of scoliosis. Scoliosis 2011,

Di Benedetto A, Vinciguerra A, Pennestri' E, Aulisa L: Biomechanics of Scoliosis Using a New Type of Brace. *Proceedings of the 8th Canadian Congress of Applied Mechanics,* Moncton, N.-B., Canada, 7-12 June, (1981), 785-786.

Dolan LA, Weinstein SL: Surgical rates after observation and bracing for adolescent idiopathic scoliosis: an evidence based review. *Spine,* 32(19 Suppl) (2007), S91-S100.

Grivas TB, Vasiliadis E, Chatziargiropoulos T, Polyzois VD, Gatos K: The effect of a modified Boston brace with anti-rotatory blades on the progression of curves in idiopathic scoliosis: aetiologic implications. Pediatr Rehabil 2003, 6(3-4):237-242.

Grivas TB, Vasiliadis E, Malakasis M, Mouzakis V, Segos D: Intervertebral disc biomechanics in the pathogenesis of idiopathic scoliosis. Stud Health Technol Inform 2006, 123:80-83.

Grivas TB, Vasiliadis ES, Rodopoulos G, Bardakos N: The role of the intervertebral disc in correction of scoliotic curves. A theoretical model of idiopathic scoliosis pathogenesis. Stud Health Technol Inform 2008a, 140:33-36.

Grivas TB and Vasiliadis ES. Cosmetic outcome after conservative treatment of idiopathic scoliosis with a dynamic derotation brace. *Stud Health Technol Inform*. 135, 2008b, 387-392.

Grivas T, Vasiliadis ES, Triantafyllopoulos G, Kaspiris A: A comprehensive model of idiopathic scoliosis (IS) progression, based on the pathobiomechanics of the deforming "three joint complex". Scoliosis 2009, 4(Suppl 2):O10.

Grivas TB, Bountis A, Vrasami A, Bardakos NV (2010) Brace technology thematic series: the dynamic derotation brace. Scoliosis 2010, 5:20

Hopf C, Heine J. Long-term results of the conservative treatment of scoliosis using the Chêneau brace. *Z Orthop Ihre Grenzgeb.* 123(3) (1985), 312-322.

Kotwicki T, Cheneau J. Biomechanical action of a corrective brace on thoracic idiopathic scoliosis: Cheneau 2000 orthosis. *Disabil Rehabil Assist Technol*. 3(3), (2008), 146-53.

Kotwicki T, Cheneau J. Passive and active mechanisms of correction of thoracic idiopathic scoliosis with a rigid brace. *Stud Health Technol Inform*. 135 (2008), 320-326.

Landauer F, Wimmer C, Behensky H: Estimating the final outcome of brace treatment for idiopathic thoracic scoliosis at 6-month follow-up. *Pediatr Rehabil* 6(3–4) (2003), 201-207.

Lupparelli S, Pola E, Pitta L, Mazza O, De Santis V, Aulisa L: Biomechanical factors affecting progression of structural scoliotic curves of the spine. Stud Health Technol Inform 2002, 91:81-85.

Maruyama T, Kitagawa T, Takeshita K, Mochizuki K, Nakamura K: Conservative treatment for adolescent idiopathic scoliosis: can it reduce the incidence of surgical treatment? *Pediatr Rehabil* 6(3–4), (2003), 215-219.

Negrini S, Grivas TB, Kotwicki T, Maruyama T, Rigo M, Weiss HR, the members of the Scientific society On Scoliosis Orthopaedic and Rehabilitation Treatment (SOSORT): Why do we treat adolescent idiopathic scoliosis? What we want to obtain and to avoid for our patients. SOSORT 2005 Consensus paper. *Scoliosis,* (2006a), 1:4.

Negrini S, Marchini G, Tomaello L: The Sforzesco brace and SPoRT concept (Symmetric, Patient-oriented, Rigid, Three-dimensional) versus the Lyon brace and 3-point systems for bracing idiopathic scoliosis. Stud Health Technol Inform 2006b, 123:245-249.

Negrini S, Marchini G. Efficacy of the symmetric, patient-oriented, rigid, three-dimensional, active (SPoRT) concept of bracing for scoliosis: a prospective study of the Sforzesco versus Lyon brace. *Eura Medicophys.* 43(2) (2007), 171- 181.

Negrini S, Atanasio S, Negrini F, Zaina F, Marchini G. The Sforzesco brace can replace cast in the correction of adolescent idiopathic scoliosis: A controlled prospective cohort study. Scoliosis. (2008a) Oct 31;3:15.

Negrini S, Atanasio S, Zaina F, Romano M, Parzini S, Negrini A. End-growth results of bracing and exercises for adolescent idiopathic scoliosis. Prospective worst-case analysis. Stud Health Technol Inform. 135 (2008b), 395-408.

Negrini S, Grivas TB, Kotwicki T, Rigo M, Zaina F and the international Society on Scoliosis Orthopaedic and Rehabilitation Treatment (SOSORT). Guidelines on "Standards of management of idiopathic scoliosis with corrective braces in everyday clinics and in clinical research": SOSORT Consensus 2008. Scoliosis 2009a, 4:2 doi:10.1186/1748-7161-4-2.

Negrini S, Atanasio S, Fusco C and Zaina F. Efficacy of conservative treatment of adolescent idiopathic scoliosis: end-growth results respecting SRS and SOSORT criteria. Scoliosis, 2009b, 4(Suppl 2):O48.

Negrini S, Atanasio S, Fusco C and Zaina F. Efficacy of bracing immediately after the end of growth: final results of a retrospective case series. Scoliosis 2009c, 4(Suppl 2):O49.

Negrini S, Atanasio S, Fusco C and Zaina F. Efficacy of bracing in worst cases (over 45°) end-growth results of a retrospective case-series. Scoliosis 2009d, 4(Suppl 2):O50

Negrini S, Atanasio S, Fusco C, Zaina F. Effectiveness of complete conservative treatment for adolescent idiopathic scoliosis (bracing and exercises) based on SOSORT management criteria: results according to the SRS criteria for bracing studies - SOSORT Award 2009 Winner. Scoliosis. 2009e Sep 4;4:19.

Negrini S, Minozzi S, Bettany-Saltikov J, Zaina F, Chockalingam N, Grivas TB, Kotwicki T, Maruyama T, Romano M, Vasiliadis ES. Braces for idiopathic scoliosis in adolescents. Cochrane Database of Systematic Reviews 2009f, Issue 1.

Negrini S, Minozzi S, Bettany-Saltikov J, Zaina F, Chockalingam N, Grivas TB, Kotwicki T, Maruyama T, Romano M, Vasiliadis ES: Braces for idiopathic scoliosis in adolescents. Cochrane Database Syst Rev 2010a, 1: CD006850.

Negrini S, Minozzi S, Bettany-Saltikov J, Zaina F, Chockalingam N, Grivas TB, Kotwicki T, Maruyama T, Romano M, Vasiliadis ES (2010b): Braces for Idiopathic Scoliosis in Adolescents. Spine (Phila Pa 1976). 2010b Jun 1;35(13):1285-93. Review.

Negrini S and Grivas TB (2010) Introduction to the "Scoliosis" Journal Brace Technology Thematic Series: increasing existing knowledge and promoting future developments Scoliosis 2010, 5:2

Odermatt D, Mathieu PA, Beausejour M, Labelle H, Aubin CE: Electromyography of scoliotic patients treated with a brace. J Orthop Res 2003, 21(5):931-936.

Périé D, Sales De Gauzy J, Sévely A, Hobatho MC. In vivo geometrical evaluation of Cheneau-Toulouse-Munster brace effect on scoliotic spine using MRI method. Clin Biomech (Bristol, Avon). 16(2) (2001), 129-137.

Richards BS, Bernstein RM, D'Amato CR, Thompson GH: Standardization of criteria for adolescent idiopathic scoliosis brace studies: SRS Committee on Bracing and Nonoperative Management. Spine 2005, 30(18):2068-2075, discussion 2076-2067.

Rigo M, Quera-Salvá G, Puigdevall N, Martínez M. et al 2002 (Retrospective results in immature idiopathic scoliotic patients treated with a Chêneau brace. *Stud Health Technol Inform.* 88 (2002), 241-245.

Rigo M, Reiter C, Weiss HR: Effect of conservative management on the prevalence of surgery in patients with adolescent idiopathic scoliosis. *Pediatr Rehabil* (2003), 6(3–4):209-214.

Rigo M, Negrini S, Weiss H, Grivas T, Maruyama T, Kotwicki T: SOSORT consensus paper on brace action: TLSO biomechanics of correction (investigating the rationale for force vector selection). Scoliosis 2006, 1:11.

Rigo MD, Weiss HR 2008 The Chêneau concept of bracing-biomechanical aspects. *Stud Health Technol Inform.* 135 (2008), 303-319

Rigo M and Gallo D. 2009 A new RSC brace design to treat single long thoracic scoliosis. Comparison of the in-brace correction in two groups treated with the new and the classical models. *Oral Presentation at Lyon 2009 SOSORT Meeting.*

Rigo M, Villagrasa M and Gallo D (2010a) A specific scoliosis classification correlating with brace treatment: description and reliability. *Scoliosis* 2010a, 5:1

Rigo M and Grivas TB (2010b) 'Rehabilitation schools for scoliosis' thematic series: describing the methods and results. Scoliosis 2010b, 5:27 doi:10.1186/1748-7161-5-27

Rowe DE. The Scoliosis Research Society Brace Manual. Milwaukee, WI: *Scoliosis Research Society*; (1998), 1–9.

Sankar WN, Albrektson J, Lerman L, Tolo VT, Skaggs DL: Scoliosis in-brace curve correction and patient preference of CAD/CAM versus plaster molded TLSOs. J Child Orthop 2007, 1(6):345-349.

Schiller JR, Thakur NA, Eberson CP Brace Management in Adolescent Idiopathic Scoliosis. *Clin Orthop Relat Res.* 2010 Mar; 468(3):670-8. Epub 2009 May 30PMID: 19484317 .

Smania N, Picelli A, Romano M, Negrini S: Neurophysiological basis of rehabilitation of adolescent idiopathic scoliosis. Disabil Rehabil 2008, 30(10):763-771.

Stokes IA, Burwell RG, Dangerfield PH: Biomechanical spinal growth modulation and progressive adolescent scoliosis - a test of the 'vicious cycle' pathogenetic hypothesis: Summary of an electronic focus group debate of the IBSE. Scoliosis 2006, 1:16.

Stokes IA: Mechanical modulation of spinal growth and progression of adolescent scoliosis. Stud Health Technol Inform 2008, 135:75-83.

The Scoliosis Research Society Brace Manual. Introduction. http://www.srs.org/professionals/bracing_manuals/section1.pdf.

Thompson GH, Richards BS III: Inclusion and assessment criteria for conservative scoliosis treatment. Stud Health Technol Inform 2008, 135:157-163.

Valavanis J, Bountis A, Zachariou C, Kokkonis D, Anagnostou D, Giahos D, Daskalakis E. Three-Dimensional Brace Treatment for Idiopathic Scoliosis. In: *Three Dimensional Analysis of Spinal Deformities* M D'Amico et al (Eds.) IOS Press, (1995), 337-341.

Vasiliadis E and Grivas TB. Quality of life after conservative treatment of adolescent idiopathic scoliosis. *Stud Health Technol Inform.* 135 (2008), 409-413.

Vasiliadis E, Grivas TB, Gkoltsiou K. Development and preliminary validation of Brace Questionnaire (BrQ): a new instrument for measuring quality of life of brace treated scoliotics. *Scoliosis.* (2006) May 20;1:7.

Vasiliadis E, Grivas TB, Savvidou O, Triantafyllopoulos G. The influence of brace on quality of life of adolescents with idiopathic scoliosis. *Stud Health Technol Inform.* 123 (2006), 352-356.

Veldhuizen AG. Idiopathic Scoliosis. A Biomechanical and Functional Anatomical Study. *Thesis, University of Groningen.* Groningen, the Netherlands: (1985).

Veldhuizen AG, Cheung J, Bulthuis GJ, Nijenbanning G. A new orthotic device in the non-operative treatment of idiopathic scoliosis. *Medical Engineering & Physics,* 24 (2002), 209–218.

Weinstein SL, Dolan LA, Spratt KF, Peterson KK, Spoonamore MJ, Ponseti IV. Health and function of patients with untreated idiopathic scoliosis: a 50-year natural history study. *JAMA,* 289 (2003), 559–567.

Weinstein SL, Ponseti IV. Curve progression in idiopathic scoliosis. *J Bone Joint Surg Am.* 65 (1983), 447–455.

Weinstein SL, Zavala DC, Ponseti IV. Idiopathic scoliosis: long-term follow-up and prognosis in untreated patients. *J Bone Joint Surg Am.* 63 (1981), 702–712.

Weiss HR, Hawes MC (2004): Adolescent idiopathic scoliosis, bracing and the Hueter-Volkmann principle. Spine J 2004, 4(4):484-485.

Weiss HR, Werkmann M, Stephan C. Correction effects of the ScoliOlogiC® "Chêneau light" brace in patients with scoliosis. Scoliosis. 2007 Jan 26;2:2.

Weiss HR, Goodall D (2008): The treatment of adolescent idiopathic scoliosis (AIS) according to present evidence. A systematic review. *Eur J Phys Rehabil Med* 44(2) (2008), 177-193.

Weiss HR, Rigo M (2008). The Chêneau concept of bracing--actual standards. *Stud Health Technol Inform.* 135 (2008), 291-302.

Weiss HR: Best Practice in conservative scoliosis care. 3rd. edition, Pflaum, Munich, 2010.

Weiss HR and Werkmann M (2010). "Brace Technology" Thematic Series - The ScoliOlogiC® Chêneau light™ brace in the treatment of scoliosis. *Scoliosis* 2010, 5:19doi:10.1186/1748-7161-5-19.

Weiss RH 2010: "Brace technology" thematic series – the Gensingen brace™ in the treatment of scoliosis. Scoliosis 2010, 5:22

Wong MS, Cheng JC, Lam TP, Ng BK, Sin SW, Lee-Shum SL, Chow DH, Tam SY: The effect of rigid versus flexible spinal orthosis on the clinical efficacy and acceptance of the patients with adolescent idiopathic scoliosis. Spine 2008, 33(12):1360-1365.

Wong MS, Cheng JC, Wong MW, So SF: A work study of the CAD/CAM method and conventional manual method in the fabrication of spinal orthoses for patients with adolescent idiopathic scoliosis. Prosthet Orthot Int 2005a, 29(1):93-104.

Wong MS, Cheng JC, Lo KH: A comparison of treatment effectiveness between the CAD/CAM method and the manual method for managing adolescent idiopathic scoliosis. Prosthet Orthot Int 2005b, 29(1):105-111.

Zaborowska-Sapeta K, Kowalski IM, Kotwicki T, Protasiewicz-Faldowska H, Kiebzak W.(2011) Effectiveness of Cheneau brace treatment for idiopathic scoliosis: prospective study in 79 patients followed to skeletal maturity. Scoliosis. 2011 Jan 25;6(1):2.

Zaina F, Negrini S, Fusco C, Atanasio S. How to improve aesthetics in patients with Adolescent Idiopathic Scoliosis (AIS): a SPoRT brace treatment according to SOSORT management criteria. *Scoliosis.* (2009) Sep 1;4:18.

Zeh A, Planert M, Klimas S, Hein W, Wohlrab D The flexible Triac™-Brace for conservative treatment of idiopathic scoliosis. An alternative treatment option ? *Acta Orthop. Belg*, 74 (2008), 512-521.

3

Motion Preservation and Shock Absorbing in Cervical and Lumbar Spine: A New Device for Anterior Cervical Arthroplasty, for Anterior or Posterior Lumbar Arthroplasty

Giuseppe Maida
Division of Neurosurgery,Departement of Neurosciences and Reabilithation
S.Anna Hospital
School of Medicine, Ferrara University
Italy

1. Introduction

In spine surgery, developing an anatomical, artificial disc prosthesis is one of the most difficult technological goals.
After surgical intervertebral disc removal, many pathologies are possible: to perform a vertebral fusion or a vertebral non fusion, depending on pathology that was treated, and in general, spinal condition in which these pathology were collocated.
While non fusion is the surgical option preferred, disc prosthesis is the preferred device.
Motion preservation, shock absorbing, biocompatibility, minimally invasive surgery for placement, and magnetic resonance imaging (MRI) compatibility, are essential technological aspects to satisfy.
Until now it was very difficult, if not impossible, to guarantee shock absorbing.
Shock absorbing, for these kinds of devices, is probably the most important biomechanical aspect.
A new device was recently ideated to resolve these problems.
Titanium and peek, materials of the devices, guarantee biocompatibility and MRI compatibility, much more resistance.
The particular shape of the device, reproducing a "molla a tazza" ("spring-cup"), the collocation and the alternation of the different materials in the device's construction, guarantee the motion preservation and, most importantly, shock absorption characteristics.
All these elements reduce device dimensions.
Disc prosthesis, in fact, until now was implanted by an anterior surgical approach.
For lumbar surgery, much more than cervical surgery, the anterior approach is potentially a very invasive surgical way.
The possibility to perform a surgical lumbar disc prosthesis placement by a posterior approach is, potentially, a "revolution" in spine surgery, for the minimally invasive spine surgery goal.
After about 40 years of research and development, artificial disc technology is considered a real option in spine surgery.

Motion preservation, load sharing and cushioning are three of the most important aspects of the technology. They make the artificial device similar to the natural intervertebral disc becoming, with spinal kinematics and histologic osseointegration at the prosthetic-bone interface, a very hard research challenge.
Two vertebral bodies and the corresponding intervertebral disc are called Spinal Functional Unit (SFU): is a three-articular-complex, in which translation and rotation are allowed by X-Z-Y axis.
In flexion, extension and lateral bending, there is a variation of the SFU's instantaneous centre of rotation that, by an elliptical instantaneous axis, falls in the posterior half and inferior discal margin (picture 1).

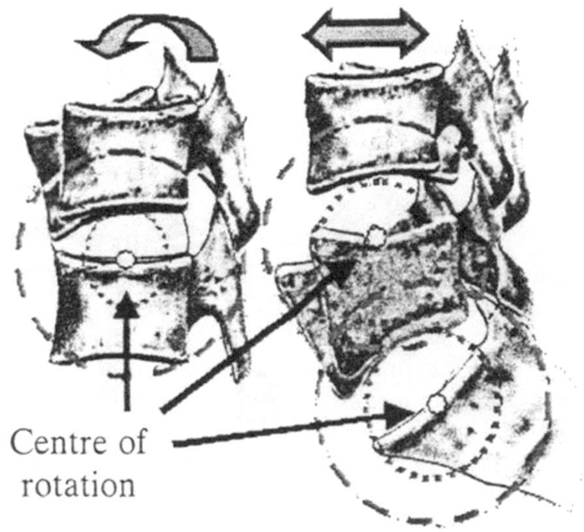

Picture 1. SFU's instantaneous centre of rotation in X-Y-Z axis motion.

The artificial intervertebral disc, in a stand alone way, under ordinary and extraordinary load, has to perform and guarantee all these.
The Kinematical classification of the intervertebral disc prosthesis, based on the own different free motion degree, organizes the devices in 3 categories.

- unconstrained
- constrained, and,
- semiconstrained

Devices in each of the above 3 groups have advantages and disadvantages.
The first have a variable centre of rotation.
They permit anterior/posterior/lateral translation, rotation, guarantee a more physiological centre of rotation, but give more stress to the articular joints.
Prefer the preservation of the posterior longitudinal ligament (PLL) during their collocation, "forgiven" minimally positional imperfections.
The second have a fix centre of rotation.
They permit only rotation, have a less physiological centre of rotation, and give less stress to the articular joints.
Another example of classification is reported in the picture 2.

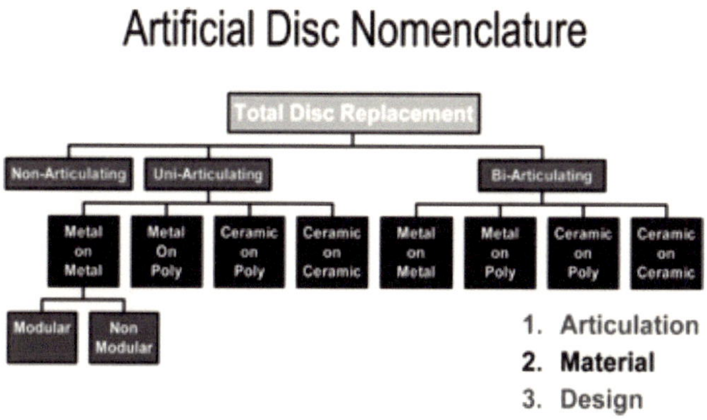

Picture 2. Cervical spine study group, 2004

2. State of art

The surgical approach used, until now, was anterior arthroplasty.
For cervical spine, the anterior extrapharingal-presternocleidomastoideo approach is a very tested and safe way, in reoperation as well (pictures 3, 4).
The incidence of cervical intraoperative complications (Haematoma, dysphagia, dysphonia), are 6.2% ; late complications (ossification, dislocation) about 5.2%.
In cervical spine it is not difficult to convert, if needed, even in long follow-ups, an arthrtoplasty to fusion.

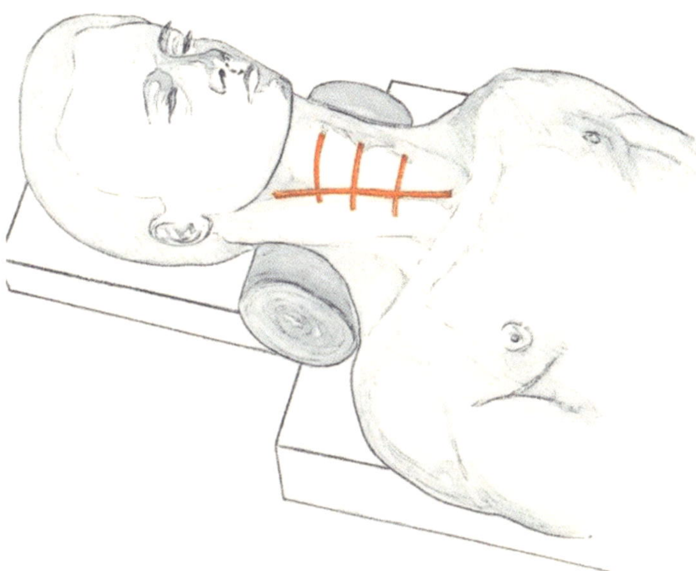

Picture 3. Cervical approach, position and skin incisions

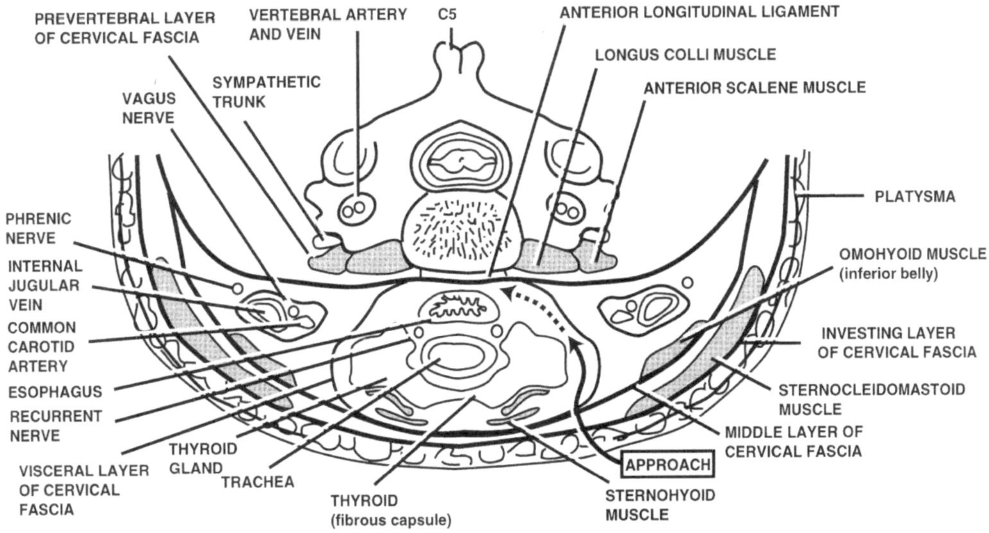

Picture 4. Cervical approach, surgical way

For lumbar spine, the anterior trans-peritoneal and retroperitoneal approach (most used) are less safe (more in reoperations), because of gross vesselles and and nervous plexis, attached to the spine: even slight dislocation, sometime, may causes neurological damages or important vascular bleeding.

This needs a long learning curve, doing the operations with the collaboration of general or vascular surgeons.

The conversion arthroplasty-fusion may be very hard in lumbar spine.

The heterotopic ossification, with consequent block of the prosthesis and intervertebral fusion, increases from 12% to 17.8%: perioperative non steroid anti-inflammatory therapy (FANS) and intraoperative "wash-out", works out to be the best prevention.

The titanium alloys and the ceramics guarantee a better MRI compatibility than cobalt-chromium alloys or steel.

The biocompatibility is very important for post-operative controls, better if MR fast spin echo imaging (than T2 imaging)

These indications in disc herniation, after surgical disc removal, in alternative to anterior intervertebral fusion, preferring one single level, no more than two levels, from C4 to C6, from L4 to S1, not responding to conservative treatments, if there is a good pre-operative range of motion (ROM) to ward after the operation, for a long time and articular joints preservation (testing by MRI, bone CT scan and dynamic RX film), to avoid adjacent-segment degeneration in a long post-operative follow-up, without pre-operative bone fractures, bone tumours, bone deformities or bone infections.

Degenerative discopathy, about ten years ago, in USA, was estimated $50 billion of annual health costs.

Arthroplasty, as alternative to fusion, was said before (pictures 5, 6).

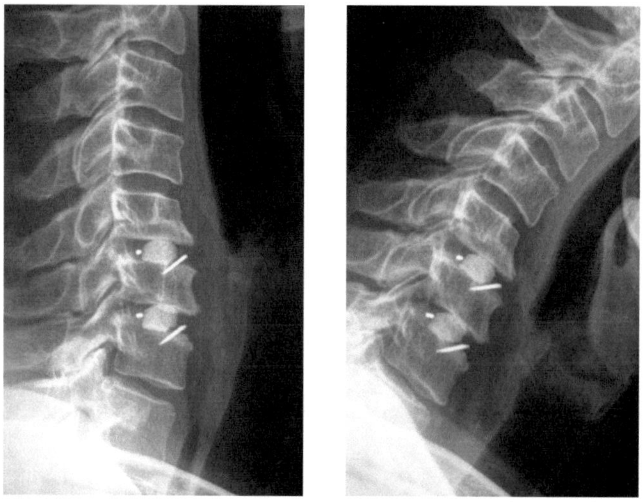

Picture 5. Cervical fusion

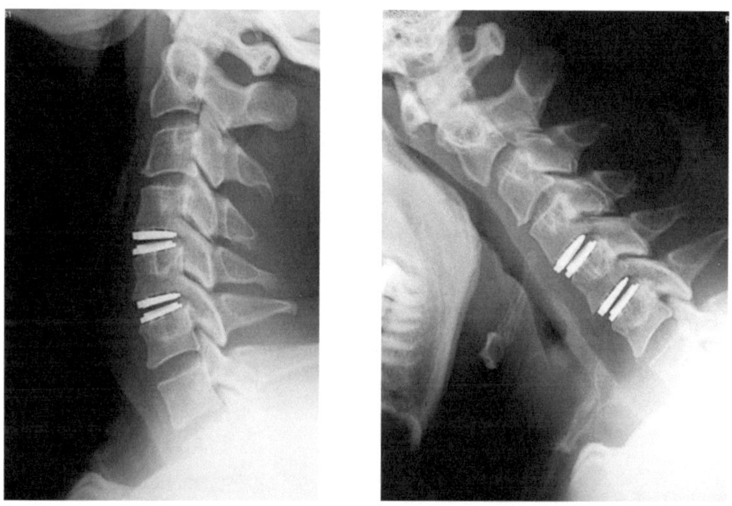

Picture 6. Cervical arthroplasty

Fusion after discectomy is a consolidate surgical treatment for disc herniation.
A lot of papers, in literature, confirm long term follow-up, its effectiveness and safeness.
In last few years, some authors have reported it the other way.
Motion preservation and adjacent -segment degeneation's prevention are the Key words of arthroplasty.
The first argument is intuitive: arthroplasty, through non fusion, preserves the motion in the operated segment .
For the second question, biomechanics and kinematics, teach that in a series of mobile segments, when one of them is blocked or unable to shock absorb or cushioning functions, the caudal and the cranial segments suffer a mechanical stress, expressing in a motion alteration.

These alteration may cause structural and anatomical alterations, as well to produce clinical failures.
From a review of literature, these are the most important elements "pro" arthroplasty vs fusion:

1. motion maintenance at treated segment (stability of range of motion data: 7.4° one year post-operation and 7.9° five years post operation)
2. reduction of adjacent-discal-segment degeneration
3. reduction of intradiscal-pressure in cranial and caudal intervertebral disc
4. no aggravations of pathologies in cranial and caudal disc, if presents before the operation
5. more patient's satisfaction, for both non post-operative orthosis dressing and early return to work.

Recently, only one paper about the not unequivocal significant utility of arthroplasty vs fusion has been reported.
In my personal surgical experience, consists of 60 patients treated with cervical microsurgical discectomy (MD group) vs 60 patients treated with cervical microsurgical discectomy and fusion (using titanium or peek devices) (MDF group) vs 60 patients treated with microsurgical discectomy and arthoplasty (MDA group), using "Discover" device ® (picture 7).

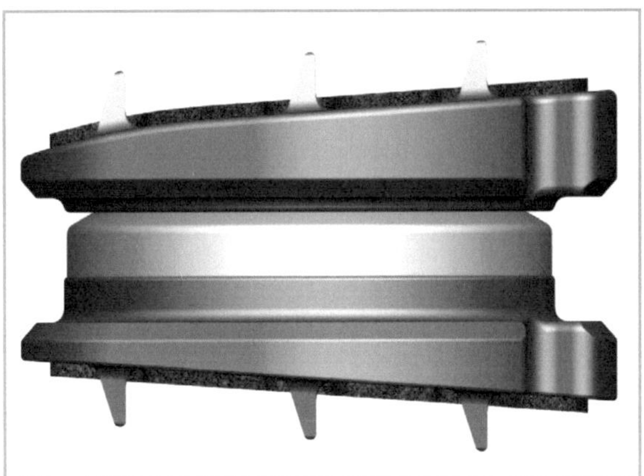

Picture 7. "Discover" device®.

The clinical and radiological follow-up was for a maximum period of 48 months.
The preliminary extrapolated data seems to confirm what had been reported before, from 1 to 5 points, encouraging and authorizing cervical arthroplasty vs cervical fusion, in discal pathology: rigorous and careful indication for patient selection, more long, shareable and verifiable follow-ups are imperative.
11.6%of clinical failures in MD group.

3. The new device (patented, all rights reserved)

One of the most important limitations of discal prosthesis (cervical and lumbar) is the axial compression.

In a non pathological lumbar disc, for example, it ranges from 0.5 to 1.5 mm: this is a very important element for shock absorbing and cushioning.
Standing to the literature data, until now, no disc prosthesis has these characteristic.
Another significant limitation, only for lumbar prosthesis is, until now, the impossibility to place it by a posterior surgical approach, more handy, and safer than anterior approach.
These new devices (picture 8) try to satisfy the two conditions just described.
For the second question, was designed a particular way, standard sized, not more bulky (thanks to particular material used) so to introduce the device, after a bilateral posterior discectomy, after a bilateral laminar reduction, by a monolateral way into the intervertebral emptied and gently distracted space.

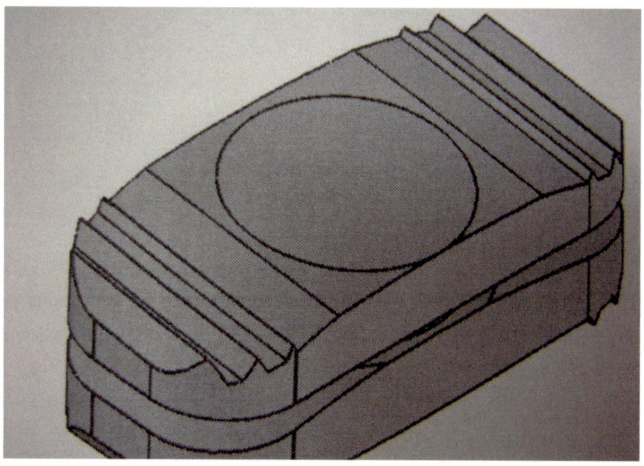

Picture 8. The new device

The device is introduced perpendicular to the intervertebral space, by a dedicated instrument and, at the anterior third of the space (verified by intraoperative RX film), by the same instruments, rotated in 180° according to the intervertebre space's axis (picture 9).

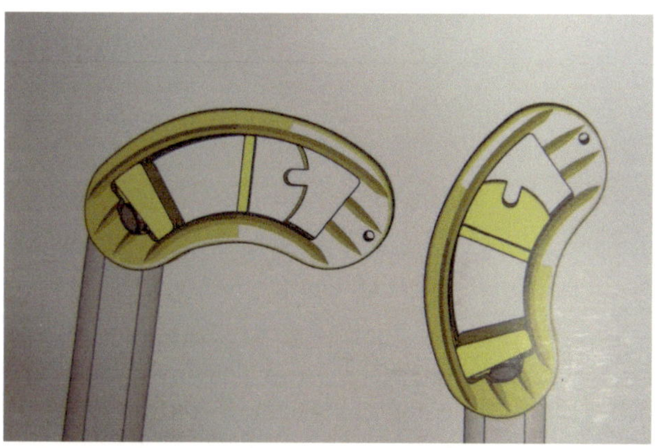

Picture 9. The device and its instruments . During the introduction (on right), after the 180° rotation (on left)

Then, unhooked the dedicated instruments, the devices remain anchored in situ after a gentle intervertebral space compression, even through its superior and the inferior tops, rough and teeth fitted, that favour a fusion, with the corresponding vertebral end plates (picture 10).

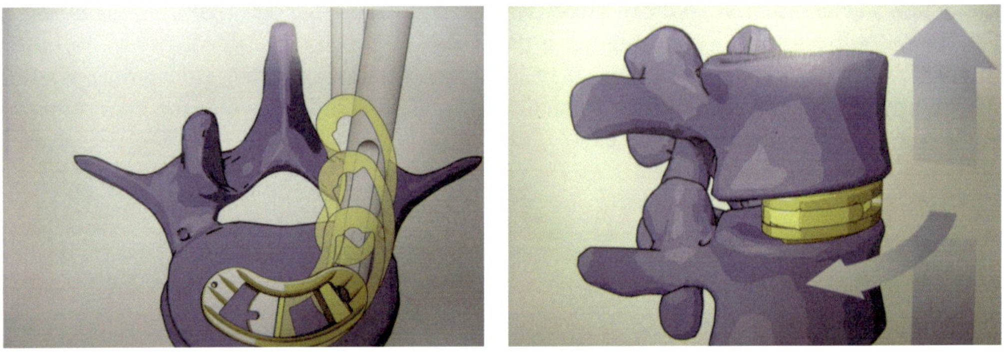

Picture 10. The device and its instruments during the introduction and rotation (on left) the device after its definitive collocation (on right)

To the second purpose, particular attention was given to the structural design, to the materials and to the sequence of the components in their assembly, in order to give to the device features similar to the non pathological intervertebral disc: bone and biocompatibility, MRI compatibility, strength to static load for maximum 330 kg , 4°-5° of maximum inclination on all side and, mostly (picture 11), features not held by other devices until now, shock absorbing, load sharing and chushioning .

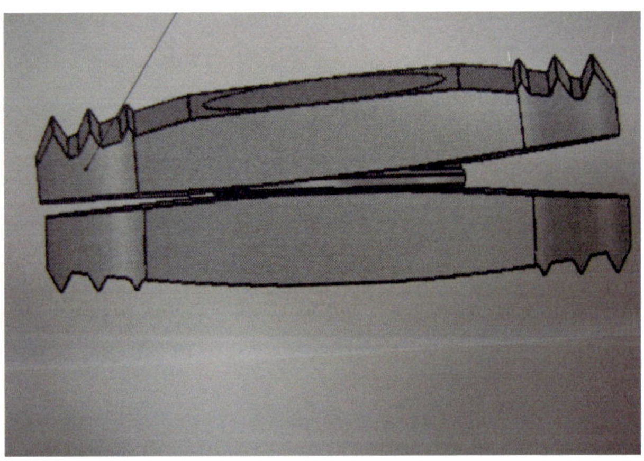

Picture 11. The device's inclination in a front view.

The use of two rough teeth fitted titanium tops (superior and inferior) , opposite to the corresponding vertebral and plates ("A" in picture 12), guarantee bone and biocompatibility, MRI compatibility, strength to static loads for maximum 330 kg.
Their mutual articulation, by a titanium spherical node (fitted of a security self-stoppage system), behaving like a "spring -cup" ("B" in picture 12), create a metal/metal antifriction interface securing strength, motion preservation and shock absorbing.

Finally, the use of an organic polymer thermoplastic (Polyether-Ether-Ketone, "PEEK") to shape a "bed" in which is collocated the titanium spherical node ("C" in picture 12) complete the shock absorbing function, giving load sharing and cushioning features, elastic response to the loads.

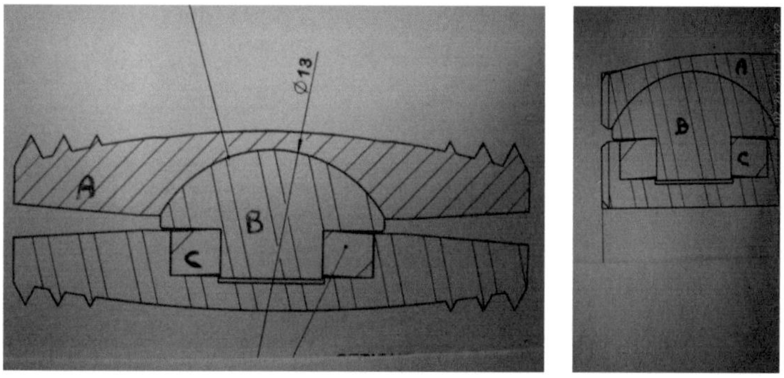

Picture 12. Structural devices view, frontal (left) and lateral (right)

4. Conclusions

The literature data, until now, had cautiously encouraged the non fusion versus the fusion, in the surgical treatment of degenerative discal pathology (in cervical and lumbar spine).
However, the elements that prefer arthroplasty over fusion were, in my opinion, mainly three:

1. a more long follow-up for fusion cases
2. the impossibility of a posterior surgical approach in lumbar spine
3. the impossibility to have a really shock absorbing, load sharing and chushioning function, but only motion preservation and strength to static load (both in cervical and in lumbar spine).

These studies, that led to the design and the creation of these device, aim to fill the gaps just emphasized above, giving a new prosthesis both in cervical spine (in order to really shock absorb, load share and cushioning functions) and in lumbar spine (in order both in these same features and in the really possibility to perform a posterior surgical approach, certainly more safe for the patients and experienced for the spine surgeons.
Only a long follow-up may validate or not these work.

5. Acknowledgment

Andrea Bedeschi, engineer in Ferrara (Italy)

6. References

Bartles, R.H.M.A et al., 2010. No justification for cervical disk prostheses in clinical practice:a meta-analysis of randomized controlled trials. Neurosurgery, Vol.66, No.6, (June 2010), pp.1-8

Brunner, H.J. et al., 2010. Biomechanics of polyaryletherketone rod composites and titanium rods for posterior lumbosacral instrumentation.J Neurosurg Spine, Vol.13, (December 2010), pp.766-772

Coric , D.et al., 2010.Prospective study of cervical arthroplasty in 98 patients involved in 1 of 3 separate investigational site with a minimum 2-year follow-up.J Neurosurg Spine, Vol 13, (December 2010), pp. 715-721

Fehlings, M.G. et al., 2010.Motion preservation following anterior cervical discectomy.J Neurosurg Spine, Vol 13, (December 2010), pp.297-298

Logroscino, C.A.et al., 2007. Artroplastica vertebrale: stato dell'arte della chirurgia protesica cervicale e lombare G.I.O.T., Vol 33 (suppl.1), pp.155-161

Maida, G.et al., 2008. Artroplastica cervicale: la nostra esperienza.Rivista Medica, Vol 14, No.3, (September 2008), pp.53-56, ISSN 1127-6339, ISBN 978-88-8041-012-6

Maida, G.et al., 2009.60 patients follow-up, from 6 months to 2 years, in patients treated with discover arthroplasty after discectomy and preliminary comparison with discectomy and fusion and discectomy alone. Eur.Spine.J, Vol 18, (May 2009), pp757

Ryu, K.S. et al., 2010.Radiological changes of operated and adjacent segments following cervical arthroplasty after a minimum 24-month follow-up: comparison between the bryan and Prodisc-C-devices. J Neurosurg Spine, Vol 13, (September 2010), pp.299-307

Part 2

Musculoskeletal Biomechanics

4

Biomechanical Characteristics of the Bone

Antonia Dalla Pria Bankoff
University Of Campinas
Brazil

1. Introduction

The bone tissue is strong and one of the most rigid structures of the body due to its combination of inorganic and organic elements. The minerals calcium and phosphate, together with collagen, constitute the organic element of the bone being responsible for approximately 60 to 70% of the bone tissue. Water constitutes approximately 25 to 30% of the bone tissue weight. (Alberts et al., 1994; Junqueira & Carneiro, 1997, 1999).
The bone tissue is a viscous-elastic material whose mechanical properties are affected by its deformation grade. The flexibility properties of the bone are provided by the collagen material of the bone. The collagen content gives the bone the ability to support tense loads. The bone is also a fragile material and its force depends on the load mechanism. The fragility grade of the bone depends on the mineral constituents that give it the ability to support compressive loads. (Alberts. et al., 1994; Junqueira & Carneiro, 1997, 1999).
Re-absorption and Bone Deposit - Bone is a highly adaptive material and very sensitive to disuse, immobilization or vigorous activity and high load levels. The bone tissue can be separated and may change its properties and setting in response to the mechanical demand. It was determined at first by the German anatomist, Julius Wolff, that gave us the theory on the bone development named Wolff Law, that says: "Each change in the form and function of a bone or only its function is followed by certain definitive changes in its internal architecture, and secondary changes equally definitive in its external compliance, in accordance to the mathematics law". (Alberts et al., 1994; Junqueira & Carneiro, 1997, 1999).

2. Bone strength and hardness

The behavior of any material under different load conditions is determined by its strength and hardness. When an external force is applied in a bone or in any other material, there is an internal reaction. The strength may be assessed by checking the relation between the load imposed (external force) and the quantity of deformation (internal reaction) that takes place in the material, known as load-deformation curve. (Holtrop, 1975).
Anisotropic Characteristics Bone tissue -Is an anisotropic material, indicating that the bone behavior will change depending on the direction of the load application. In general, the bone tissue may lead to higher loads in the longitudinal direction and a lesser quantity of load when applied over the bone surface. The bone is strong to support loads in the longitudinal direction because it is used to receive loads in this direction. (Holtrop, 1975).
Viscoelastic Characteristics - The bone is also viscoelastic, which means that it responds differently depending on the speed to which the load is applied and the length of the load.

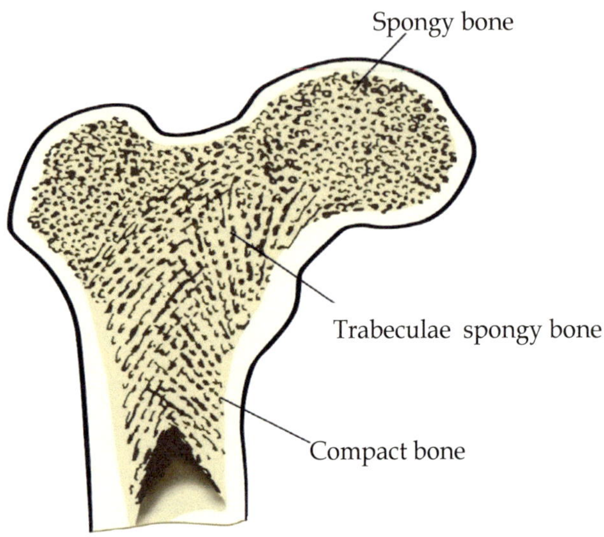

Fig. 1.1. This mean section of the proximal tip of the femur shows both the compact bone and the sponge bone. The dense compact bone covers the external part of the bone, going downside in order to form the bone body. The sponge bone is found in the tips and is identified by its truss appearance. Watch for the curvature in the trabeculae, which is formed to support the stresses. Bankoff (2007, p. 122).

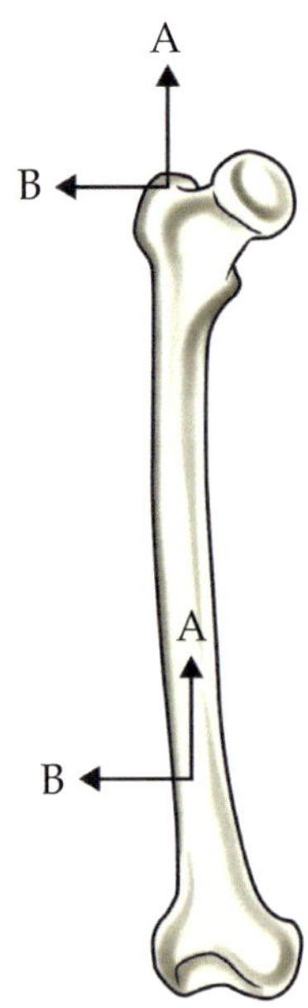

Fig. 1.2. The bone is considered anisotropic because it responds differently when the forces are applied in different directions. (A) The bone can lead to great forces applied in the longitudinal direction. (B) The bone is strong when it leads with forces applied transversally crossing its surface. Bankoff (2007, p. 123).

In very fast speeds of load placement, the bone can lead to higher loads before it fails or breaks. As showed in FIGURE 1.3, the bone that receives the load slowly breaks with a load that is approximately half of that it could support if the load was more quickly applied.

Elastic Response - When the load is firstly applied, a bone is deformed by a change in the extent or angular format. The bone is deformed up to 3%. This is considered the elastic amplitude of the load-deformation curve because, when the load is removed, the bone is recovered and goes back to the original format or extent. The stress-distension or load-deformation curve is presented in FIGURE 1.4. An exam of this curve may be used in order to determine if a material is hard, flexible, fragile, strong or weak. The curve showed could represent a material that is strong and flexible. (Hay, 1982; Holtrop, 1975).

Plastic Response - With the continuous placement of load on the bone tissue, its deformation point is reached, after which the external fibers of the bone tissue will start to cede, experiencing micro-breaks and disconnection of the material within the bone. (Hay, 1982; Holtrop, 1975).

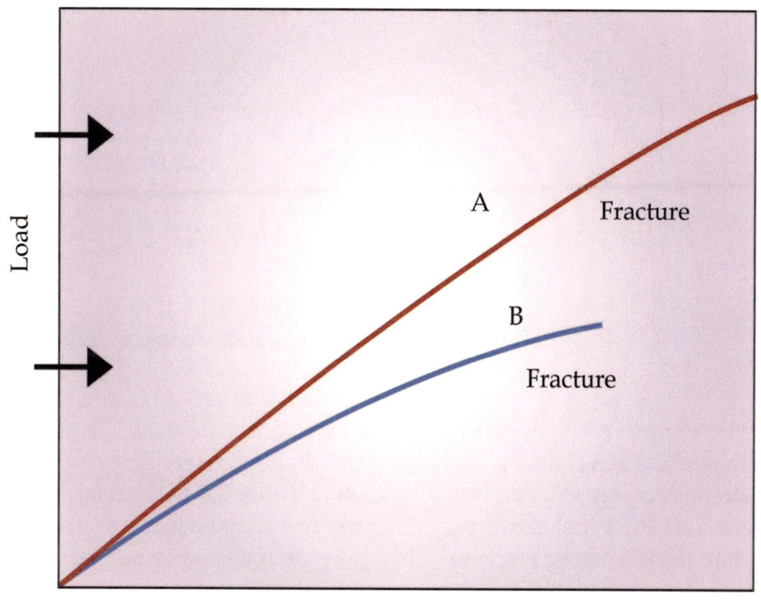

Fig. 1.3. The bone is considered viscoelastic because it responds differently when it receives loads in different speeds. (A) When it receives the load quickly, the bone responds more rigidly, and may handle a higher load before it breaks. (B) When it receives the load slowly, the bone is not so rigid or strong, breaking under lesser loads. Bankoff (2007, p.123).

To which we name plastic or non-elastic phase in the load-deformation curve. The bone tissue starts to deform permanently and eventually breaks if the load continues in the non-elastic phase. Thus, when the load is removed, the bone tissue does not retake the original extent and is permanently elongated. (Hay, 1982; Holtrop, 1975; Bankoff, 2007).

Strength - The strength of the bone or any other material is defined by the point of failure or by the load sustained before the failure. The strength may also be analyzed in terms of storage of energy, the area under the load-deformation or stress-distension curve. (Hay, 1982; Holtrop, 1975; Bankoff, 2007).

Hardness - The hardness, or elasticity module of a material, is determined by the decrease of the load-deformation curve (FIGURE 1.4) during the amplitude of the elastic response and is represented by the resistance of the material to the load as the structure is deformed. This response occurs in many materials, including bones, tendons and ligaments. The stress-distension curve for flexible, fragile materials and for the bone is represented in FIGURE 1.5. (Choi & Goldstein, 1992).

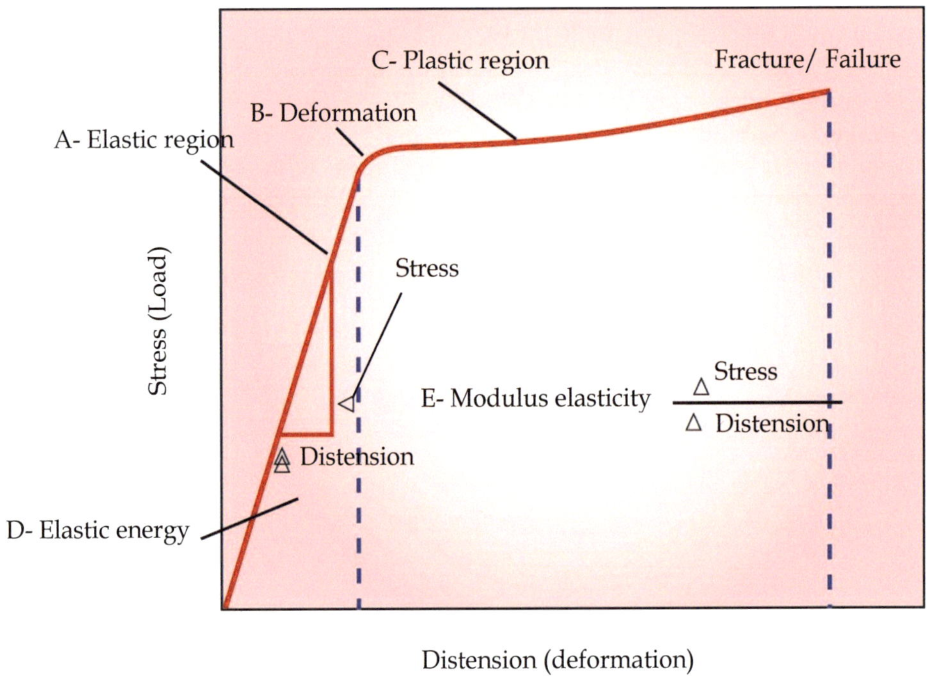

Fig. 1.4. The stress-distension curve and the load-deformation curve illustrate the performance strength characteristic of a material when subjected to the load. When the load is applied, there is an (A) initial elastic response that eventually reaches a (B) deformation point, getting into the (C) plastic response when the material is deformed permanently or is broken. The strength of the material is determined by the (D) energy or area under the curve. The hardness of a material, called elasticity module is determined by the (E) inclination of the curve during the elastic response phase. Bankoff (2007, p. 123).

A hard material will respond with a minimum deformation to the load increase. When the material fails in the end of the elastic phase, it is considered a fragile material. The glass is an example of fragile material. The bone is not so hard as the glass or metal, and, differently of those materials, it does not respond linearly, because it cedes and deforms not uniformly during the load placement phase. The higher the load imposed to the bone, the higher the deformation. In addition, if the load exceeds the elastic limits of the material, there will be a permanent deformation and failure of the material. If a material continues to over-elongate and over-deform in the plastic phase, it is known as flexible material. (Choi & Goldstein, 1992). The skin is an example of material that is deformed considerably before the failure. The bone is a material that has properties that respond in both the fragile and the flexible mode. (Choi & Goldstein, 1992).

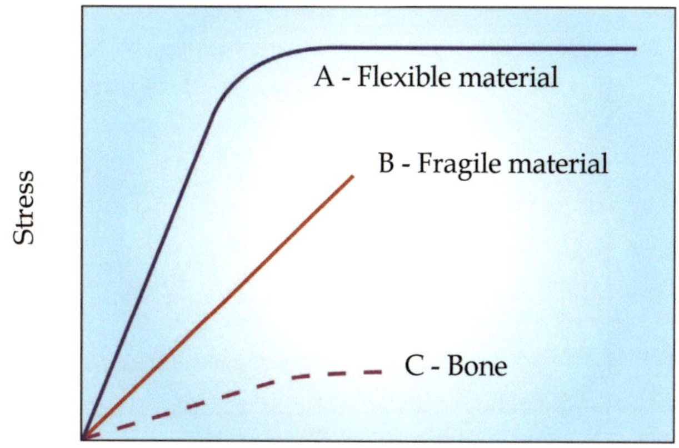

Fig. 1.5. Such stress-distension curves illustrate the differences of behavior among (A) flexible material, (B) fragile material and (C) bone, that has both fragile and flexible properties. When the load is applied, a fragile material responds linearly and fails or breaks before experiences any permanent deformation. The flexible material will get into the plastic area and will be considerably deformed before the failure or break. The bone is slightly deformed before the failure. Bankoff (2007, p. 124).

A plurality of materials was signaled in a graph of FIGURE 1.6 in accordance to its strength and hardness. Examples of materials considered hard and weak are glass and copper, hard and strong materials are steel, iron and gold. Flexible and strong materials are fiberglass and silk, and flexible and weak materials are oak-wood, lead and a spider web. The bone is considered a flexible and weak material. (Choi & Goldstein, 1992; Cook et al., 1987).

Stress and Distension – Another way to assess the behavior of the bone or any other material when subjected to the load is to measure the stress, or the load by area of transversal section and the distension, or deformation, regarding the original extent of the material. A stress-distension curve may be produced on such a way that, as the load-deformation curve, illustrates the mechanical behavior of the material and may be used in order to check the strength and hardness of the material (FIGURE 1.4). (Choi & Goldstein, 1992; Cook et al., 1987).

The load-deformation curve of a material in particular seems exactly to the stress-distension curve for the same material and is interpreted on the same way described previously using the load-deformation curve. The only difference among the curves is in the units used to represent each one of them. The load-deformation curve is represented by absolute values of load and deformation, while the stress-distension by relative values regarding the material extent and transversal section. The benefit of producing a stress-distension curve is that the standardization regarding the unit of the area and extent allows for the comparison of different materials. (Cook et al., 1987; Choi & Goldstein, 1992).

Stress or Normal Distension and Shear – The stress and distension may occur perpendicularly to the transversal section plan of the object that receives the load, known as normal stress and distension, or parallel to the transversal section plan, known as shear stress and distension. For example, a normal distension involves a change in the extent of an object, while distension with shear is characterized by a change in the original angle of the object. (Cook. et al., 1987; Schaffler & Burr, 1988; Choi & Goldstein, 1992).

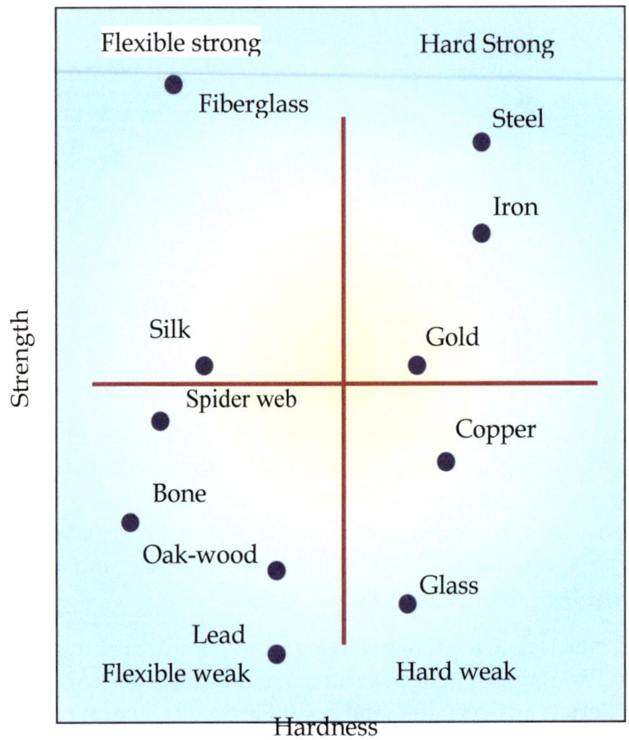

Fig. 1.6. The strength and stiffness of a variety of different materials are represented by four quadrants representing flexible and weak materials which is (A); flexible and weak; (B) hard and weak (C) hard and strong; and (D) flexible and strong. Watch that the bone is classified as flexible and weak together with other materials such as spider web and oak wood. (Adapted from Hamil and Knutzen, 1999 p.44) Bankoff (2007, p. 124).

3. Load types

The skeletal system is subjected to a variety of different types of forces on such a way that the bone receives loads in different directions. There are loads produced by the weight sustentation, by the gravity, by muscle forces and by external forces. The loads are applied in different directions producing forces that may vary from five different types: compression, tension, shear, curvature or torsion (Shipman, Walker & Bichell, 1985).
The skeletal system injury can be produced by applying a high-magnitude single strength of one these types of load or by repeated application of low-magnitude loads over a long period. The second type of injury in the bone is called stress fracture; fatigue fracture or bone distension. FIGURE 1.8 shows the X-ray photo of a stress fracture on the metatarsus. These fractures occur because of cumulative microtrauma imposed on the skeletal system, when the placement of loads on the system is so frequent that the process of bone repair cannot be equal to the breakdown of the bone tissue. (Shipman, Walker & Bichell, 1985; Egan, 1987).

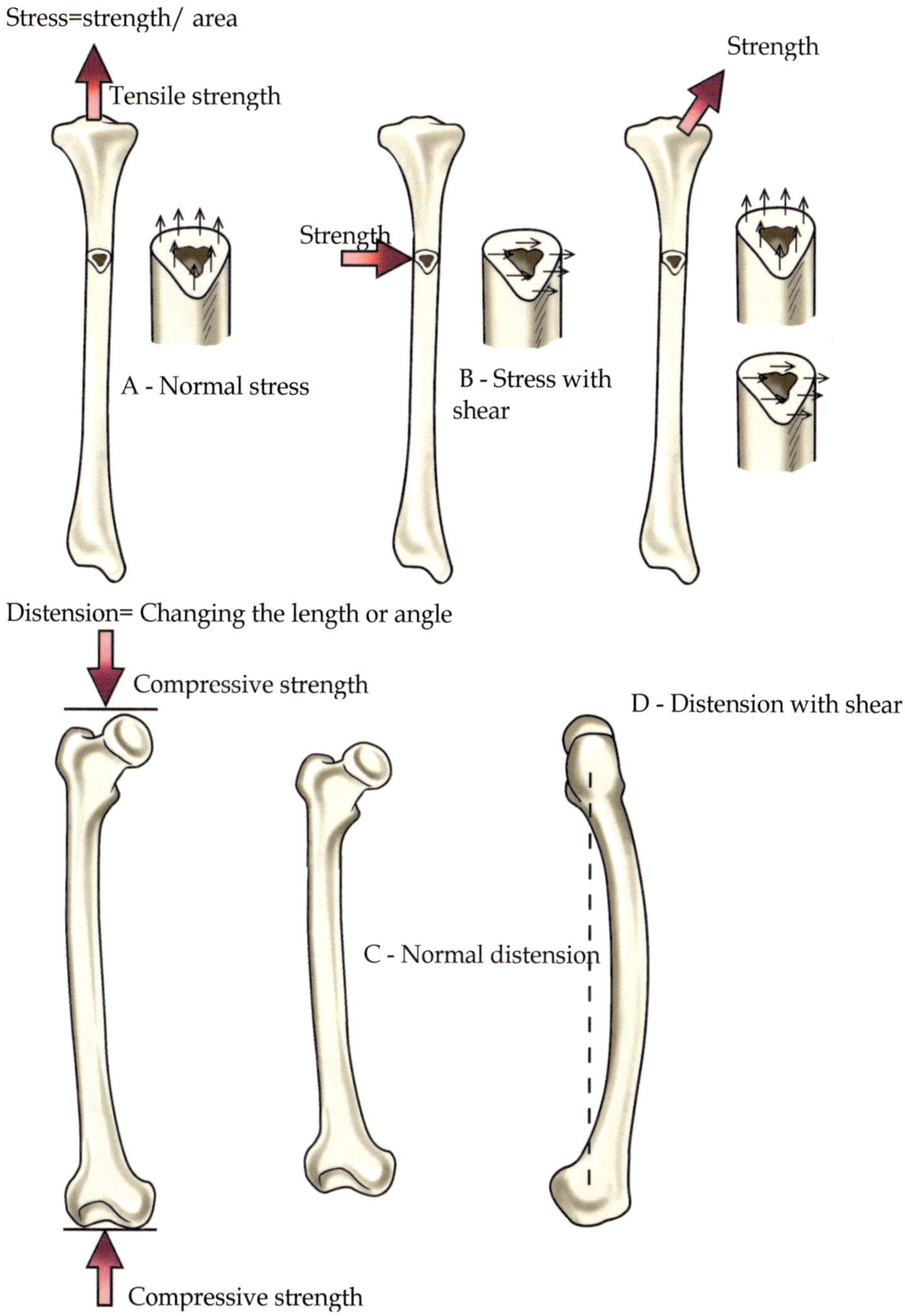

Fig. 1.7. The stress, that is the force by area unit, may occur perpendicularly to the plan (normal stress) as represented in (A), or in parallel to the plan (stress with shear) represented in (B). The distension, that is the deformation of the material, is represented in (C) normal distension, in which the extent varies and (D) distension with shear, in which the angle is changed. Bankoff (2007, p. 125).

Compressive Strengths – A compressive strength presses the edges of the same bone at the same time; and is produced by muscles, weight support, gravity or some external load that come down the length of the bone. Compressive stress and distension within the bone causes bone shortening and extension and bone absorbs maximum stress in a plane perpendicular to the compressive strength (Figure 1.9 -A and B). Compressive strengths are necessary for the development and growth of the bone. The stresses and distensions produced by the compressive strengths and other strengths are responsible for facilitating the deposition of the bone material. (Egan, 1987).

If a large compressive strength is applied and if the loads exceed the structure stress limits, fracture occurs. There are numerous places in the body prone to fracture or compressive injuries. A compressive strength is responsible for patellar pain by softening and destruction of cartilage under the patella, known as chondromalacia patellar chondromalacia. As knee articulation moves in amplitude of movement, the patella moves up and down in its sulcus. The load between the patella and femur increases and decreases until a point where the patellofemoral compressive strength is greater than about 50 degrees of flexion and lower in full extension or hyperextension of knee articulation. High compressive strength in flexion, primarily on the lateral patellofemoral surface, is the source of the destructive process that breaks down cartilage and underlying surface of the patella. (Egan, 1987; Schaffler & Burr, 1988; Hoffman & Grigg, 1989).

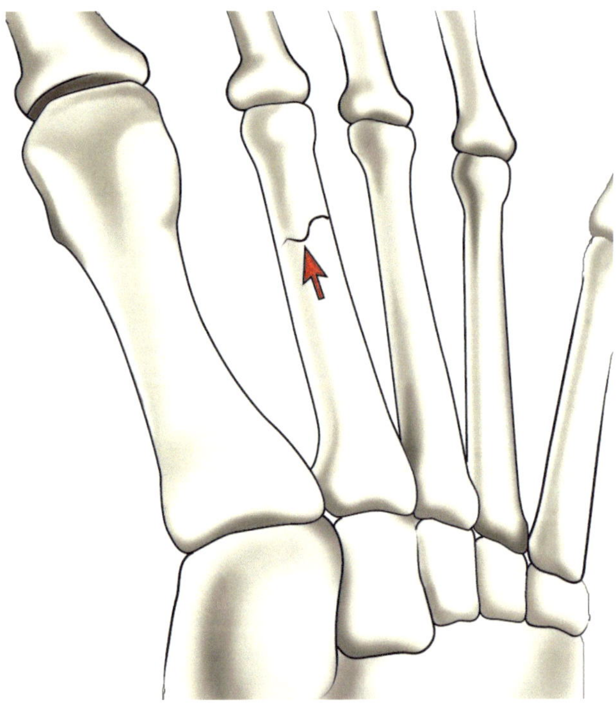

Fig. 1.8. Stress fractures occur in response to the excessive loads on the skeletal system so that cumulative microtrauma occurring in the bone. A stress fracture in the second metatarsal, as shown above, is caused by running on hard surfaces or using hard shoes. It is also associated with people with high arches and can be created by fatigue of neighbors muscles. Bankoff (2007, p. 126).

Compression is also a source of fractures in the vertebrae. Fractures have been reported in the cervical area in activities like water sports, gymnastics, wrestling, rugby, ice hockey and American football. Normally, the cervical spine is slightly extended with a convex curvature previously, if the head is lowered, the cervical spine is rectified to approximately 30 degrees of flexion. If strength is applied against the top of the head when you are in this position, the cervical vertebrae get a load to down in its extension caused by a compressive strength, creating a dislocation or fracture-dislocation of the vertebral facets. When spearing or butting (throw of the player by other of the team) with the head in flexion was banned in American football, the number of cervical spine injuries has been drastically reduced. (Cook et al., 1987; Fine et al., 1991; Halpbern & Smith, 1991).

There are also reports of compression fractures in lumbar vertebrae of weightlifters, defenders of American football or gymnasts who had overwhelmed the vertebrae in the spine being with hyperlordotic or lordotic position. Finally, compression fractures are common in subjects with osteoporosis. (Matheson et al., 1987; Fine et al., 1991; Halpbern & Smith, 1991).

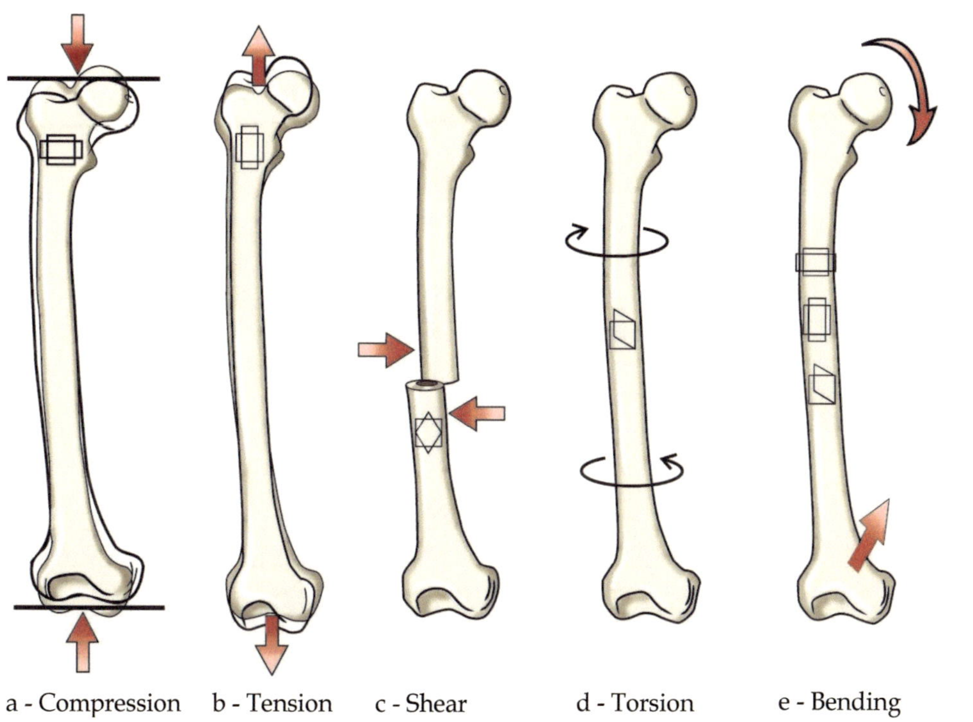

Fig. 1.9. A The skeletal system is subject to a variety of loads that alternate within the bone. The solid layout in the femur above indicates the original state of the bone tissue. The colored area shows the effect of the applied strength on the bone. (A) A compressive strength causes shortening and extension, (B) a tensile strength causes narrowing and elongation (C) shear strength and (D) torsional strength creates an angular distortion, and (E) the bending strength includes all changes seen in compression, tension and shear. Bankoff (2007, p. 126).

Spondylolysis may occur and a stress fracture of the vertebrae interarticular part. Specific weight lifting that have high incidence of this fracture are the clean and jerk (direct weight lifting from the ground up above the head). It also occurs in gymnasts and is associated with positions of extreme extension of the lumbar vertebrae region. (Matheson et al., 1987).

A compressive strength on the hip joint can increase or decrease the potential of femoral neck injury. The hip joint needs to absorb compressive strengths of approximately 3-7 times the body weight during walking. Compressive strengths are over 15 to 20 times the body weight in the jump. In a normal standing posture, the hip joint takes about one third of body weight if the two members are on the ground. This creates large compressive strengths on the lower portion of the femoral neck and a large traction strength or tension on the upper portion of femur neck. FIGURE 1.12 shows how this happens as the body pushes down the femoral head, pushing at the base of the femoral neck and tractioning the top of the femoral neck out while creating a bending (Matheson et al., 1987, Jackson, 1990).

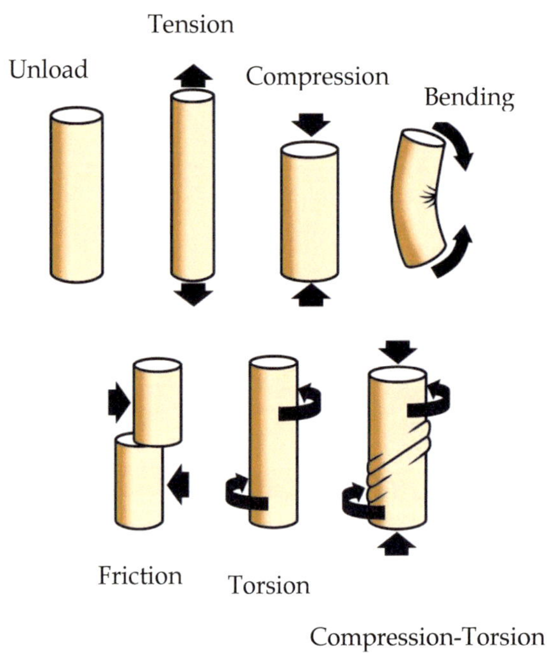

Fig. 1.9. B. Types of loads that can be applied in a tissue such as the bone. Bankoff (2007, p. 126).

The abductors of the hip, specifically the medial gluteus, constrict to interpose the weight of the body during support. They also produce a compressive load on the upper femoral neck that reduces the tensile strength and the potential of injury in the femoral neck, since the bone will breaks generally earlier with a tensive strength (FIGURE 1.12). It was reported that runners have developed femoral neck fractures because the medial gluteus fatigue, failing to act in reduction of high tensive strength that is on the upper neck, producing fracture. A fracture of femoral neck can also be produced by a strong co-contraction of the hip muscles, specifically the abductors and adductors, creating excessive compressive strengths on the upper femur neck. (Matheson et al., 1987; Marks & Popoff, 1988).

We can also cite as compressive strength on the temporomandibular joint (TMJ) the role of sternocleidomastoid muscles when accommodated and shortened to a particular situation. They exert this kind of strength on the mandibular condyles, thus changing the whole mastication and mandibular morphology and may cause damage to joints and headache. This happens due to the insertion of the sternocleidomastoid be in the mastoid process of the temporal bone, exactly in the bone where the jaw is articulated, and because these are anti-gravity muscles in relation to its origins and insertions. (Halpbern et al., 1987; Matheson et al., 1987).

Tensive Strengths – A tensive strength is usually applied on the bone surface and it pulls or elongates the bone, tending it to extend and narrow the bone (FIGURE 1.9 A and B). The maximum stress, as in compression is perpendicular to the applied load. The source of tensive strength is usually the muscle. When the muscle applies a tensive strength to the system by the tendon, the collagen in the bone tissue is aligned with the tensive strength of the tendon. (See FIGURE 1.13 – an example of alignment of collagen in the tibial tuberosity). This figure also illustrates the influence of tensive strengths in development of apophyses, showing how the tibial tuberosity is formed by tensive strengths. The failure of the bone usually occurs at the site of muscle insertion. Tensive strengths can also create ligament avulsions that occur more frequently in children. In addition, the ligament avulsions are common in lateral ankle because of ankle sprain. (Choi & Goldstein, 1992; Cook et al., 1987).

Besides the bone tissue, here are other examples of biomechanical properties of connective tissue, represented by tendons and ligaments. Biomechanical properties of tendons and ligaments are often characterized as a relation load versus deformation in response to a tensive load (FIGURE 1.10). In these experiments, a sample (e.g., ligament, tendon, and ligament-bone) is obtained from a corpse, and is assembled on a device that elongates the tissue to a prescribed speed (distension speed) until the tissue is broken, and measures the displacement (elongation) and strength. The clinical observations on the disruption of the connective tissue suggest that tissular breakdown is more common than avulsion of the bone. (FIGURE 1.10) shows the variation in peak strength and elongation in different samples. For example, a sample medial patellar tendon-bone was elongated by 10 mm and a tensive strength exerted peak ($\sigma_{rupture}$) of about 3 kN before starting to fail, while a sample of anterior cruciate ligament-bone was elongated by 15 mm and exerted a tensive strength of peak ($\sigma_{rupture}$) of approximately 1.5 kN before failing. The data shown in Figure 1.10 were taken from a study comparing the mechanical properties from several collagen tissues for use in the reconstruction of the articular cartilage of the knee joint. The gracilis tendon is the tissue between the muscle and the tibial insertion. (Lakes et al., 1990).

A sample of the fascia lata had 70 to 10 cm wide and was taken from the middle of the thigh near the lateral femoral condyle. The data indicate that the patellar tendon-bone sample was stronger than the sample anterior cruciate ligament-bone, but both were stronger than the samples of the gracilis tendon and fascia lata (FIGURE 1.10) When the load and deformation are normalized so that the load is expressed per unit of cross-sectional area and deformation is described as a percentage of the initial length, the biomechanical properties of tissues can be compared to overload versus distension relations. FIGURE 1.11 represents an overload relation versus distension idealized for collagen tissues, like tendons and ligaments. The overload relation versus distension comprises three regions: tip, linear and rupture. The region of the tip corresponds to the initial part of the relationship, in which the collagen fibers are elongated and rectified from the standard of rest in a zigzag. The linear region represents the ability of elastic tissue; the inclination of the relation in this region is called the elastic modulus and is more pronounced in tissues that are more rigid. Outside of linear region, the inclination decreases, since some fibers are broken in the region of rupture.

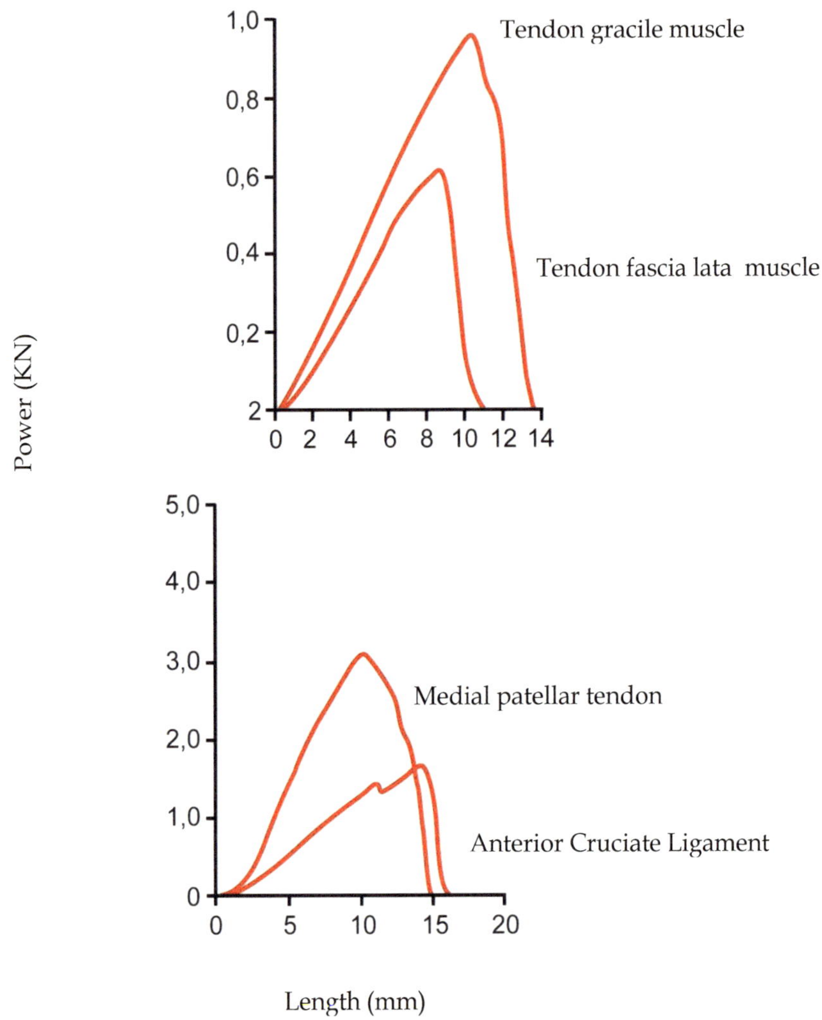

Fig. 1.10. The load versus deformation relationships for the sample of connective tissue elongated up to rupture. Bankoff (2007, p. 127).

When the connective tissue experiences a distension of this magnitude, the tissue undergoes plastic changes and there is a change in its resting length. From of overload relation versus distension, the tissue can be characterized by measures of final overload (σ final) of final distension (σ final) of the elastic modulus and the energy absorbed (area under curve; overload versus distension). These properties tend to decline in the face of conditions such as reduced use (e.g., immobilization, bed rest), aging and steroid use, but increases with long-term exercise. Moreover, the properties of tendon may vary with the muscle function. (Lakes et al., 1990; Matheson et al., 1987).

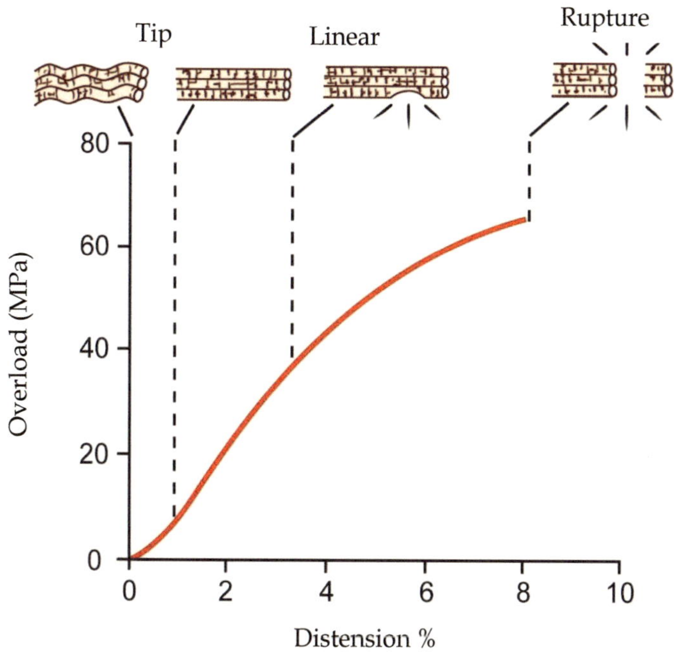

Fig. 1.11. Relation of overload versus distension idealized for collagen tissue. The tissue may experience only a small change in length before being damaged. Bankoff (2007, p. 128).

The avulsion fractures occur when the tensive strength of the bone is not sufficient to prevent fracture. This is typical in some injuries that occur in movements of high-speed pitch, as in the arm of throw, sore of basketball of junior players. The avulsion fracture in this case is usually in the medial epicondyle due to the tension generated in the wrist flexors. (Lakes et al., 1990).

Two other fractures produced by common tension, are in the fifth metatarsal due to tensive strengths, generated by fibular muscle group, and in the calcaneus where the triceps surae muscle generates the strengths. The tensive strength on the calcaneus can also be produced in the support phase of walking to the extent that the arc is depressed and the plantar fascia that covers the plantar surface of the foot is tensioned, exerting a tensive strength on the calcaneus. Some sites of avulsion fractures in the pelvic region, shown in Figure 1.14, include the upper and lower spines, the lesser trochanter, the ischial tuberosity and the pubic bone. (Lakes et al., 1990; Mundy et al., 1995).

Tensive strengths are mostly responsible for distensions and sprains. For example, a typical ankle sprain in inversion occurs when the foot, rolls to the side, elongating the ligaments. Tensive strengths are also identified with canelite when the anterior tibial pulls its insertion site and the interosseous membrane. (Bechtol, 1954). Another body part exposed to high tensive strength is the tibial tuberosity that transmits very high tensive strength when the quadriceps femoris muscle group is active. This tensive strength, under sufficient magnitude and duration, may create a condition of tendinitis in senior participant. In the youngest participant, however, the damage usually occurs at the insertion site of tendon-bone and may result in inflammation, bony deposits or avulsion fracture of tibial tuberosity.

Osgood Schlatter disease is the name of a condition characterized by inflammation and formation of bony deposits in the tendon-bone junction. (Boume, 1976).

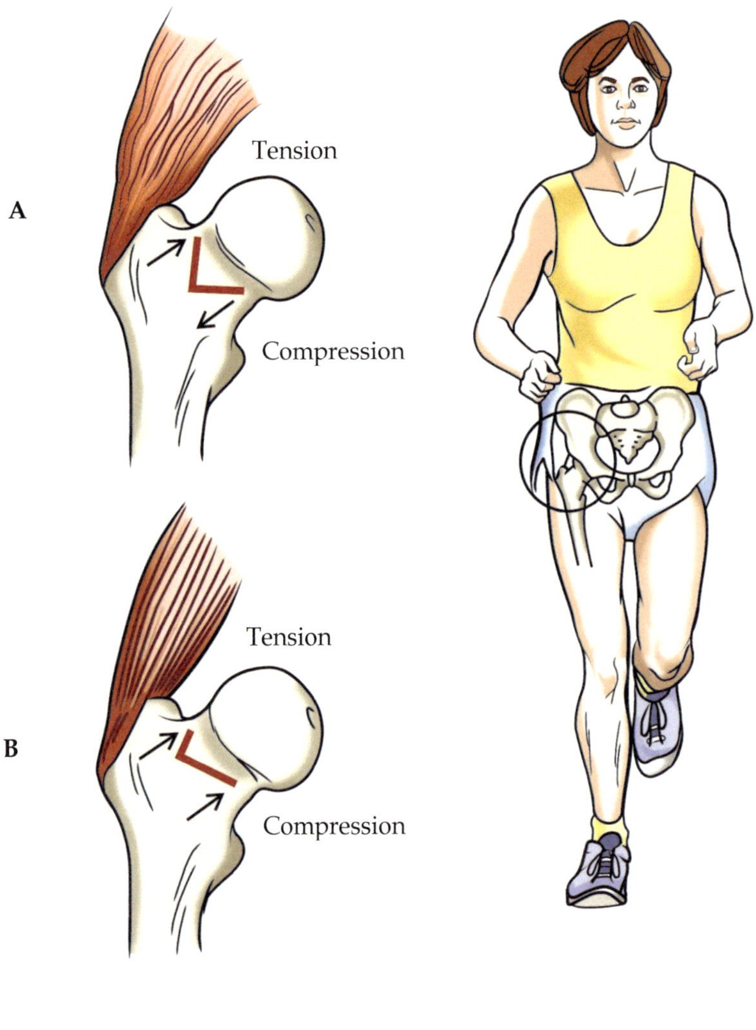

Fig. 1.12. When standing or in the stance phase of walking or running, there is a bending strength applied on the femoral neck. This strength creates an intense compressive strength on the lower femoral neck and a tensive strength on the upper femoral neck (see A above). When the medial gluteus constricts, the compressive strength is increased; and the tensile strength is decreased (B above). That reduces the potential for injury, since there is greater probability of injury with tension. Bankoff (2007, p. 128).

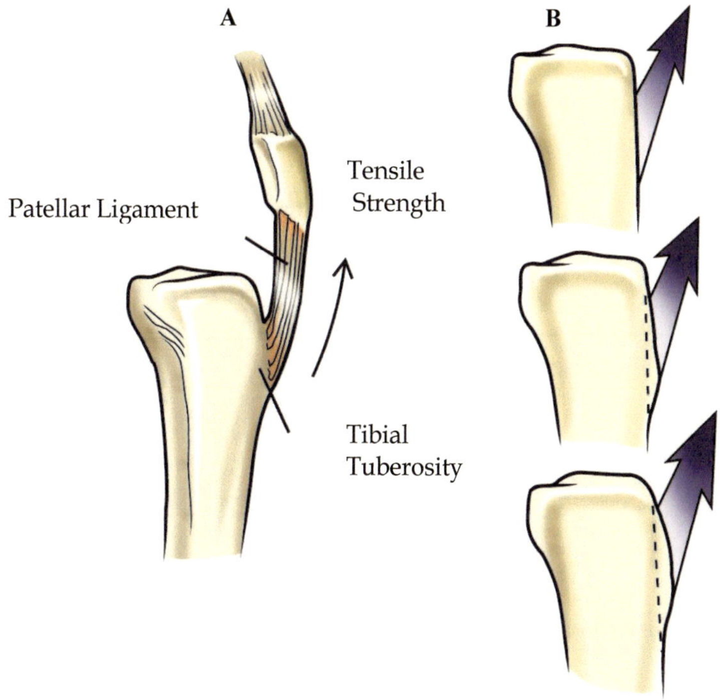

Fig. 1.13. Fig. (A) When tensive strengths are applied on the skeletal system, the bone is strengthened in the direction of traction while the collagen fibers align with the traction of the tendon or ligament. (B) Tensive strengths are also responsible for the development of apophyses, which are bony growths such as processes, tubers or tuberosities. Bankoff (2007, p. 128).

The bone responds to the demands placed on it as described by Wolff's Law already mentioned. Thus, different bones and different sections in a bone will respond differently to compressive and tensive strengths. For example, tibia and femur participates in support weight in lower limb and are strongest when the load is coming from a compressive strength. The fibula does not participate significantly in supporting weight, but it is a muscle insertion site, is stronger when tensive strengths are applied. (Hamil & Knutzen, 1999; Bankoff, 2007).

An assessment of the differences that can be found in the femur showed higher tensive strength capacity of the middle slope of the body that is loaded by a bending strength in the supporting weight. In the femur neck, the bone may withstand large compressive strengths, and in the insertion sites of muscles, there is great tensive strength. (Hamil & Knutzen, 1999; Bankoff, 2007).

Shear strengths – A shear strength is applied parallel to the surface of an object, creating internal deformation in an angular direction (FIGURE 1.9 A and B). Maximum shear stresses act on the surface parallel to the plane of applied strength. The shear stresses are created when a bone is subjected to compressive strengths, tensive strength or both. FIGURE 1.16 shows how a shear stress is developed by applying a compressive or tensile strength. Observe and change the shape of the diamond. As the diamond undergoes distortion by compression or tension, a shear strength applied to the surface occurs. (Riegger, 1985; Bankoff, 2007).

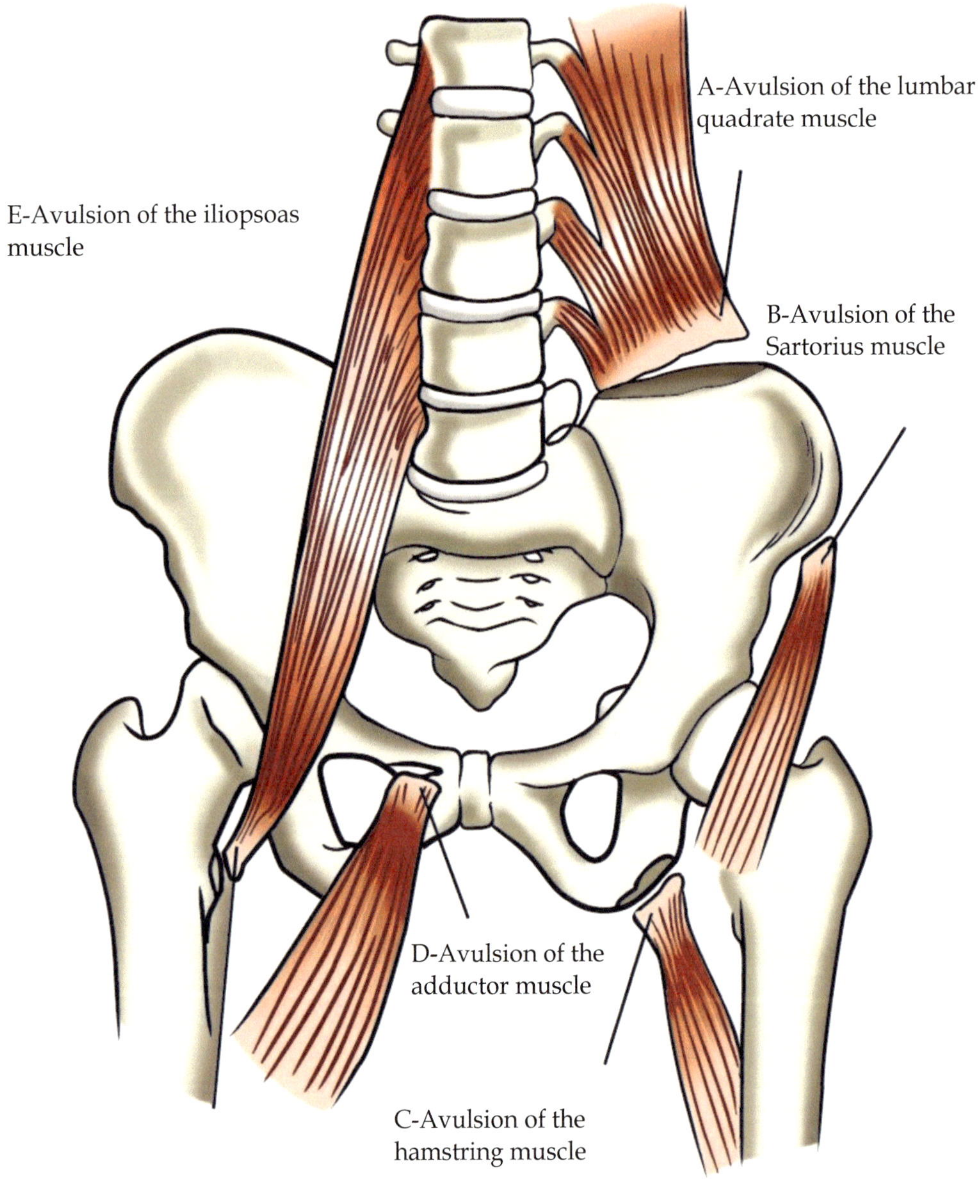

Fig. 1.14. Fractures with avulsion may occur because of tension applied by a tendon or ligament. The sites of injury in which the fracture with avulsion occur in the pelvic region, are shown above and include: (A) anterosuperior spine, (B) anteroinferior spine, (C) ischial tuberosity, (D) pubic bone, and (E) lesser trochanter. Bankoff (2007, p. 129).

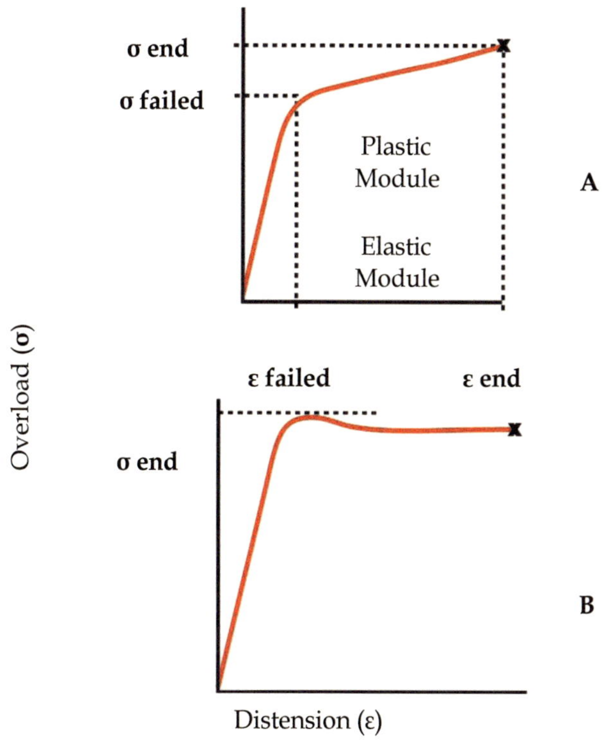

Fig. 1.15. Relation of overload versus distension idealized of a cortical bone subjected to tension loads (a) and compression (b). Bankoff (2007, p. 129).

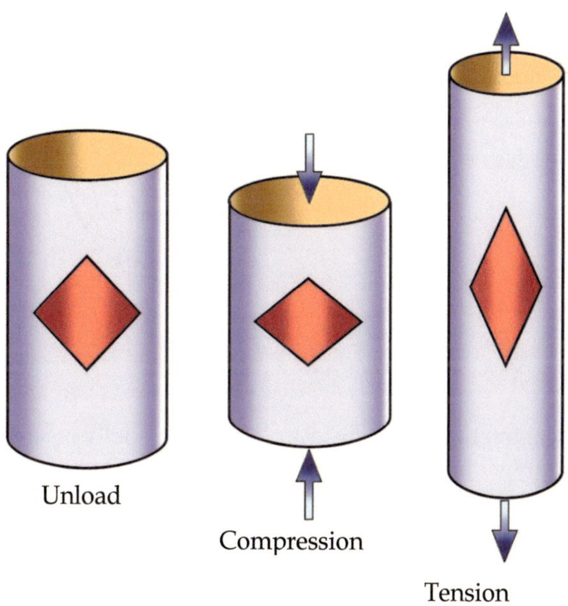

Fig. 1.16. Shear stress and distension accompanies both, tensive and compressive loads. Bankoff (2007, p. 130).

The bone fails more quickly when exposed to a shear strength rather than a compressive or tensive strength. This is because the bone is anisotropic and responds differently when it receives loads of different directions. (Riegger, 1985; Bankoff, 2007).

The shear strengths are responsible for problems in the vertebral discs. A shear strength may produce spondylolisthesis, in which one vertebra slips over another previously. In the lumbar spine, shear strength by vertebrae, increases with increasing lordosis and with hyperlordosis. The pull of muscle on the lumbar vertebrae also creates an increasing shear strength on the vertebrae. (Bankoff, 2007).

Examples of fractures due to shear strengths are frequently found in the femoral condyles or tibial plateau. The injury mechanism of both is usually a hyperextension of the knee with some fixing of the foot and a valgus strength or medial on the thigh or shin. In adults, this shear strength may create a fracture or injury in the collateral or crossed ligaments. In the developing child, this shear strength may create epiphyseal fractures, such as the distal femoral epiphysis. The mechanism of injury and resultant epiphyses injury.

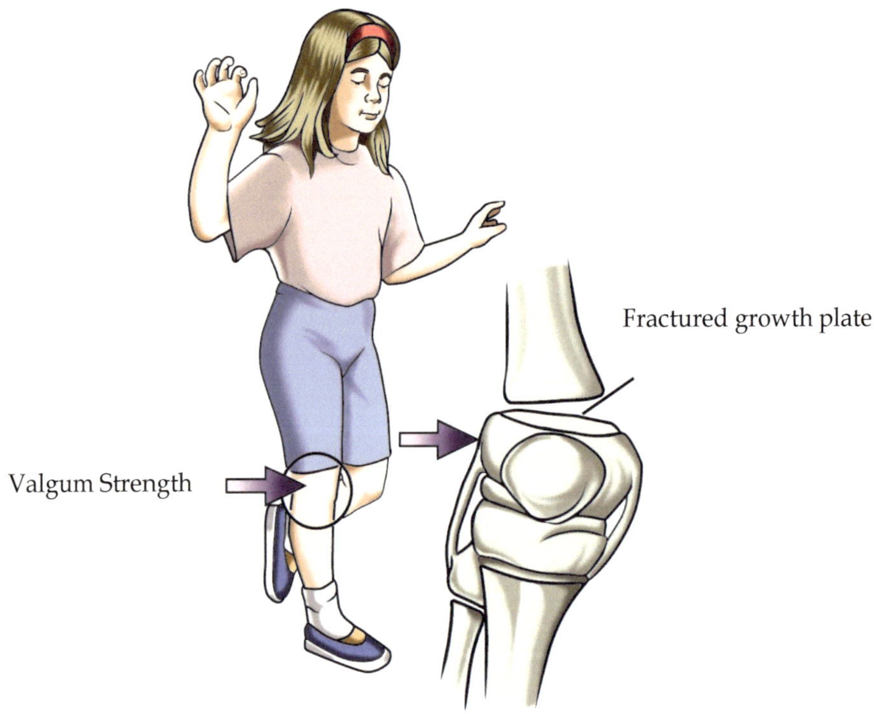

Fig. 1.17. An epiphyseal fracture of the distal epiphysis is usually created by a shearing strength. A strength applied in valgus on the thigh or shin with the foot fixed and hyperextended knee is commonly produced. Bankoff (2007, p. 130).

are shown in FIGURE 1.17. The effects of such a fracture can be quite significant since that epiphysis is the fastest growing in the body and is responsible for approximately 37% of bone growth in the leg. (McConkey & Meeuwisse, 1988; Holich, 1998; Bankoff, 2007).

It is usual the bone is loaded with different types of strength at the same time. FIGURES 1.18 and 1.19 contain an examination of multiple loads absorbed by the tibia during walking and running, respectively. In the walking, there is a compressive stress on the heel contact, created by the weight-bearing, ground contact and muscle contraction. Tensive stress dominates in the middle phase of support because of muscle contraction. It develops a compressive stress in preparation for propulsion, as it increases the strength on the ground and muscle contractions. A shear strength is also present in the propulsive phase of support, and is believed to be related to torsion created by external rotation of the tibia. (Holich, 1998) In the running, stress increases substantially, and stress patterns are different from those seen in the walking. There are similarities in support phase of foot, since it creates a compressive strength due to contact with the ground, body weight and muscle contraction. This is followed by a great tensive continuous stress throughout the withdrawal phase of the toes and balancing phase. (Holich, 1998)

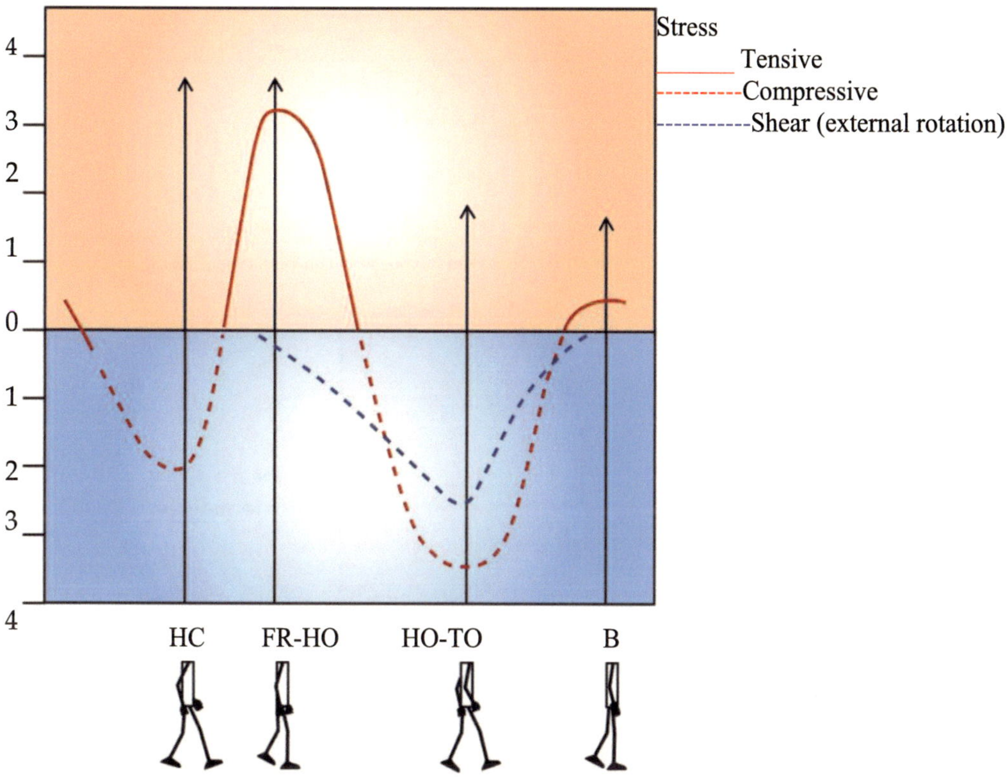

Fig. 1.18. Tensive stress, compressive stress and with shear on the tibia of the adult during the walking - HC = heel contact; FR = foot rectification; HO = heel output; TO = toes output; B = balancing. Bankoff (2007, p. 131).

The pattern of shear stress is also different, and is representative of the twist created in response to internal and external rotation of the tibia. Compressive strengths, tensive and shear applied simultaneously on the bone are important in the development of the bones strength. FIGURE 1.20 illustrates both the compressive stress lines as tensive in the tibia and femur during the running. The bone strength is developed along these lines of stress. (Holich, 1998).

Bending Strengths – A bending strength is the strength applied to an area that has no support offered by the framework. When a bone is subjected to a bending strength and deformation occurs, one side of the bone will form a convexity in which will have tensive strengths, and the other side of the bone, will form a concavity in which compressive strengths are present (FIGURE 1.9 A and B). Typically, the bone will fail and break on the convex side in response to high tensive strengths since the bone may withstand greater compressive strengths than tensive. The magnitude of tensive and compressive strengths produced by bending becomes larger the farther away one is the axis of the bone, so they are larger in the outer portions of the bone. (Holich, 1998; Bankoff, 2007).

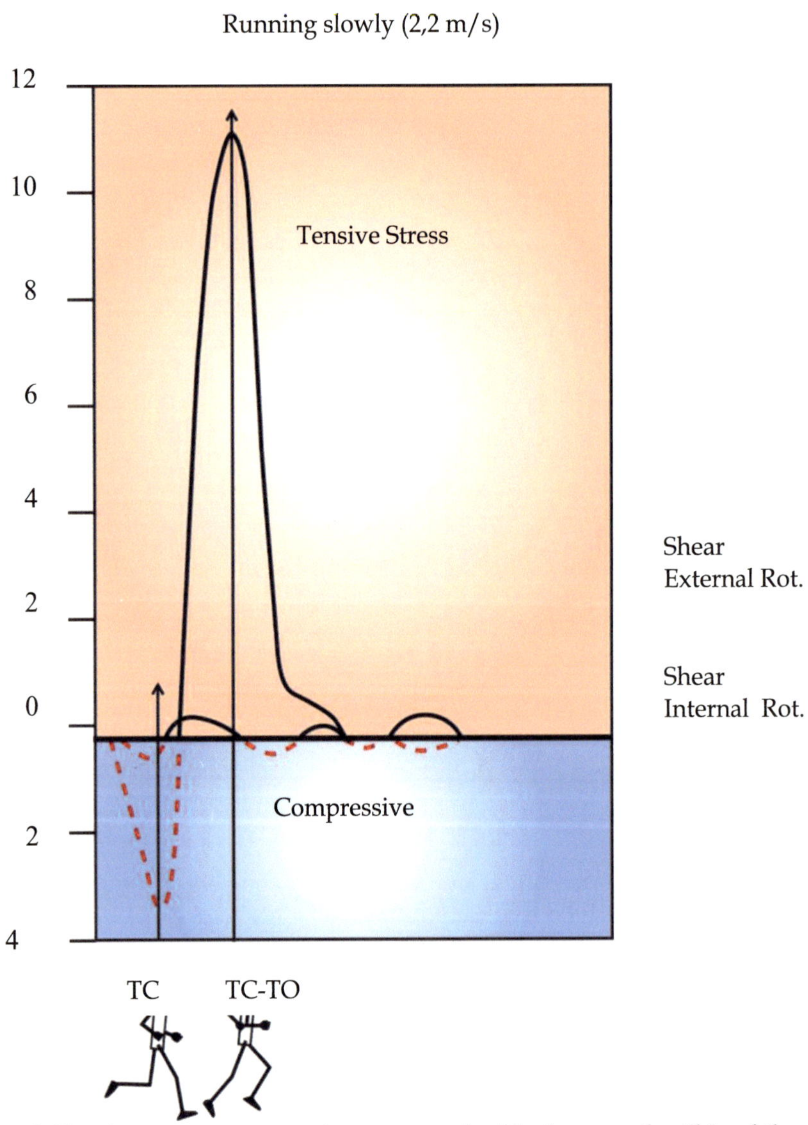

Fig. 1.19. Tensive stress, compressive stress and with shear on the tibia of the adult during the running - TC = toes contact; TO = toes output. Bankoff (2007, p. 131).

During the regular support, there is bending produced both in the femur and in the tibia. The femur is tilted both anteriorly and laterally due to the format and mode of transmission of strength by the supporting weight. The support weight produces an anterior bending in the tibia. Although these bending strengths are not producers of injury, when one examines the strength of the tibia and femur, the bone is stronger in those regions in which the bending strength is greater. (Keller & Spengler, 1989; Jackson, 1990).

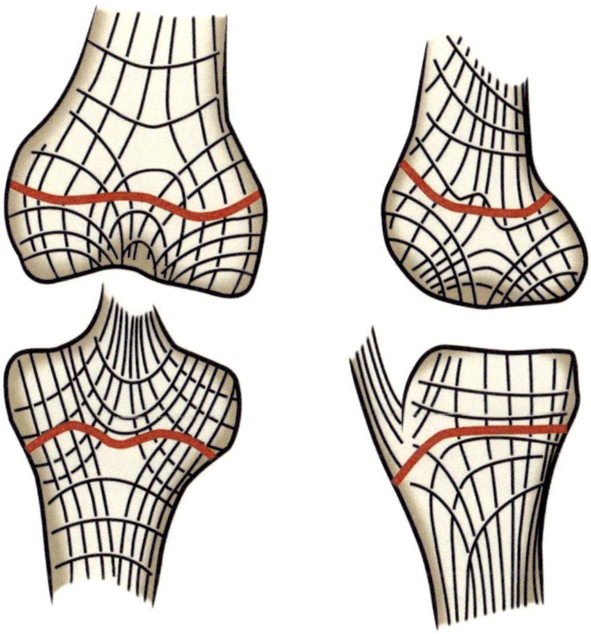

Fig. 1.20. Lines of compressive stress (solid line) and Tensive stress (dashed line) are represented for the distal femur and proximal tibia during the support phase of the running. Bankoff (2007, p. 131).

Bending loads, generators of injury are produced by application of strength in three or four points. The application of strength at three points usually involves strengths applied perpendicularly to the bone at the ends of the bone, with a strength applied in the opposite direction in the middle of the bone. The bone will break in half as occurs in the fracture in ski boot shown in FIGURE 1.21. That fracture is produced when the skier falls on top of the boot with the ski, and the boot pulling the other direction. The bone will break generally in the back because that is where the convexity is given and where are applied the tensive strengths. (Keller & Spengler, 1989; Jackson, 1990).
The bending strength in three points is also liable for injuries to the finger, which is squeezed and forced in hyperextension and knee injuries or lower limb, when the foot is fixed on the ground and lower body bends. Just eliminating the long supports in footwear of American football players, and playing in fields in good condition, this type of injury may be reduced by half. (Keller & Spengler, 1989; Jackson, 1990).
The application of bending strengths at three points can also be used in orthoses. FIGURE 1.22 shows two applications of orthoses using the application of strength at three points for

a correct postural deviation or stabilize a region. A bending load is applied at four points with the application of two equal and opposite pairs of strength in each end of the bone. In the case of four-point bending, the bone will break at the weakest point. This is illustrated in Figure 1.23 with the application of a bending strength of four points on the femur. The femur breaks at its weakest point. (Jackson, 1990).

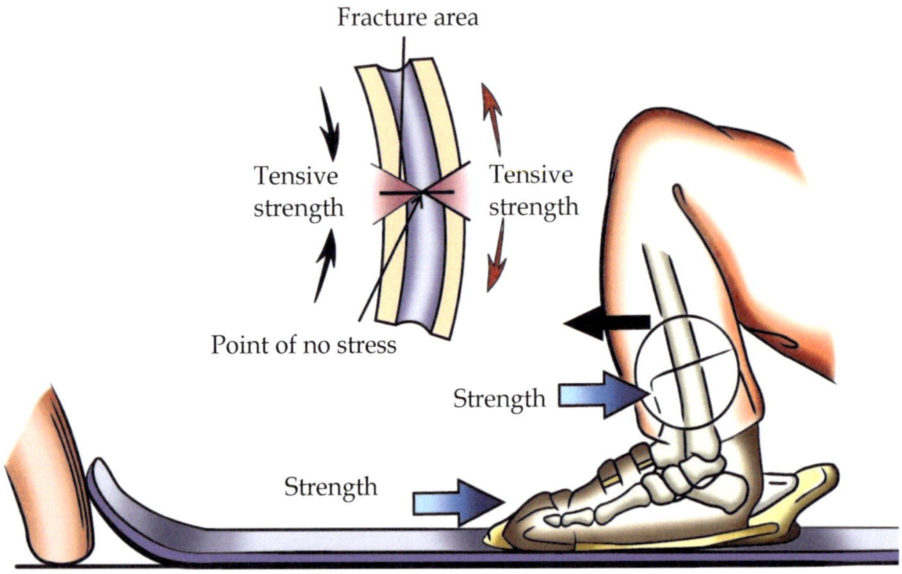

Fig. 1.21. A bending load at three points creates fracture in ski boots and occurs when the ski is detained abruptly. A compressive strength is created in the anterior tibia and a tensile strength is created in the posterior tibia. The tibia fracture is usually on the back. Bankoff (2007, p. 132).

Torsion Strengths – A torsion strength applied to the bone is a rotational strength, creating a stress with shear on the material (FIGURE 1.9 A and B). The magnitude of stress increases with distance from the axis of rotation, and the maximum shear stress acts both perpendicular as parallel to the axis of the bone. A torsion load also produces tensive and compressive strengths at the angle through the structure. (Cook et al., 1987; Choi & Goldstein, 1992).
Gregerson, 1971, described that fractures that result from torsion strength occur at the humerus when imperfect launching techniques create a twist of the arm and lower limb when the foot is planted and the body changes direction. A spiral fracture is produced because of applying a torsion strength. An example of the mechanism of a spiral fracture at the humerus is what happens to a pitcher as shown in FIGURE 1.24. The fracture usually starts on the outside of the bone and parallel to the middle of the bone. The torsion load on the lower limb is also responsible for injuries at the cartilage and ligaments in the knee joint.
Injury vs. Load – If a bone will or not suffers an injury because of an applied strength, it depends on the limits of critical strength of the material and the history of loads received by the bone. These limits are influenced primarily by the load of the bone can be increased or decreased by physical activity and conditioning by immobilization and skeletal maturity of the individual. The speed with which the load is placed is also important because the response and tolerance are sensitive to it. Loads placed very quickly, when the bone tissue is unable to deform at the same speed, can cause injury. (Pirnay, 1987).

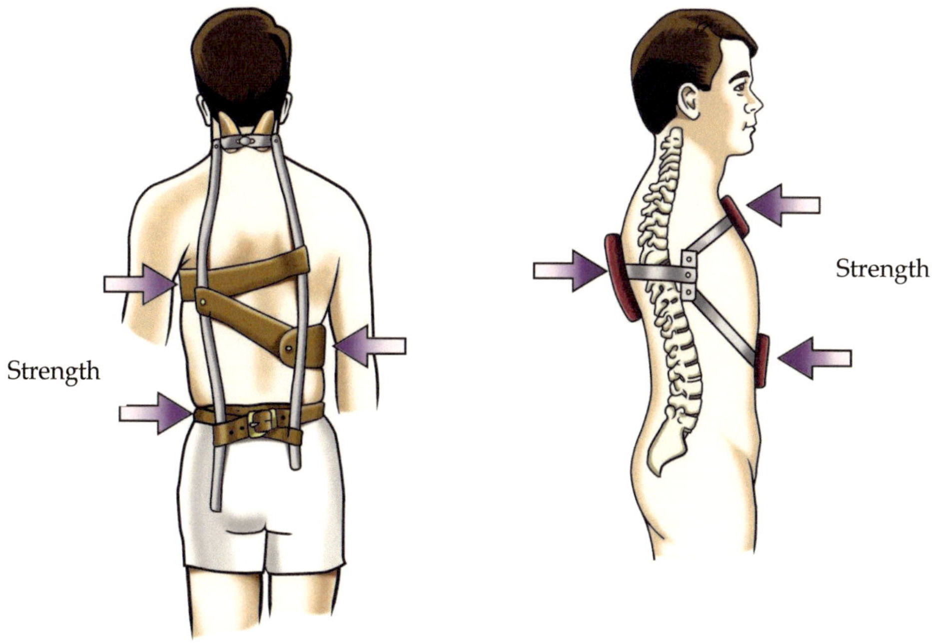

A - Collect Milwaukee B - Collect Jewett

Fig. 1.22. Bending load is used at three points in many types of orthoses. (A) The Milwaukee brace was used for correction of lateral curvature of the spine and was applied a bending strength at three points in the column. (B) The Jewett orthosis applies a bending strength at three points in the thoracic spine to create extension of column in the region. Bankoff (2007, p. 132).

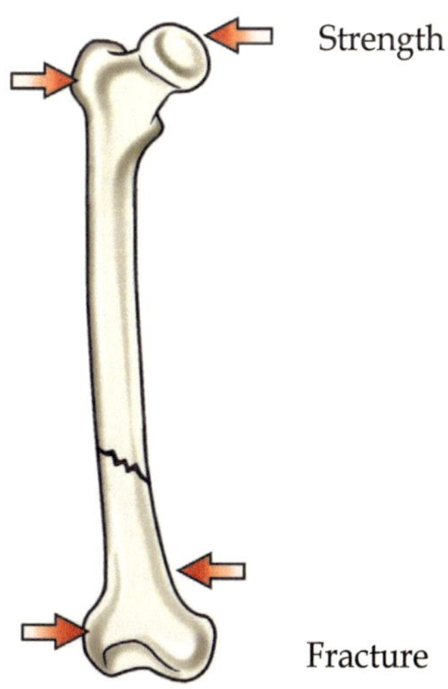

Fig. 1.23. A bending load in four points applied in a structure will create a break or failure at the weakest point. Above, it is a hypothetical example using the femur. Bankoff (2007, p. 132).

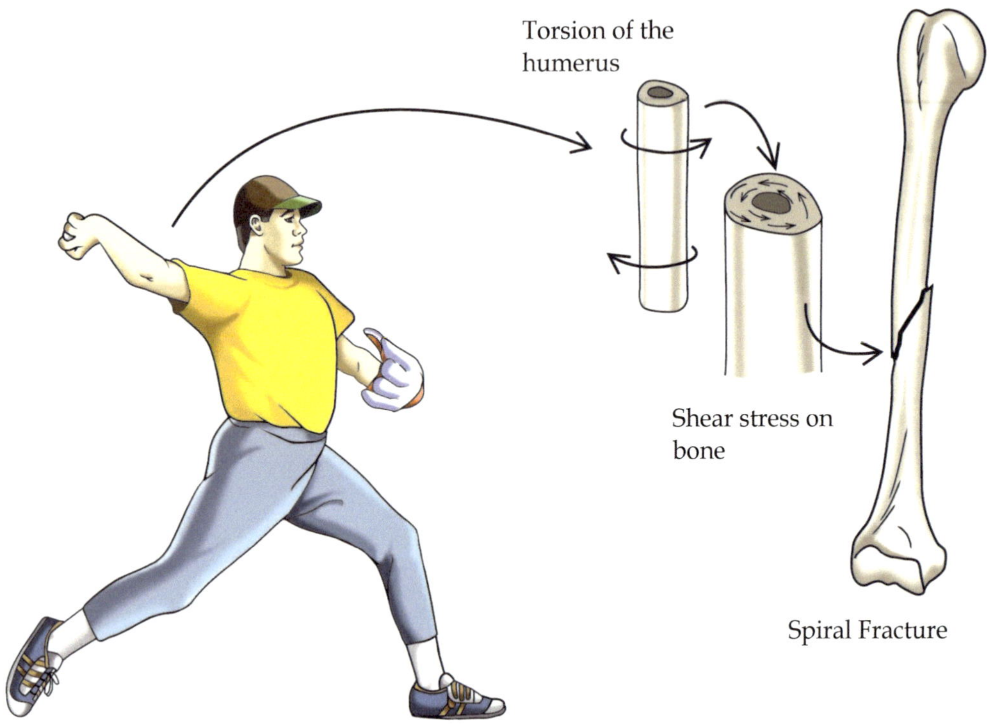

Fig. 1.24. A torsion strength applied to the bone creates a shear stress to the surface. An example of torsion applied to the humerus is shown above. Bankoff (2007, p. 133).

Muscular activity vs. Load – Muscle activity can also influence in loads that can be managed by the bones. The muscles change the strengths applied at the bone creating tensive and compressive strengths. These muscle strengths can reduce tensive strengths or redistribute the strengths on the bone. Since most bones can withstand large amounts of compressive strengths, the total amount of load may increase due to the contribution of muscles. However, if the muscle fatigues during a series of exercises, this decreases its ability to lighten the load on the bone. The altered distribution of stress or increase in tensile strengths makes the athlete or player, prone to injury. (Pirnay, 1987).

Stress Fracture – The typical stress fracture occurs during load application, which produces a shear distension or tension, resulting in lacerations, fractures, ruptures or avulsions. The bone tissue can also develop a stress fracture in response to compressive or tensive loads that overwhelm the system, either by a magnitude of excessive strength applied to one or a few times, or by applying strength in a low or moderate level, but with an excessive frequency. The relation between the magnitude and frequency of load applications on the bone . Tolerance of the bone for the injury is a function of load and cycles of load placement. (McCue, 1970; Matheson, 1987).

Stress fracture occurs when the bone resorption weakens the bone too much and bone deposit does not occur quickly enough to strengthen the area. The cause of stress fractures at the lower limb can be attributed to muscle fatigue, which reduces the shock absorption and allows the redeployment of strengths to specific focal points in the bone. In the upper limb, the stress fractures are created by repetitive muscular strengths that exert traction on the bones. This type of fracture responds for 10% of all injuries in athletes. (McCue, 1970; Matheson, 1987).

4. Conclusion

The research in bone biomechanics mentioned in this section contributed to show the importance of this area of study and brought brief discussions on the bone tissue and its incorporation in the biomechanical aspect of human skeletal and locomotor system. The information contained in this study by the authors was a cited research and placed the bone tissue (histology, anatomy, biomechanics and kinesiology) as a material adaptive level of loads.

5. Acknowledgments

To the researchers cited in this section for scientific contributions on bone biomechanical considerations:
To the *Espaço da Escrita* from the University of Campinas for their contribution in the process of translating the text.
To the Graduate School of Physical Education in particular to Prof. Antonio Carlos de Moraes.
Prof. Dr. Carlos Aparecido Zamai aid in the development and technical preparation of the text.

6. References

Alberts, B. et al. (1994). Molecular biology of the cell. Garland Press, 3rd ed.

Bankoff, A.D.P. (2007). Morfologia e Cinesiologia Aplicada ao Movimento Humano. Editora Guanabara Koogan, Rio de Janeiro- Brasil.

Bechtol, C.O. (1954). Grip test. J. Bone Joint Surg., 36-A, 820-824.

Boume, G.H. (editor). (1976). The biochemistry and physiology of bone. 2nd ed. 4 vols. Academic Press.

Choi, K. & Goldstein, S.A. (1992). A comparison of the fatigue behavior of human trabeculae and cortical bone tissue. Journal Biomechanics, 25: 1371.

Cook, S.D. et al. (1987). Trabeculae bone density and menstrual function in women runners. The American Journal of Sports Medicine.15: 503.

Egan, J.M. (1987). A constitutive model for the mechanical behavior of soft connective tissues. Journal of Biomechanics. 20: 681-692.

Fine, K.M.; Vegso, J.J.; Sennett, B., & Torg, J.S. (1991). Prevention of cervical spine injuries in football. The Physician and Sports Medicine.Vol 19 (10): 54-64.

Gregerson, H.N. (1971). Fractures of the Humerus from Muscular Violence. Acta Orthop. Scand., 42, 506-512.

Halpbern, B.C., & Smith, A. D. (1991). Catching the cause of low back pain. The Physician and Sports Medicine. Vol. 19(6): 71079.

Halpbern, B., et al. (1987). High school football injuries: Identifying the risk factors. The American Journal of Sports Medicine. 15: 316.

Hamill, J.; & Knutzen, K.M. (1999). Bases biomecânicas do movimento humano. São Paulo: Manole

Hay, E.D. (editor). (1982). Cell biology of extracellular matrix. Plenum.

Hoffman, A.H.; & Grigg, P. (1989). Measurement of joint capsule tissue loading in the cat knee using calibrated mechano-receptors. Journal of Biomechanics. 22: 787-791.

Holtrop, M.E. (1975). The ultra structure of bone. Ann Clin Lab Sci, 5:264.

Holick, M.F. (1998). Perspective on the impact of weightlessness on calcium and bone metabolism. Bone, New York. v.22, n.5, p.105-111.

Jackson, D.L. (1990). Stress fracture of the femur. The Physician and Sports Medicine. v. 9, (7), pp. 39-44

Junqueira, L.C.; & Carneiro, J. (1999). Histologia básica. 9 ed, Rio de Janeiro: Guanabara Koogan.

Junqueira, L.C.; & Carneiro, J. (1997). Biologia celular e molecular. 6 ed. Rio de Janeiro: Guanabara Koogan.

Keller, T.S.; Spengler, D.M. (1989). Regulation of bone stress and strain in the immature and mature rat femur. Journal of Biomechanics. 22:1115-1127.

Lakes, R.S.; Nakamura, S.; Behiri, J.C. E.; & Bonfield, W. (1990). Fracture mechanics of bone with short cracks. Journal of Biomechanics. 23:967-975.

Marks Jr, S.C.; & Popoff, S.N. (1988). Bone cell biology the regulation of development structure, and function in the skeleton. Amer J Anat 183:1.

McConkey, J.P., & Meeuwisse, W. (1988). Tibial plateau fractures in alpine skiing. The American Journal of Sports Medicine. 16: 159-164.

Matheson, G.O. et al. (1987). Stress fractures in athletes. The American Journal of Sports Medicine. 15:46-58.

McCue, F.C. (1970). Athletic Injuries of the Proximal Interphalangeal Joint Requiring Surgical Treatment. J. Bone Joint Surg., 52-A, 937-956.

Mundy, G.R. et al. (1995). The effects of cytokines and growth factors on osteoblastic cells. Bone. 17:71.

Pirnay, F.M. et al. (1987). Bone mineral contend and physical activity. International Journal Sport Medicine, 8: 331.

Riegger, C.L. (1985). Mechanical properties of bone. Ln: Orthopaedic and Sports Physical Therapy. Edited by J.A. Gouldand G.J. Davies. St. Louis, C.V. Mosby Co, 3-49.

Schaffler, M.B.; & Burr, D.B. (1988). Stiffness of compact bone: Effects of porosity and density. Journal of Biomechanics. 21:13-16.

Shipman, P., Walker, A.; & Bichell, D. (1985). The Human Skeleton. Cambridge, Harvard University Press.

5

Biomechanical Studies on Hand Function in Rehabilitation

Sofia Brorsson
Halmstad University, School of Business and Engineering, Sweden

1. Introduction

Hand function requires interaction of muscles, tendons, bones, joints and nerves. The unique construction of the hand provides a wide range of important functions such as manipulation, sense of touch, communication and grip strength (Schieber and Santello 2004). The hand is used in many ways, and in many different situations in our daily lives; so injuries, diseases or deformities of the hand can affect our quality of life. Several of our most common injuries and diseases affect hand function. Therefore, it is very important to understand how healthy and diseased hands work in order to be able to design optimal rehabilitation strategies pursuant to hand injury or disease.

There are many different methods used today for evaluating hand and finger functions. One widely accepted method that provides an objective index of the hand and finger functions is hand force measurement (Balogun, Akomolafe et al. 1991; Innes 1999; Incel, Ceceli et al. 2002). There is also a potential for using modern non-invasive methods such as ultrasound and finger extension force measurements, but these have not been completely explored so far.

An important factor in developing grip force is the synergy between the flexor and extensor muscles. The extensor muscles are active when opening the hand, which is necessary for managing daily activities (Fransson and Winkel 1991). Even though the extensor muscles are important for optimal hand function, surprisingly little attention has been focused on these muscles. It has, however, been difficult to evaluate hand extension force, since there is no commercially available measurement instrument for finger extension force. In addition, because of the lack of a device to assess extension force, there is limited basic knowledge concerning different injuries and how diseases affect the static and dynamic forearm muscle architecture or/and muscle interaction.

Impaired grip ability in certain diseases such as Rheumatoid Arthritis (RA) could be caused by dysfunctional extensor muscles leading to inability to open the hand (Neurath and Stofft 1993; Vliet Vlieland, van der Wijk et al. 1996; Bielefeld and Neumann 2005; Fischer, Stubblefield et al. 2007). Deformities of the MCP-joints are common, and may lead to flexion contractures and ulnar drift of the fingers. Weak extensor muscles may play a role in the development of these hand deformities. Furthermore, knowledge concerning how the muscles are influenced by RA and the mechanism of muscle force impairments is not fully understood for RA patients. This group of patients would benefit from further hand/finger evaluation methods for evaluation of rehabilitation and interventions. There is also a need for further knowledge of the dynamic action of skeletal muscle and the relation between muscle morphology and muscle force. The force that can be generated is dependent on the muscle architecture; these architectural parameters can be studied non-invasively with US. By using US it is possible to obtain detailed, dynamic information on the muscle architecture. In order to assess how disease influences muscle morphology and function, it is necessary to establish baseline knowledge concerning normal forearm muscles. The general aim of this book chapter was to further our knowledge about biomechanics of the hand, RA patient, non-invasive evaluation methods used for evaluation of rehabilitation interventions and muscle biomechanics will be further presented.

2. Biomechanics of the hand

It is important to understand the biomechanics of the hands and fingers as well as the muscle architecture and structure in order to develop new evaluation methods for finger extension force. The construction of the hand is quite complicated, including 29 joints, 27 bones and more than 30 muscles and tendons working together for range of motion (ROM), performing perception and force production.

2.1 The construction of the hand

The metacarpophalangeal (MCP) joints II-V are condyloid joints that allow for movement in two planes, flexion/extension or adduction/abduction. The ROM in the joints is approximately 30–40 degrees extension, 70–95 degrees flexion and 20 degrees adduction/abduction. Ligaments connect the bones and provide stability of the joints; in the hand there are numerous ligaments that stabilize the joints. To provide stability to the metacarpal bones, there are ligaments working in conjunction with a thick tissue located in the palm (the palmar aponeurosis). Muscles that control the hand and have their origin located near the elbow are called the extrinsic muscles. The tendons of these muscles cross the wrist and are attached to the bones of the hand. The large muscles that bend (flex) the fingers originate from the medial aspect of the elbow. The large muscles that straighten (extend) the fingers originate from the lateral aspect of the elbow. The extrinsic muscles are responsible for powerful grip ability. In addition to these large muscles, there are smaller muscles in the hand, intrinsic muscles, that flex, extend, abduct (move outwards) and adduct (move inwards). The agonist for extension in fingers II–V is the muscle extensor digitorum communis (EDC). This muscle originates at the lateral epicondyle of humerus; the muscle is connected to phalanges II–V by four tendons, which glide over the MCP-joints articulations. The tendons divide into three parts. The main part is attached to the extensor hood and two collateral ligaments are attached at the lateral and medial parts of the fingers. The extensor hood covers the whole phalange and is formed from the extensor digitorum tendon and fibrous tissue. The extension ability in the MCP-, proximal interphalangeal-, and distal interphalangeal joints are produced by EDC, interossei and lumbricales muscles (Smith 1996; Marieb 1997). Finger extension force is dependent on the wrist position. However, at the present time there is no consensus for the optimal wrist angle for finger extension force measurement. Researchers believe that a wrist position between 10-30 degrees is suitable for finger extension measurements (Li 2002).

2.2 Muscle force

The forces a muscle can produce depend on many factors such as the muscles' structure, muscle architecture, muscle-nerve interaction and physiological aspects. This thesis focuses mainly on how the muscle structure, at macro level, affects the forces produced. A brief overview of the micro architecture level and muscle control is described in this chapter.

The skeletal muscles have four behavioral properties, extensibility, elasticity, irritability and the ability to develop tension. Extensibility and elasticity provide muscles the ability to stretch or to increase in length and to return to normal length after stretching and these properties provide a smooth transmission of tension from muscle to the bones. The muscle's ability to respond to stimuli, irritability, provides the capability to develop tension. The tension that muscles provide has also been referred to as contraction, or the contractile component of muscle function. The tension that a muscle can develop affects the magnitude of the force generated, the speed, and length of time that the force is maintained; all these parameters are influenced by the muscle architecture and function of the particular muscle. The manner in which the muscles are constructed and controlled contributes to muscle force production. The force that a muscle generates is also related to the velocity of muscle shortening, such as the force-velocity relationship, length-tension relationship, stretch-shortening cycle and electromechanical delay (Wickiewicz, Roy et al. 1984; Brand 1993; Fitts and Widrick 1996; Kanehisa, Ikegawa et al. 1997; Debicki, Gribble et al. 2004; Hopkins, Feland et al. 2007).

2.2.1 Macro-architecture

Muscle architecture has been studied by muscle-imaging techniques such as magnetic resonance imaging and ultrasound (US), and research has shown that there are numerous variations in the muscle architecture (i.e. fibre length, pennation angle, cross-sectional area (CSA), muscle volume etc.) within and between species. The architecture of a skeletal muscle is the macroscopic arrangement of the muscle fibres. These are considered relative to the axis of force generated (Otten 1988; Blazevich and Sharp 2005). The arrangements of muscle fibres affect the strength of muscular contraction and the ROM which a muscle group can move a body segment. It is important to understand the impact of muscle architecture parameters in order to design effective interventions for disease, injury rehabilitation, as well as for athletic training and exercise, especially considering the results of adaptation to physical training. The pennation angle is the angle between the muscle fibre and the force generating axis (Figure 1). Early researchers have reported greater pennation angles in subjects that practice weight training compared to untrained subjects. It has been claimed that increase in pennation angle is biomechanically important since more tissue can attach to a given area of tendon, and slower rotation of the muscle fibre during contraction is possible through a greater displacement of the tendon, thus generating more force (Aagaard, Andersen et al. 2001; Kawakami, Akima et al. 2001).

Fascicle length (muscle fibre) can be of importance for the biomechanics of the muscles, the change in fascicle length has been reported to have impact on high-speed force generation (Fukunaga, Ichinose et al. 1997). The fascicles containing a greater number of sarcomeres in series and generate force over longer ranges of motion and longer fibres also possess greater shortening speeds. From experimental studies, it has been claimed that the physiological cross-sectional area (PCSA) of a muscle is the only architectural parameter that is directly proportional to the maximum tetanic tension generated by the muscle. Theoretically, the

PCSA represents the sum of all CSA of the muscle fibres inside the muscle. The design of the muscles in terms of pennation angle, fibre length and PCSA reflects the muscles' capacity to develop force. Although each muscle is unique in architectural design, a number of generalizations have been made on the lower extremity muscles. For example quadriceps muscles are designed with high pennation angles, large PCSA and short muscle fibres, and this design is suitable for large force production. The same design pattern can be observed in the upper extremity, and the flexor muscles structure predicts that they generate almost twice the force as the extensor muscles (Lieber and Friden 2000). To summarize: the research about muscle architecture and adaptation to speed and strength exercises shows that muscle architecture is plastic and can respond to exercise, although more research is required to fully understand the impact of varying methods of strength and speed training. To fully understand the adaptation of muscle architecture to all forms of interventions would require a formidable research effort. Surprisingly little research has described changes of muscle architecture when aging, despite that aging is associated with significant sarcopenia.

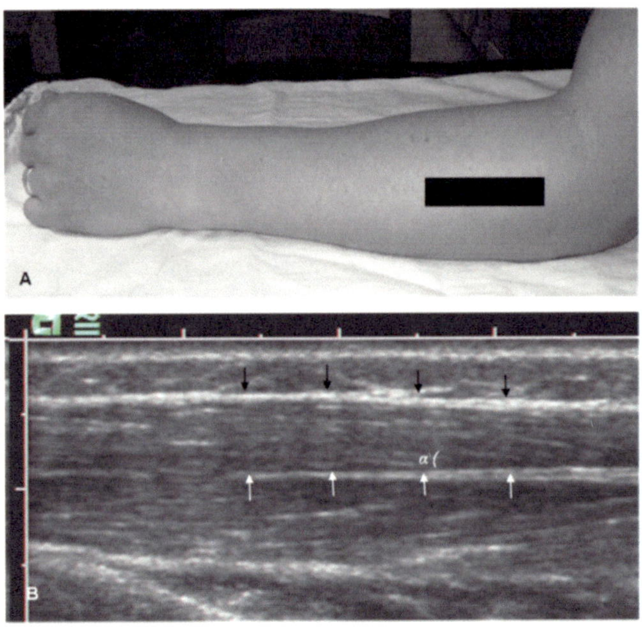

Fig. 1. (A) The black rectangle shows the position of the US probe during pennation angle measurements of the m.EDC. (B) The longitudinal US image showing the superficial aponeurosis (black arrows), the deep aponeurosis (white arrows) and the pennation angle (α). ©Sofia Brorsson

Previous research has claimed that pennation angle and fascicle length were significantly smaller in older than younger individuals in some muscles such as m. soleus, m. gastrocnemius medialis and lateralis (Kubo, Kanehisa et al. 2003; Narici, Maganaris et al. 2003; Morse, Thom et al. 2005), but there were no age related changes in m. triceps brachii and m. gastrocnemius medialis concerning pennation angles for women (Kubo, Kanehisa et al. 2003).Furthermore, little research has been done concerning how muscle architecture adapts to disuse or diseased muscles, which is very important from a rehabilitation perspective. Kawakami et al. (2000) investigated changes in the muscle parameters fascicle length, pennation angle and CSA in m.triceps brachii and m. vastus lateralis after 20 days of bed rest. They found no significant changes in fascicle length and pennation angle even though there was a significant reduction of the CSA (Kawakami, Muraoka et al.2000). Other researchers have reported decreased muscle size, muscle strength and decreased pennation angles after bed rest (Akima, Kuno et al.1997; Narici and Cerretelli 1998; Kawakami, Akima et al. 2001). It has been claimed that one explanation for the different adaptations of muscle architecture in different disused muscles (due to bed rest) is that the changes depends on the individual muscle actions.

2.2.2 Micro-architecture

The skeletal muscles have a wide range of variations in size, shape, and arrangement of fibres. Skeletal muscles are composed of muscle fibres that are bundled together in fascicles, the fascicles are composed of about 200 muscle fibres. Each muscle fibre is surrounded by the endomysium, which is connected to muscle fascia and tendons. The muscle fibres are formed by myofilaments, comprised of myofibrils. A contractile myofibril is composed of units, sarcomeres (Smith 1996; Marieb 1997). By using electron microscopy researchers have observed the muscle structure (ultra-structure) and structures such as sarcomeres, actin and myosin were analysed (Alberts 2002). These structures have become the basis of the theory of sliding filaments during muscle contraction and later to the Cross-bridge theory, which has become the accepted paradigm for muscle force production (Huxley 1954; Huxley 1957; Huxley and Simmons 1971).

2.2.3 Muscle control

Muscles allow us to move our joints, to apply force and to interact with our world through action. Muscles are important for us because they have the unique ability to shorten, and to do that with enough force to perform movements. Muscle fibres are arranged into functional groups; there, all fibres are innervated by one single motor neuron; these groups are called motor units. Movements that are precisely controlled such as the finger movements are produced by motor units with small numbers of fibres (Kandel, Schwartz et al. 1991). When a muscle fibre is activated by a motor nerve impulse, the actin and myosin filaments in the sarcomere connect strongly to each other, pulling the filaments together. Sarcomeres are arranged in long chains that build up the muscle fibre, so when the sarcomeres contract, become shorter, the whole fibre becomes shorter. To be able to produce force the muscle must be innervated by a motor neuron, and the excitation-contraction coupling is along the whole fibre length simultaneously through the T-tubule system. This leads to rapid release of calcium ions from the sarcoplasmic reticulum. When the contraction signal ends, the calcium is driven back to the sarcoplasmic reticulum through ATP-driven calcium pumps (Kandel, Schwartz et al. 1991). Increase in neuromuscular function and muscle strength is attained when the load intensity exceeds that of the normal daily activity of the individual muscles (Hellebrandt and Houtz 1956; Karlsson, Komi et al. 1979). Increase in muscle performance at the beginning of strength training can be explained by physiological and neural adaptation, such as effective recruitment of motor units and reduction of inhibitory inputs of the alpha motor neurons (Hakkinen, Malkia et al. 1997). Several researchers have reported that muscle hypertrophy occurs after 6–8 weeks of strength training and that a certain level of muscle strength is needed to prevent a decline in functional capacity (Nygard, Luopajarvi et al. 1988; Sale 1988; Kannus, Jozsa et al. 1992). Inactivity or decrease in physical activity leads to loss of muscle strength and a decrease in neuromuscular performance, this has been observed for patients with arthritis (Hakkinen, Hannonen et al. 1995). Some researcher claim that, during the early phase, muscle force production after exercise is more related to improved innervations than increased CSA (Blazevich, Gill et al. 2007).

3. Non-invasive evaluation methods in rehabilitation

In this thesis, the effect of both the static and dynamic muscle architecture and the ability to produce force is studied in the extensor muscle EDC in healthy subjects and RA patients; either as physical performance or self-reported function. There are different evaluation methods available to evaluate muscle architecture, force production and hand function in rehabilitation.

3.1 Grip force measurements

Hand force is an important factor for determining the efficiency of interventions such as physiotherapy and hand surgery. Hand force/grip strength is widely accepted as providing an objective measure of the hand function (Balogun, Akomolafe et al. 1991; Incel, Ceceli et al. 2002) and measurements of grip force have been used to evaluate patients with upper extremity dysfunction. However, measurements have mainly been made of the flexion force and pinch force. Even though flexion forces represent only 14 % and tripod pinch grip only 10 % of all daily hand grip activity (Adams, Burridge et al. 2004). Surprisingly little measurements have been made of the finger extension force, despite the fact that extension force is important in developing grip force. Furthermore, it has been difficult to evaluate hand extension force impairment, since no commercially available measurement instrument for finger extension force exists. Some research instruments have been designed. However they are complicated, with little clinical potential and do not have the ability to measure both whole hand extension force and single finger extension forces as the new force measurement device, EX-it, has (Brorsson 2008 a, Kilgore, Lauer et al. 1998; da Silva 2002; Li, Pfaeffle et al. 2003). Hand grip measurements have been seen to be a responsive measure in relation to hand pain and correlate well with patients' overall opinion of their hand ability; these measurements provide a quick evaluation of patient's progress throughout treatment (Incel, Ceceli et al. 2002; Adams, Burridge et al. 2004). Grip force is influenced by many factors including fatigue, time of day, hand dominance, pain, sex, age and restricted motion. Interestingly, the synergistic action of flexor and extensor muscles is an important factor for grip force production (Richards, Olson et al. 1996; Incel, Ceceli et al. 2002). It is widely accepted that grip and pinch force measurements provide an objective index of the functional integrity of the upper extremity. Today there are devices for measuring some grips, such as Jamar™, Grippit™, MIE digital power and pinch grip analyser™ and Pinchmeter ™ (Nordenskiold and Grimby 1993; Lagerstrom and Nordgren 1998; Mitsionis, Pakos et al. 2008). Severe weaknesses in RA patients' grip forces have been reported by several authors. Nordenskiöld et al. (1993), reported reduced flexion force for RA women compared to healthy controls using the Grippit device. Furthermore, Nordenskiöld (1997) reported a relationship between significant grip force and daily activities (Nordenskiold and Grimby 1993; Nordenskiold 1997). The activity limitations in relation to grip force and sex after 3 years of RA has been claimed to be lower for women than for men. The authors concluded that this result may be explained by reduced grip force rather than sex (Thyberg,

Hass et al. 2005). Fraser et al. (1999) reported weakness in three different grip types using an MIE digital power and pinch grip analyser. They measured flexion force, pinch force and tripod force. They also measured forearm parameters which they expected to be relevant for producing forces, such as hand and forearm volume. They could however not find any significant differences between healthy and RA parameters (Fraser, Vallow et al. 1999). Buljina et al. (2001) reported the effectiveness of hand therapy for RA patients. They evaluated grip strength with the measuring device called Jamar 1113 (Sammons-Preston, Jackson, MI), then they analysed the tip-to-tip pinch, palmar pinch, key pinch, range of motions in the MCP-joints while pain in the hands was measured by a visual analog scale (VAS). They reported the effectiveness of therapy and that the RA patients significantly increased their hand force (Buljina, Taljanovic et al. 2001). Jones et al. (1991) reported that RA patients hand force was 75 % lower than healthy subjects (Jones, Hanly et al. 1991). Even though hand exercises are used frequently for keeping and preventing loss of grip force for RA patients, only few studies have evaluated the result of grip improvement (Hoenig, Groff et al. 1993). Adams et al. (2004) reported flexion and tripod force recorded by an MIE digital grip analyser, hand function was evaluated with the Grip ability test (GAT) and the patient's questionnaire Disability Arm Shoulder Hand (DASH). They concluded that grip force was significantly correlated to self-reported assessment and hand function (Adams, Burridge et al. 2004). Brorsson etal. (2008 a,b) showed that the extension force was significantly reduced in the RA group (men, $p < 0.05$, and women $p < 0.001$) compared to the control group. Furthermore, they showed that there was a significant difference between the finger extension force for healthy men and women ($p < 0.001$), the finger extension force and flexion force in the dominant hand for healthy subjects and RA patients are presented in Figure 2.

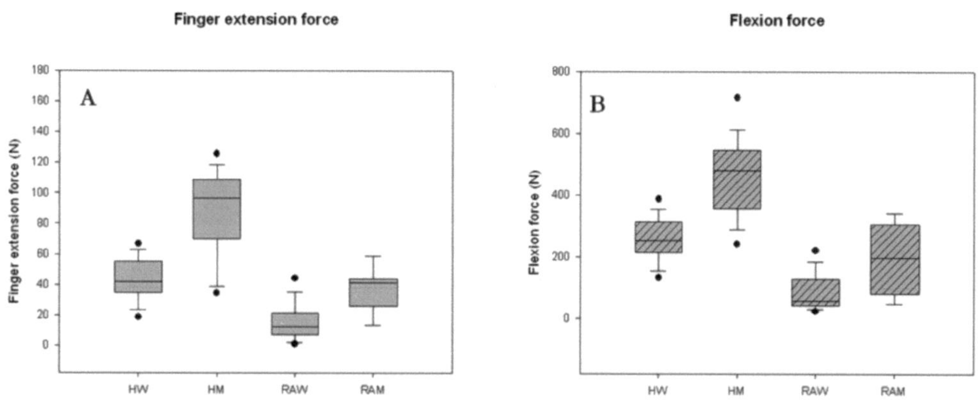

Fig. 2. (A) Finger extension force in dominant-hand. (B) Flexion force in dominant-hand. The box-plots represent healthy women (HW), healthy men (HM), women with RA (RAW) and men with RA (RAM). The results are from participants in all papers (n=80 HW, n=47 HM, n=65 RAW and n=12 RAM).

3.2 Ultrasound examination in skeletal muscle architechture

Ultrasound technology provides new and exciting possibilities to non-invasively access physiological mechanisms inside the living body, both at rest and during muscle contraction. Ultrasonic devices collect sound waves that are emitted by a probe after reflecting off the body's internal tissues; this provides detailed images of the body structures. The recent developments of the probes have enabled the use of US to examine the joint and surrounding soft tissues such as the muscles. The increasing interest for US among rheumatologists contributes to the understanding of the natural history of rheumatic diseases, and US is today important in the early diagnosis of RA (Kane, Balint et al. 2004; Grassi, Salaffi et al. 2005) . US has been used in several studies to provide *in vivo* information about the muscle architecture of different muscles. Zheng et al. (2006) combined US with surface electromyography for evaluating changes in muscle architecture after using prosthetics (Zheng, Chan et al. 2006). US has also been used to study the differences between men and women regarding muscle parameters such as muscle pennation angles and muscle fascicle length (Kubo, Kanehisa et al. 2003). US allows for dynamic studies of muscle architecture, Fukunaga et al. (1997) have developed a method to study the fascicle length during contraction (Fukunaga, Ichinose et al. 1997). Furthermore, US has been used to analyse the muscle architecture's response to age, the authors concluded that some muscles in the lower extremities decreased in thickness with aging but the fascicle length did not decrees with aging (Kubo, Kanehisa et al. 2003). Loss of muscle mass with aging has been reported to be greater in the lower extremities than in the upper extremities. Decreases in CSA of the muscles have been reported to be 25-33 % lower in young compared to elderly adults (Narici, Maganaris et al. 2003). However, several researchers have reported decreased muscle strength but not decreased CSA, so the force, expressed per unit of muscle CSA, has been reduced in older individuals (Young 1984; Macaluso, Nimmo et al. 2002; Narici, Maganaris et al. 2003). US has been applied to the rotator cuff muscles to analyse the dynamic contraction pattern of these muscles to confirm the neuromuscular intensity (Boehm, Kirschner et al. 2005). Fukunaga et al. (1997) used US to measure muscle architecture and function in human muscles. They pointed out that the use of cadavers for studies of architecture and modelling of muscle functions would result in inaccurate and, in some cases, misleading results (Fukunaga, Kawakami et al. 1997). Aagaard et al. (2001) used US to measure the response to strength training and the changes in muscle architecture. They concluded that the quadriceps muscle increased both its CSA and the pennation angle after heavy resistance training (Aagaard, Andersen et al. 2001). Rutherford and Jones (1992) did not find any increased pennation angles after resistance training, even though they reported increased CSA and muscle force in the quadriceps muscle (Rutherford and Jones 1992). Brorsson et al. (2008) showed that there was a significant difference between the muscle anatomy of healthy men and women. The results of the ultrasound measurements and the differences in muscle architecture parameters between healthy men and women, and healthy women and RA women are summarised in Table 1.

The overall shape changes in muscle CSA during contraction were more pronounced for men than for women, ($p < 0.01$). US studies have also been performed on human skeletal muscles to explore the changes in muscle architecture that occur during dynamic contractions. The authors found that at a constant joint angle, the fascicle length and the pennation angles changed significantly during muscle contraction (Reeves and Narici 2003).

3.3 Function test evaluation, patients' questionnaires and visual analogue scale in hand rehabilitation

The Grip Ability Test (GAT) is designed for individuals with RA; it measures ADL ability. The test is based on three items chosen to represent different daily grip types. The test is performed following a standardized protocol consisted of three items: to put a "sleeve"

(Flexigrip™ stocking) on their non-dominant hand, place a paper clip on an envelope and pour 200 ml into a cup from a 1 litre water jug. GAT is a reliable, valid and sensitive ADL test (Dellhag and Bjelle 1995). Hand function has been assessed by GAT for measuring grip ability and activity limitations in several studies. Dellhag et al. (1992) reported that RA patients have improved their hand function after just 4 weeks of hand exercise (Dellhag, Wollersjo et al. 1992). Bjork et al. (2007) showed significant differences in activity limitations between healthy controls and RA patients in there study using GAT (Bjork, Thyberg et al. 2007). The relationship between self-reported upper limb function and grip ability was studied in an early rheumatoid population by Adams et al. (2004). They reported correlation between GAT and the questioner DASH (Adams, Burridge et al. 2004). Dellhag et al. (2001) reported in their study that patients with RA that have good hand function, low GAT score, displayed normal or increased safety margin during precision grip-lift compared to healthy controls (Dellhag, Hosseini et al. 2001).

Muscle parameter	Healthy men (n=20)	Healthy women (n=20)	RA women (n=20)
Thickness (cm)	1.2 (1.0-1.6)**	1.0 (0.7-1.2)*	0.8 (0.6-1.2)
CSA (cm²)	2.5 (1.6-3.3)**	1.8 (1.0-2.6)*	1.7 (0.4-2.5)
Fascicle length (cm)	6.6 (3.8-9.5)**	4.8 (3.9-7.0)*	4.4 (2.4-6.7)
Pennation angle (degree)	6.7 (3.3-8.5)*	5.3 (4.0-8.5)	5.6 (3.8-6.5)
Volume (cm³)	27.5 (18.6-43.1)**	16.7 (9.7-28.9)**	12.5 (3.1-23.5)

Muscle parameters are presented as median (range)
*$p < 0.05$, ** $p < 0.01$ (significant differences between healthy men - healthy women and between healthy women - RA women).

Table 1. Muscle architechture of EDC

Self-administered questionnaires are recommended for evaluating functional disability from the patients' perspective (Guillemin 2000; Liang 2000). The hand function is affected early on in RA and can be evaluated with different methods. One widely used selfadministrated extremity-specific questionnaire is the Disability of the Arm, Shoulder and Hand (DASH) that is been reliable and validated for assessing upper limb functional ability in the RA population (Atroshi, Gummesson et al. 2000). DASH has been used for evaluating the effectiveness of patient-oriented hand rehabilitation programmes, and has shown significant differences between two rehabilitation programmes and surgery (Gummesson, Atroshi et al. 2003; Harth, Germann et al. 2008). Furthermore, DASH has been used by Solem et al. (2006) for evaluation of long-term results of arthrodesis (Solem, Berg et al. 2006). Adams et al. (2004) showed in their study that DASH was useful to evaluate the relationship between upper limb functional ability and structural hand impairment (Adams, Burridge et al. 2004). Another commonly used generic questionnaire for evaluating functional disability in people is the Short Form 36-item Health Survey (SF-36), there a validated Swedish version has been developed (Sullivan, Karlsson et al. 1995). Generic healthy status measurements are commonly used for evaluation of RA patients. SF-36 has been used to detect the treatment effect in the study outcomes. Furthermore, use of SF-36 permits comparisons of physical and mental aspects in the RA population, as well as comparison between patients with RA, other patients groups and the general population (Tugwell, Idzerda et al. 2007). SF-36 has been used in several studies to evaluate the clinical outcome and quality of life after arthroplasty, and concluded the health status and the overall physical functions with significant improvements for RA patients (Angst, John et al. 2005; Ringen, Dagfinrud et al. 2008; Uhlig, Heiberg et al. 2008).

Visual analog scale (VAS) pain is a method frequently used to measure perceived pain level and the impact that high pain levels have on functional disability. Decreased functional ability in patients with RA has been reported correlated with on disease activity, disease duration, age, grip force and high pain level (Oken, Batur et al. 2008). Hand disabilities were detected in 81 % of RA patients and strongly correlated to pain level, grip force and clinical and laboratory activity. Female RA patients have reported more pain and worse disability than men (Bodur, Yilmaz et al. 2006; Hakkinen, Kautiainen et al. 2006). Brorsson et al. (2008) reported that neither the RA group nor the controls showed any significant improvement in DASH score after 6 weeks of hand exercise therapy. However, after 12 weeks of hand exercise the RA group showed a significant improvement in the DASH score, while there was still no improvement in the control group. Neither group showed any significant improvement in the SF-36 score after the hand exercises (Figure 3). However, some of the RA patients reported "tiredness" in their hands after the exercise.

The exercises caused no significant change in the pain level (Table 2).

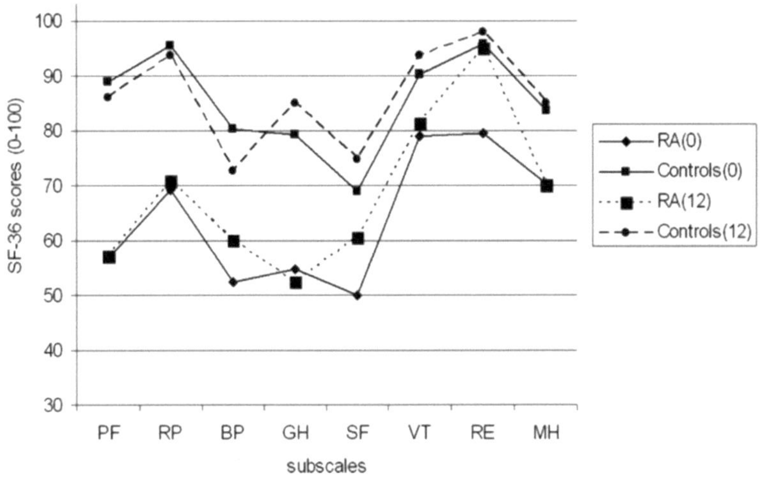

Fig. 3. SF-36 score pre- and post hand exercise therapy
Results of the SF-36 questionnaire, before (0) and after 12 weeks (12), of hand exercises. The scale is 0–100, from worst to best. The questionnaire is designed for measuring the generic health in the general population but is also useful for different patient groups. SF-36 is divided into eight health profiles scales; physical function (PF), role physical (RP), bodily pain (BP), general health (GH), vitality (VT), social functioning (SF), role emotional (RE) and mental health (MH). All dimensions are independent of each other.

4. The hand in rheumatoid arthritis

RA is our most frequent autoimmune inflammatory disease, with prevalence of nearly 1 %. RA is found throughout the world and affects all ethnic groups. It may strike at any age, but its prevalence increases with age; the peak incidence being between the fourth and sixth decades. The prevalence is about 2½ times higher in women than in men. The onset of symptoms usually involves symmetrical joints in hand and feet, but RA is a systemic disease and might affect any organ such as vessels, pleura or skin. There is often involvement of multiple joints and surrounding tissues. It's estimated that 80-90 % of the RA patients suffer from decreased hand function (Maini 1998; O'Brien, Jones et al. 2006). The hand in most patients may develop some typical pattern of deformity. These deformities are influenced by several factors, such as inflammation in the joint with distension of the joint capsule and ligament attenuation. Inflammation in and around tendons might distend tendon sheaths and cause tendon ruptures. The influence of disease by the characteristic MCP-joint deformity of ulnar drift (Figure 4), results of local joint forces (Smith and Kaplan 1967; McMaster 1972; Tan, Tanner et al. 2003; Bielefeld and Neumann 2005). Muscle involvement can lead to weakness and contractures. RA patients are frequently affected by pain, weakness and restricted mobility: the deformities of the hand, in various degrees, leads to limitation in activities of daily living (ADL) (Chung, Kotsis et al. 2004; Mengshoel and Slungaard 2005; Masiero, Boniolo et al. 2007).

		RA group (n=18)		Control group (n=18)	
	Week	Median	Range	Median	Range
GAT	0	19.8	16.5 - 51.6	13.4	8.9 - 16.9
	6	16.8	14.4 - 40.0**	11.5	6.9 - 15.8**
	12	16.1	12.1 - 30.2**	10.3	7.7 - 14.5**
DASH	0	37.3	8.8 - 62.5	2.5	0.0 - 16.3
	6	37.5	5.8 - 75.0	2.1	0.0 - 17.5
	12	39.2	6.7 - 47.5*	5.0	0.0 - 15.0
VAS	0	1.5	0.0 - 6.0	0.0	0.0 - 2.0
	6	2.5	0.0 - 7.0	0.0	0.0 - 2.0
	12	2.0	0.0 - 7.0	0.0	0.0 - 2.0

Median values of hand function tests before (week 0) and after 6 and 12 weeks of hand exercise. Median and range are given for the grip ability test (GAT), disability, of arm shoulder and hand questionnaire (DASH) and reported pain level (VAS). Number of participants (n=#)

*p < 0.05, **p < 0.01

Table 2. Hand function evaluations before and after hand exercise.

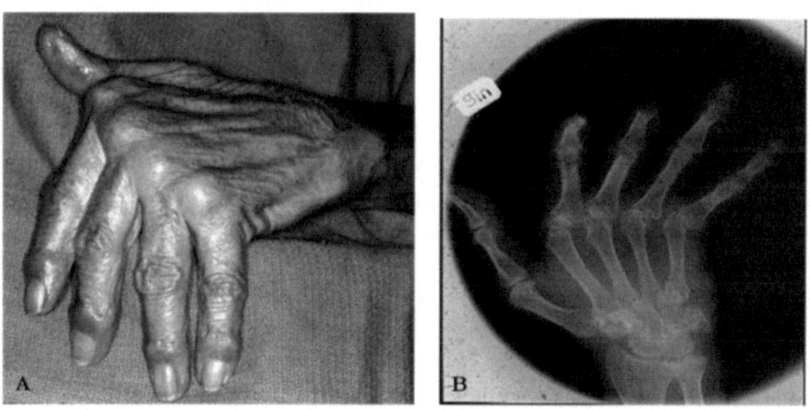

Fig. 4. The hand in most patients may develop some typical pattern of deformity; these images show the characteristic MCP-joint deformity of ulnar drift. ©Sofia Brorsson

The exact cause of RA is still unknown, however genetic, hormonal and environment factors have been reported to be involved in autoimmune diseases such as RA (Ollier and MacGregor 1995; Reckner Olsson, Skogh et al. 2001; Tengstrand, Ahlmen et al. 2004). Diagnosis of RA are based on ACR criteria which include; pain and swelling in at least three joint areas, symmetrical presentation, early morning joint stiffness for more than 1 hour, involvement of MCP joint or PIP joint or wrists, subcutaneous nodules, positive rheumatoid factor and radiological evidence of erosions. At least four of these signs or symptoms should be present for six weeks (Arnett, Edworthy et al. 1988). Pain and tenderness of the joints are well described and documented (Pearl and Hentz 1993), but there is less knowledge concerning how the muscles are influenced by the disease. The most common histological findings in RA are the pronounced muscle atrophy and nodular myositis. Magyar et al. (1973) observed changes in the muscles consistent with denervation using electron microscopy. These authors showed that the muscle changes might be due to a direct involvement of the neuromuscular system and that the pathological changes affect the contractile element in the muscles (Magyar, Talerman et al. 1973). An important part of hand function is based on the function of the muscles which are involved in finger and wrist motion and the ability to develop grip force. RA patients often report that they feel weakness, particularly when performing flexion force. There are several possible reasons for this weakness such as reduction in muscle fibre diameter, direct involvement of inflammatory processes in the muscle, joint deformity influencing muscle function and pain (Haslock, Wright et al. 1970; Leading 1984; Bruce, Newton et al. 1989). The muscle structure (ultra-structure) and changes in rheumatoid arthritis have been recognised pathologically and clinically. Although electron microscopy is valuable in investigating human skeletal muscle both in normal and RA muscles, only a few data sources document muscle ultra-structural alterations in RA patients (Haslock, Wright et al. 1970; Magyar, Talerman et al. 1973; Wollheim 2006). Furthermore, a non-invasive study on muscle architecture in RA patients appears to be poorly investigated.

4.1 Rehabilitation and intervention of the Rheumatoid Arthritis hand

Treatment of RA is focused on reducing the inflammatory activity by medication, rehabilitation and surgery (Stenstrom and Minor 2003). New disease modifying drugs for RA patients administered early after onset have made it possible for people with this disease to stay more active and more fit than 10-20 years ago (Pincus, Ferraccioli et al. 2002). Today's treatment options to increase hand function for RA patients include electrotherapy, injection therapy, manual therapy and traditional exercise prescription, but the evidence base for treatments remains weak, particularly when focusing on the hand (Weiss, Moore et al. 2004; Plasqui 2008). In 1974, Lee et al. reported in their study that immobilization and/or physical rest were beneficial in the treatment of RA, leading to a decrease in pain and joint swelling (Lee, Kennedy et al. 1974). Other groups have reported that the forces involved in using the hand lead to joint erosion and increased deformities (Ellison, Flatt et al. 1971; Kemble 1977). Despite earlier fear of aggravating symptoms, there is now scientific evidence showing that various forms of exercise are both safe and beneficial (Stenstrom and Minor 2003). However, comparatively little research has evaluated the evidence for the benefits of hand exercise in RA (O'Brien, Jones et al. 2006). Recently reviewed effectiveness on hand exercise therapy in RA patients showed that only nine eligible studies have incorporated hand exercise therapy as part of the intervention (Chadwick 2004; Wessel 2004). Hoening et al. (1993) showed in their study that a home hand exercise program was effective for increasing the grip force in the RA hand (Hoenig, Groff et al. 1993). Intensive hand exercise has previously been reported to be effective for improving grip- and pinch force for RA patients (Ronningen and Kjeken 2008). Brorsson et al. (2008) have showed that a regular home exercise programme for the RA hand, evaluated with force measurements, ultrasound examination, function test and patients questionnaires (Figure 5), is beneficial for grip (flexion and extension) force production. Furthermore, they reported that hand exercise improves the relation between flexion and extension forces as well as improved hand function. They also reported improved flexion and extension force for the RA patients after 12-weeks of hand exercise (Figure 6).

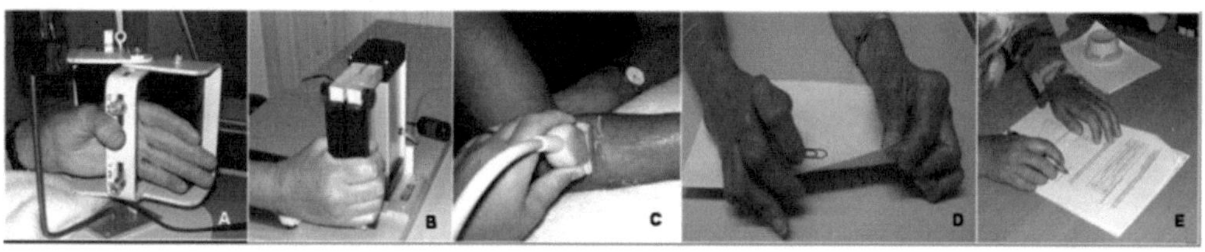

Fig. 5. The total study period was 18 weeks of home hand exercise, divided into 6-week periods. Baseline values were determined at week 0 (Occasion I) and 6 (Occasion II). Thereafter, the hand exercise programme was started, and the effects were measured after 6 weeks (Occasion III) and 12 weeks (Occasion IV). Evaluation methods used: (A) finger extension force measurements (EX-it), (B) Flexion force measurements (Grippit™), (C) US examination of the EDC muscle, (D) grip ability test, and (E) questionnaires.

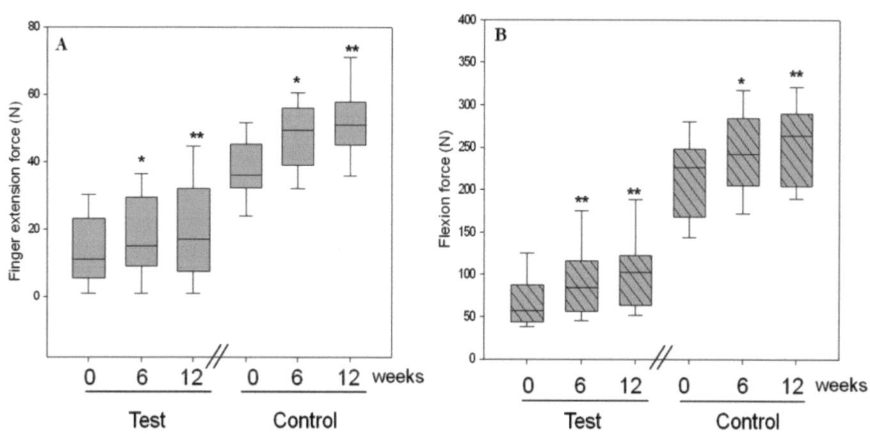

Fig. 6. Illustrates the finger extension force (A) and flexion force (B) in the two groups of participants in paper IV after 6 and 12 weeks of hand exercise. Both groups show significant improvement after 6 and 12 weeks (* $p < 0.05$, ** $p < 0.01$).

Hand surgery has been regarded as beneficial for some patients with RA. Arthroplastic procedures of the wrist and fingers have been performed since 1960. An increasing number of patients with RA receive joint replacements in the MCP joints of the hand. The purpose of these operations is to improve the patients' extension ability, extension force, and hand function as well as reduce pain (Weiss, Moore et al. 2004). At present, when the outcome of surgery is evaluated, it is impossible to objectively test if the patients' finger extension force has been improved or not, since no force measurement device for finger extension force is commercially available. It is necessary to find methods to objectively measure hand function in order to be able to evaluate the functional impairment, as well as the results of therapeutic interventions i.e. surgery or physical therapy.

5. Conclusion

To further our understanding of hand function, and specifically the extensor muscles' function and ability to produce force in rehabilitation, this book chapter describes the development and results of new non-invasive methods, a new finger extension force measurement device, EX-it, and an ultrasound imaging method (Brorsson et al. 2008 a,b). Furthermore, the results of this book chapter show that finger extension force measurements and ultrasound are effective methods for evaluating improvement after the intervention hand exercise. The effect of hand exercise on the extensor muscles could be objectively evaluated with EX-it and ultrasonic imaging. This chapter also reported the usefulness of short-term hand exercise for patients with RA and that a home exercise programme can enhance hand function.

Various methods can be used to study muscle architecture, including ultrasound, magnetic resonance imaging (Juul-Kristensen, Bojsen-Moller et al. 2000; Aagaard, Andersen et al. 2001) and laser diffraction. Laser diffraction is an invasive technique, while magnetic resonance imaging is only suitable for static measurements. Ultrasound, on the other hand, is non-invasive and clearly shows the movement of the muscle (Fukunaga, Ichinose et al. 1997). It is also harmless, can be repeated and offers the possibility of dynamic examinations. The limitations with US are the quality of the examinations, which are dependent on the investigator's ability to reproduce the imaging conditions (measurements), to find correct landmarks in both transverse and longitudinal direction and standardise the procedures. Ultrasound has been shown to be a highly valuable tool to assess *in vivo* muscle architecture for studying muscle function and relationships between muscle force and muscle size (Maughan, Watson et al. 1984; Hakkinen and Keskinen 1989; Kawakami, Abe et al. 1993; Fukunaga, Kawakami et al. 1997).

In rheumatoid arthritis, impaired finger extension is a common symptom; differences in extension muscle force capacity as well as in muscle architectural parameters, between normal and RA muscles are reported. Earlier studies have reported that RA patients also have weaker grip, pinch and tripod force than healthy controls, and it has been suggested that force assessment could be used as an accurate indicator of upper limb ability and that grip force (i.e. flexion and pinch force) should be included in the evaluation and follow-up of the patients with RA in hand rehabilitation units (Helliwell and Jackson 1994; Fraser, Vallow et al. 1999; Adams, Burridge et al. 2004; Bodur, Yilmaz et al. 2006). The decrease in force capacity could be explained by a direct effect of the disease on muscle function, disuse or impaired neuromuscular transmission, or different medications, but the decrease could also be due to the fact that the RA patients experienced more pain than the healthy subjects, a situation which could influence their maximal muscle exertion. Loss of hand grip force has been shown to result from pain, or fear of pain, or mechanical malfunction (Fraser, Vallow et al. 1999).

Ultrasound is a non-invasive and harmless method that can be used to visualise functionally important muscle parameters dynamically. Finger extension control is one of the most difficult motions to regain after disease/injury and is also very important for prehensile activities (Cauraugh, Light et al. 2000). Since both EX-it and ultrasound have been shown to be sensitive in their evaluation of hand exercise, it can be expected that these methods can be used to evaluate other interventions, such as surgical procedures, physiotherapy and/or pharmacological treatment. With these new methods, arthroplastic interventions in the MCP-joints of the fingers can objectively be evaluated. In a longer perspective it may be possible to establish more efficient rehabilitation programmes for RA patients. Furthermore, force measurements are a quick and easy measure of hand impairment and function, and are useful when evaluating hand status. EX-it in combination with other non-invasive evaluation methods (i.e.grip ability tests and health assessment questionnaires) will provide more information on hand function. Patients with rheumatoid arthritis suffer from a variety of functional deficiencies, of which impaired muscle function is a serious one. There is a recent trend towards the use of non-invasive methods in studying disease-specific changes, such as magnetic resonance imaging and ultrasound. Increased knowledge concerning muscle morphology and function in RA will allow better diagnosis and evaluation of interventions, such as surgical procedures, physiotherapy and/or pharmacological treatment. In a longer perspective it may be possible to establish a more efficient rehabilitation programme for RA patients. If combined with functional and clinical measures of disability, information on muscle architecture could then be used as an objective tool in the assessment of hand function after physical therapy and hand surgery.
In this thesis no negative effects of EX-it, ultrasound or the exercise programme on self reported pain level were reported in the RA group. It is possible that RA patients need continuous exercise to prevent loss of muscle strength and to improve the performance of activities of daily living (Stenstrom 1994; Hakkinen, Malkia et al. 1997; O'Brien, Jones et al. 2006; Masiero, Boniolo et al. 2007). However, the response to exercise from RA patients must be further evaluated to find out if longer exercise period can obliterate the differences between healthy and rheumatoid arthritis muscle strength and function; or to find out if these differences depend on a disease-specific effect on the rheumatoid arthritis muscles.

5.1 Future implications

Several questions have arisen during writing this book chapter and performed research in this area and require further research. It would be of interest to analyse how EDC responds during contraction at different locations of the muscle. Brorsson et al. (2008 a,b, 2009)reported that the inter muscle movement pattern in the muscle was observed, but were unable to measure it with the methods used for this thesis. Further knowledge about *in vivo* muscle pattern could provide information about the muscle as well as the elastic characteristics of the aponeurosis and tendon.

- Is it possible that the EDC, a muscle designed for precision tasks and grip control rather than force exertion, is constructed differently from the large force-generating muscles?
- Can US be used as a diagnostic tool for analysing muscle disease?
- Are muscle movement patterns related to force production?
- Does this muscle movement appear in other muscle groups?

RA patients significantly increased their hand force and hand function after exercise. However, the response to exercise from RA patients must be further evaluated. It would be interesting to combine invasive and non-invasive methods to be able to answer the following questions:

- Would longer periods of hand exercise obliterate the differences between healthy and rheumatoid arthritis muscle force and function?
- Do the muscle's architecture, force production and decreased function depend on disease specific effects on the rheumatoid arthritis muscles?

It would be of great interest to investigate the possibility to objectively evaluate interventions, such as surgical procedures, physiotherapy and/or pharmacological treatment with the help of finger force measurements and ultrasound evaluations.

- In a longer perspective, can it be possible to establish more efficient rehabilitation programmes for RA patients through further knowledge about the muscle biomechanics?

6. Acknowledgment

My research is a result of multidisciplinary collaboration between the School of Business and Engineering, Halmstad University, Department of Hand Surgery, Sahlgrenska University Hospital, Göteborg, Research and Development centre at Spenshult Hospital for Rheumatic Diseases, Halmstad, Department of Diagnostic Radiology and Department of Research and Education, Halmstad Central Hospital, Halmstad.

I would like to thank all the patients and healthy subjects, for your participation in the studies, for performing the tests and answering the questioners. I extend my thanks to Professor Marita Hilleges, for your inspiration, critical comments and clever suggestions.

7. References[1]

Aagaard, P., J. L. Andersen, et al. (2001). "A mechanism for increased contractile strength of human pennate muscle in response to strength training: changes in muscle architecture."J Physiol 534(Pt. 2): 613-23.

Adams, J., J. Burridge, et al. (2004). "Correlation between upper limb functional ability and structural hand impairment in an early rheumatoid population." Clin Rehabil 18(4): 405-13.

Akima, H., S. Kuno, et al. (1997). "Effects of 20 days of bed rest on physiological cross-sectional area of human thigh and leg muscles evaluated by magnetic resonance imaging." J Gravit Physiol 4(1): S15-21.

Ashford, R. F., S. Nagelburg, et al. (1996). "Sensitivity of the Jamar Dynamometer in detecting submaximal grip effort." J Hand Surg [Am] 21(3): 402-5.

Atroshi, I., C. Gummesson, et al. (2000). "The disabilities of the arm, shoulder and hand (DASH) outcome questionnaire: reliability and validity of the Swedish version evaluated in 176 patients." Acta Orthop Scand 71(6): 613-8.

Balogun, J. A., C. T. Akomolafe, et al. (1991). "Grip strength: effects of testing posture and elbow position." Arch Phys Med Rehabil 72(5): 280-3.

Berntson, L. and E. Svensson (2001). "Pain assessment in children with juvenile chronic arthritis: a matter of scaling and rater." Acta Paediatr 90(10):1131-6.

---

[1] *Since many references have several authors, only the two first authors are mentioned (in alphabetical order) in this book chapter.*

Bielefeld, T. and D. A. Neumann (2005). "The unstable metacarpophalangeal joint in rheumatoid arthritis: anatomy, pathomechanics, and physical rehabilitation considerations." J Orthop Sports Phys Ther 35(8): 502-20.

Bjork, M. A., I. S. Thyberg, et al. (2007). "Hand function and activity limitation according to health assessment questionnaire in patients with rheumatoid arthritis and healthy referents: 5-year followup of predictors of activity limitation (The Swedish TIRA Project)." J Rheumatol 34(2): 296-302.

Blazevich, A. J., N. D. Gill, et al. (2007). "Lack of human muscle architectural adaptation after short-term strength training." Muscle Nerve 35(1): 78-86.

Blazevich, A. J. and N. C. Sharp (2005). "Understanding muscle architectural adaptation: macro- and micro-level research." Cells Tissues Organs 181(1): 1-10.

Bodur, H., O. Yilmaz, et al. (2006). "Hand disability and related variables in patients with rheumatoid arthritis." Rheumatol Int 26(6): 541-4.

Boehm, T. D., S. Kirschner, et al. (2005). "Dynamic ultrasonography of rotator cuff muscles." J Clin Ultrasound 33(5): 207-13.

Brand, W. (1993). Clinical Mechanics of the Hand. St. Louis, Missouri.

Brorsson, S., A. Nilsdotter, et al. (2008a). "A new force measurement device for evaluating finger extension function in the healthy and rheumatoid arthritic hand." Technol Health Care 16(4): 283-92.

Brorsson, S., A. Nilsdotter, et al. (2008b). "Ultrasound evaluation in combination with finger extension force measurements of the forearm musculus extensor digitorum communis in healthy subjects." BMC Med Imaging 3:8:6.

Bruce, S. A., D. Newton, et al. (1989). "Effect of subnutrition on normalized muscle force and relaxation rate in human subjects using voluntary contractions." Clin Sci (Lond) 76(6): 637-41.

Buljina, A. I., M. S. Taljanovic, et al. (2001). "Physical and exercise therapy for treatment of the rheumatoid hand." Arthritis Rheum 45(4): 392-7.

Chadwick, A. (2004). "A review of the history of hand exercises in rheumatoid arthritis." Musculoskeletal Care 2(1): 29-39.

Chadwick, E. K. and A. C. Nicol (2001). "A novel force transducer for the measurement of grip force." J Biomech 34(1): 125-8.

Chung, K. C., S. V. Kotsis, et al. (2004). "A prospective outcomes study of Swanson metacarpophalangeal joint arthroplasty for the rheumatoid hand." J Hand Surg [Am] 29(4): 646-53.

Debicki, D. B., P. L. Gribble, et al. (2004). "Kinematics of wrist joint flexion in overarm throws made by skilled subjects." Exp Brain Res 154(3): 382-94.

Dellhag, B. and A. Bjelle (1995). "A Grip Ability Test for use in rheumatology practice." J Rheumatol 22(8): 1559-65.

Dellhag, B., N. Hosseini, et al. (2001). "Disturbed grip function in women with rheumatoid arthritis." J Rheumatol 28(12): 2624-33.

Dellhag, B., I. Wollersjo, et al. (1992). "Effect of active hand exercise and wax bath treatment in rheumatoid arthritis patients." Arthritis Care Res 5(2): 87-92.

Doebelin, E. (1990). Measurements Systems, Application and Design. New York, Mc Graw-Hill.

Ekdahl, C. and S. I. Andersson (1989). "Standing balance in rheumatoid arthritis. A comparative study with healthy subjects." Scand J Rheumatol 18(1): 33-42.

Ellison, M. R., A. E. Flatt, et al. (1971). "Ulnar drift of the fingers in rheumatoid disease. Treatment by crossed intrinsic tendon transfer." J Bone Joint Surg Am 53(6): 1061-82.
Fess, E. (1992). "Grip strength." American Society of Hand Therapists. Clinical Assment Recommendations Vol. 2nd: 41-45.
Fischer, H. C., K. Stubblefield, et al. (2007). "Hand rehabilitation following stroke: a pilot study of assisted finger extension training in a virtual environment."Top Stroke Rehabil 14(1): 1-12.
Fitts, R. H. and J. J. Widrick (1996). "Muscle mechanics: adaptations with exercisetraining." Exerc Sport Sci Rev 24: 427-73.
Flatt, A. E. (1974). "Letter: Shoulder-hand syndrome." Lancet 1(7866): 1107-8.
Fransson, C. and J. Winkel (1991). "Hand strength: the influence of grip span and grip type." Ergonomics 34(7): 881-92.
Fraser, A., J. Vallow, et al. (1999). "Predicting 'normal' grip strength for rheumatoid arthritis patients." Rheumatology (Oxford) 38(6): 521-8.
Freilich, R. J., R. L. Kirsner, et al. (1995). "Isometric strength and thickness relationships in human quadriceps muscle." Neuromuscul Disord 5(5): 415-22.
Fukunaga, T., Y. Ichinose, et al. (1997). "Determination of fascicle length and pennation in a contracting human muscle *in vivo*." J Appl Physiol 82(1): 354-8.
Fukunaga, T., Y. Kawakami, et al. (1997). "Muscle architecture and function in humans." J Biomech 30(5): 457-63.
Fukunaga, T., M. Miyatani, et al. (2001). "Muscle volume is a major determinant of joint torque in humans." Acta Physiol Scand 172(4): 249-55.
Fung, Y. (1993). Biomechanics Mechanical properties of living tissues. New York, Springer-Verlag.
Grassi, W., F. Salaffi, et al. (2005). "Ultrasound in rheumatology." Best Pract Res Clin Rheumatol 19(3): 467-85.
Guillemin, F. (2000). "Functional disability and quality-of-life assessment in clinical practice." Rheumatology (Oxford) 39 Suppl 1: 17-23.
Gummesson, C., I. Atroshi, et al. (2003). "The disabilities of the arm, shoulder and hand (DASH) outcome questionnaire: longitudinal construct validity and measuring self-rated health change after surgery." BMC Musculoskelet Disord 4: 11.
Hakkinen, A., P. Hannonen, et al. (1995). "Muscle strength in healthy people and in patients suffering from recent-onset inflammatory arthritis." Br J Rheumatol 34(4): 355-60.
Hakkinen, A., H. Kautiainen, et al. (2006). "Muscle strength, pain, and disease activity explain individual subdimensions of the Health Assessment Questionnaire disability index, especially in women with rheumatoid arthritis." Ann Rheum Dis 65(1): 30-4.
Hammer, A. and B. Lindmark (2003). "Test-retest intra-rater reliability of grip force in patients with stroke." J Rehabil Med 35(4): 189-94.
Harth, A., G. Germann, et al. (2008). "Evaluating the effectiveness of a patient oriented hand rehabilitation programme." J Hand Surg Eur Vol.
Haslock, D. I., V. Wright, et al. (1970). "Neuromuscular disorders in rheumatoid arthritis. A motor-point muscle biopsy study." Q J Med 39(155): 335-58.
Helliwell, P. S. and S. Jackson (1994). "Relationship between weakness and muscle wasting in rheumatoid arthritis." Ann Rheum Dis 53(11): 726-8.
Hoenig, H., G. Groff, et al. (1993). "A randomized controlled trial of home exercise on the rheumatoid hand." J Rheumatol 20(5): 785-9.

Hopkins, J. T., J. B. Feland, et al. (2007). "A comparison of voluntary and involuntary measures of electromechanical delay." Int J Neurosci 117(5): 597-604.

Huxley, A.F., R. Niedergerke (1954). "Structural Changes in Muscle During Contraction. Interference Microscopy of Living Muscle Fibrers " Nature 173: 971.

Huxley, A. F. and R. M. Simmons (1971). "Proposed mechanism of force generation in striated muscle." Nature 233(5321): 533-8.

Ichinose, Y., H. Kanehisa, et al. (1998). "Morphological and functional differences in the elbow extensor muscle between highly trained male and female athletes." Eur J Appl Physiol Occup Physiol 78(2): 109-14.

Incel, N. A., E. Ceceli, et al. (2002). "Grip strength: effect of hand dominance."Singapore Med J 43(5): 234-7.

Innes, E. (1999). "Handgrip strength testing: A review of the literature." Australian Occupational Therapy Journal 46: 120-140.

Jones, E., J. G. Hanly, et al. (1991). "Strength and function in the normal and rheumatoid hand." J Rheumatol 18(9): 1313-8.

Juul-Kristensen, B., F. Bojsen-Moller, et al. (2000). "Muscle sizes and moment arms of rotator cuff muscles determined by magnetic resonance imaging."Cells Tissues Organs 167(2-3): 214-22.

Kandel, E.R., J.H. Schwartz, et al. (1991). Principles of neural science, Appleton & Lange.

Kane, D., P. V. Balint, et al. (2004). "Musculoskeletal ultrasound - a state of the art review in musculoskeletal ultrasound in rheumatology."Rheumatology (Oxford) 43(7): 823-8.

Kawakami, Y., T. Abe, et al. (1993). "Muscle-fiber pennation angles are greater in hypertrophied than in normal muscles." J Appl Physiol 74(6): 2740-4.

Kawakami, Y., H. Akima, et al. (2001). "Changes in muscle size, architecture, and neural activation after 20 days of bed rest with and without resistance exercise." Eur J Appl Physiol 84(1-2): 7-12.

Kawakami, Y., Y. Muraoka, et al. (2000). "Changes in muscle size and architecture following 20 days of bed rest." J Gravit Physiol 7(3): 53-9.

Kemble, J. V. (1977). "Functional disability in the rheumatoid hand." Hand 9(3): 234-41.

Kubo, K., H. Kanehisa, et al. (2003). "Muscle architectural characteristics in women aged 20-79 years." Med Sci Sports Exerc 35(1): 39-44.

Lieber, R. L. and J. Friden (2000). "Functional and clinical significance of skeletal muscle architecture." Muscle Nerve 23(11): 1647-66.

Maini, R. N. (1998). "Rheumatoid arthritis. A paradigm of inflammatory disease of the musculoskeletal system." Acta Orthop Scand Suppl 281: 6-13.

Marieb, E. (1997). Human anatomy and physiology. California, Benjamin/Cummings Science.

Narici, M. and P. Cerretelli (1998). "Changes in human muscle architecture in disuseatrophy evaluated by ultrasound imaging." J Gravit Physiol 5(1): P73-4.

Nordenskiold, U. (1997). "Daily activities in women with rheumatoid arthritis. Aspects of patient education, assistive devices and methods for disability and impairment assessment." Scand J Rehabil Med Suppl 37: 1-72.

Nordenskiold, U. M. and G. Grimby (1993). "Grip force in patients with rheumatoid arthritis and fibromyalgia and in healthy subjects. A study with the Grippit instrument." Scand J Rheumatol 22(1): 14-9.

Otten, E. (1988). "Concepts and models of functional architecture in skeletal muscle." Exerc Sport Sci Rev 16: 89-137.

Pearl, R. M. and V. R. Hentz (1993). "Extensor digiti minimi tendon transfer to prevent recurrent ulnar drift." Plast Reconstr Surg 92(3): 507-10.
Qvistgaard, E., S. Torp-Pedersen, et al. (2006). "Reproducibility and inter-reader agreement of a scoring system for ultrasound evaluation of hip osteoarthritis." Ann Rheum Dis 65(12): 1613-9.
Richards, L. G., B. Olson, et al. (1996). "How forearm position affects grip strength." Am J Occup Ther 50(2): 133-8.
Ringen, H. O., H. Dagfinrud, et al. (2008). "Patients with rheumatoid arthritis report greater physical functional deterioration in lower limbs compared to upper limbs over 10 years." Scand J Rheumatol 37(4): 255-9.
Ronningen, A. and I. Kjeken (2008). "Effect of an intensive hand exercise programme in patients with rheumatoid arthritis." Scand J Occup Ther: 1-11.
Schieber, M. H. and M. Santello (2004). "Hand function: peripheral and central constraints on performance." J Appl Physiol 96(6): 2293-300.
Solem, H., N. J. Berg, et al. (2006). "Long term results of arthrodesis of the wrist: a 6-15 year follow up of 35 patients." Scand J Plast Reconstr Surg Hand Surg 40(3): 175-8.
Sollerman, C. and A. Ejeskar (1995). "Sollerman hand function test. A standardised method and its use in tetraplegic patients." Scand J Plast Reconstr Surg Hand Surg 29(2): 167-76.
Stenstrom, C. H. (1994). "Home exercise in rheumatoid arthritis functional class II: goal setting versus pain attention." J Rheumatol 21(4): 627-34.
Stenstrom, C. H. and M. A. Minor (2003). "Evidence for the benefit of aerobic and strengthening exercise in rheumatoid arthritis." Arthritis Rheum 49(3): 428-34.
Sullivan, M., J. Karlsson, et al. (1995). "The Swedish SF-36 Health Survey.Evaluation of data quality, scaling assumptions, reliability and construct validity across general populations in Sweden." Soc Sci Med 41(10): 1349-58.
Tan, A. L., S. F. Tanner, et al. (2003). "Role of metacarpophalangeal joint anatomic factors in the distribution of synovitis and bone erosion in early rheumatoid arthritis." Arthritis Rheum 48(5): 1214-22.
Tengstrand, B., M. Ahlmen, et al. (2004). "The influence of sex on rheumatoid arthritis: a prospective study of onset and outcome after 2 years." J Rheumatol 31(2): 214-22.
Thyberg, I., U. A. Hass, et al. (2005). "Activity limitation in rheumatoid arthritis correlates with reduced grip force regardless of sex: the Swedish TIRA project." Arthritis Rheum 53(6): 886-96.
Trappe, S. W., T. A. Trappe, et al. (2001). "Calf muscle strength in humans." Int J Sports Med 22(3): 186-91.
Tugwell, P., L. Idzerda, et al. (2007). "Generic quality-of-life assessment in rheumatoid arthritis." Am J Manag Care 13 Suppl 9: S224-36.
Weiss, A. P., D. C. Moore, et al. (2004). "Metacarpophalangeal joint mechanics after 3 different silicone arthroplasties." J Hand Surg [Am] 29(5): 796-803.
Wessel, J. (2004). "The effectiveness of hand exercises for persons with rheumatoid arthritis: a systematic review." J Hand Ther 17(2): 174-80.
Vliet Vlieland, T. P., T. P. van der Wijk, et al. (1996). "Determinants of hand function in patients with rheumatoid arthritis." J Rheumatol 23(5): 835-40.
Wollheim, F. A. (2006). "Aging, muscles, and rheumatoid arthritis." Curr Rheumatol Rep 8(5): 323-4.

6

Cervical Spine Anthropometric and Finite Element Biomechanical Analysis

Susan Hueston[1], Mbulelo Makola[1], Isaac Mabe[1] and Tarun Goswami[1,2]
[1]*Biomedical and Industrial Human Factors Engineering*
[2]*Department of Orthopedic Surgery, Sports Medicine and Rehabilitation*
United States of America

1. Introduction

A multidisciplinary approach to the study of the cervical spine is presented. The cervical spine provides higher levels of flexibility and motion as compared to the lumbar and thoracic spine regions. These characteristics can be attributed to the anatomy of the specific cervical vertebra. A statistical analysis of cervical vertebra anthropometry was performed in order to determine if significant relationships exist between vertebral features. The analysis was performed on a cohort of Chinese Singaporean cervical spines.

Mathematical analysis methods provide an extremely useful tool in the study of the cervical spine. Analyses can provide force displacement response characteristics of the cervical spine. Additionally, mathematical analysis methods can provide internal stress, and strain response characteristics for cervical vertebra and intervertebral discs. Mathematical analyses of the cervical spine require robust and accurate constitutive and geometric models. A review of cervical spine finite element modeling techniques and approaches is presented in order to help frame analysis and modeling best practices.

A finite element analysis study was performed focusing on vertebral endplate subsidence. Subsidence is a failure mechanism in which a vertebral endplate fails after implantation of an intra vertebral implant device. The effects of vertebral endplate morphology on stress response were analyzed in order to better understand indicators for subsidence.

1.1 Analysis of Chinese Singaporean cervical spine anthropometry

With respect to biomechanics it is important to understand the anatomy of the body. In this particular section the anatomy of the cervical spine will be presented, with investigation into the morphometry of the vertebra themselves. To accomplish this, an investigation on how the different dimensional anatomy of the cervical spine changes relates to each other will be presented.

To begin, a brief explanation of the anatomy of spine will be presented in order to aid in understanding of the anthropometry of the cervical spine. The spine consists of 5 sections: cervical, thoracic, lumbar, sacrum, and coccyx (from top to bottom) (Saladin and Miller, 2004). There are 33 vertebrae in the whole spine: 7 in the cervical spine which is located in the neck, 12 in the thoracic spine which is located in the chest, 5 in the lumbar spine which is located in the lower back, 5 in the sacrum that is located at the base of the spine, followed by the 4 small vertebrae in the coccyx (Saladin and Miller, 2004).

As stated previously there are 7 vertebrae in the cervical spine. The first two vertebrae are particularly unique and allow for movement of the head, the first is known as the Atlas (C1) and the second the Axis (C2) (Saladin and Miller, 2004). Because of their unique features analysis on correlations present in the dimensional anatomy was not completed. For the remaining 5 vertebrae from C3-C7 an investigation in the correlation in the dimensional anatomy was completed. The results of this investigation will allow for more accurate modeling of this region, in order to assist in the development of improved spinal implants as well as more efficient surgical device placement techniques. Additionally, these statistics will lead to a better understanding of cervical spine functionality and its susceptibility to failure. The different dimensional aspects that were analyzed were based on the anthropometric measurements completed from a published study by Tan on Chinese Singaporeans (Tan, Teo and Chua, 2004).

The present study involved the anthropometric measurements of linear, and angular aspects, as well as area. The linear measurements included: upper and lower end plate width (EPWu, and EPWl), upper and lower end plate depth (EPDu and EPDl), anterior and posterior vertebral body height (VBHa and VBHp), spinal canal width (SCW), spinal canal depth (SCD), left and right pedicle height (PDHl and PDHr), left and right pedicle width (PDWl and PDWr), spinous process length (SPL), and the transverse process width (TPW). The area measurements included: the upper and lower end plate area (EPAu and EPAl), spinal canal area (SCA), and the left and right pedicle area (PDAl and PDAr). Finally, the angular measurements included: upper and lower end plate transverse inclination (EPItu and EPItl), left and right pedicle sagittal inclination (PDIsl and PDIsr), and the left and right pedicle transverse inclination (PDItl and PDItr). Analysis was completed using the concepts of linear regression, ANOVA, and parameter estimation. Utilizing these results an investigation into any relationship that might be present between the previous anthropometric measurements was completed for each segment. As an example, a comparison between the EPWu of the C3 vertebra and the PDIsr of the C3 vertebra was analyzed to determine if there was any statistically significant relationship present (Tan, Teo and Chua, 2004).

Previous research, as discussed in this section, has been to provide quantitative measurements for the cervical spine. The purpose of the analysis completed was to develop any significant relationships present between the different anthropometrics of each vertebra. Of these significant relationships it was important to see why they were significant, which were significant in the opposite comparison (for example between EPWu vs. EPWl and EPWl vs. EPWu), and which were found in more than just one vertebral segment.

1.2 Materials & methods

To begin the analysis of the correlations present in the cervical spine anthropometrics, measurements were collected from Tan's study on Chinese Singaporeans. The linear, angular, and area measurements are depicted in Figure 1. In this analysis, a comparison of just one vertebral body's measurements was compared. A good example is comparing data from the C3 vertebra to other C3 vertebral data. These comparisons totaled approximately 600 for each vertebral body segment. The statistical analysis was completed using linear regression (including parameter estimation), and ANOVA with the use of SAS® 9.2 TS Level 2M0. A regression analysis is a statistical technique used to explore relationships that are present between two or more variables. In particular a linear regression analysis relates these various variables into a straight-line relationship where the slope and the y-intercept

of the line are the regression coefficients. Not all points will lie on this line, but a majority of the points will be within certain deviation of this line resulting in a model. For this particular study, a simple linear regression was used. It involves just one independent variable (x), also known as a regressor or predictor. With this linear regression analysis, parameter estimation was used. Parameter estimation is a technique of statistical inference, which is a way to make conclusions from random variation data. In this particular case, parameter estimation was used to find the y-intercept and the slope of the linear relationship between two anthropometric variables. ANOVA stands for Analysis of Variance, and can be used in order to test the significance of regression analysis. For the ANOVA, 95% confidence interval was used to test the significance between variables, while a 97.5% interval was used for the parameter estimation. Another test of significance was based off the R^2 value, which is also known as the correlation ratio. This correlation coefficient is the proportion of total variance of the dependent variable that is explained by the independent variable. Thus a higher value showing that the model is more accurate. In the case of the analysis described in this paper if the R^2 value was >0.6, the model was assumed to be a good fit (Montgomery and Peck, 1982; Gamst, Meyers and Guarino, 2008).
In the study completed by Tan on the Chinese Singaporeans, measurements of 10 cadaveric males were completed based on the measurements defined in Figure 1. The measurements mean and standard deviation found by Tan are displayed in Table's 1-3 where Table 1 displays the linear measurements, Table 2 lists the area measurements, and Table 3 illustrates the angular measurements that were taken in this study (Tan, Teo and Chua, 2004).

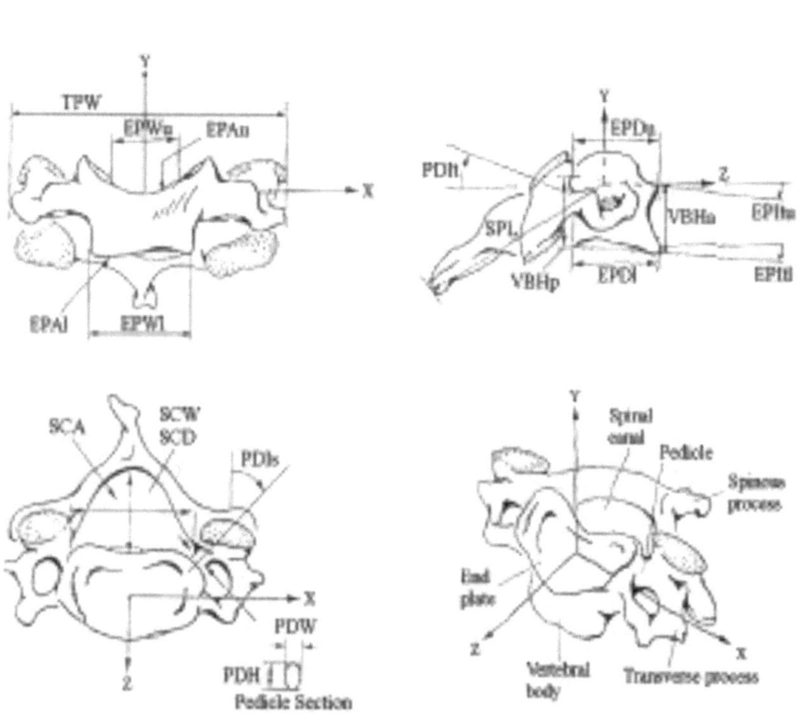

Fig. 1. Depiction of anthropometric measurements (Tan, Teo and Chua, 2004)

	C3		C4		C5		C6		C7	
	Mean	Std dev	Mean	Std dev	Mean	Std dev	Mean	Std dev	Mean	Std dev
EPWu	13.8	0.1	14.7	0.1	14.9	0.1	15.8	0.0	19.0	0.1
EPWl	14.3	0.1	15.0	0.1	15.9	0.1	19.5	0.2	20.3	0.2
EPDu	13.6	0.1	14.0	0.1	14.3	0.1	14.6	0.2	15.1	0.2
EPDl	15.1	0.2	15.2	0.4	15.1	0.3	15.7	0.3	15.6	0.3
VBHa	10.0	0.2	9.9	0.3	9.6	0.2	10.4	0.3	11.2	0.2
VBHp	11.2	0.1	11.3	0.2	11.3	0.1	11.3	0.2	11.8	0.3
SCW	19.2	0.4	19.3	0.5	20.3	0.4	20.6	0.4	19.7	0.4
SCD	10.3	0.3	10.3	0.3	10.3	0.3	10.3	0.3	11.0	0.2
PDHl	6.7	0.2	6.6	0.2	6.3	0.3	6.0	0.3	6.5	0.2
PDHr	6.8	0.2	6.7	0.2	5.9	0.2	6.0	0.1	6.1	0.1
PDWl	4.5	0.2	4.6	0.2	4.7	0.1	5.1	0.2	5.6	0.2
PDWr	4.4	0.2	4.5	0.2	4.9	0.2	5.4	0.2	5.7	0.2
SPL	25.6	0.5	30.3	0.4	33.6	1.0	40.5	1.5	46.9	1.1
TPW	41.4	0.8	44.9	0.8	47.6	1.0	48.4	0.9	53.8	1.0

Table 1. Linear Measurements from Tan study (mm) (Tan, Teo and Chua, 2004)

	C3		C4		C5		C6		C7	
	Mean	Std dev	Mean	Std dev	Mean	Std dev	Mean	Std dev	Mean	Std dev
EPAu	154.7	3.8	169.2	4.9	187.4	6.6	210.5	10.0	220.8	9.0
EPAl	216.8	10.1	241.5	10.6	286.4	10.3	316.3	7.4	340.0	10.3
SCA	149.7	9.0	159.9	8.4	166.8	8.0	163.7	10.2	167.5	6.7
PDAl	27.6	1.0	27.7	0.8	27.4	1.1	29.4	1.5	33.7	2.6
PDAr	28.5	1.0	28.8	1.0	28.5	1.1	33.0	1.3	32.1	1.6

Table 2. Surface Area measurements from Tan study (mm^2) (Tan, Teo and Chua, 2004)

Utilizing the mean and standard deviations from Tan's study, SAS® random number generation was used to create a normally distributed data set. From this random number generation, 100 observations were simulated in order to make the comparisons more robust. From this increase in sample size, linear regression analysis was completed simultaneously with the ANOVA. The results of this analysis are shown and discussed in succeeding paragraphs.

	C3		C4		C5		C6		C7	
	Mean	Std dev	Mean	Std dev	Mean	Std dev	Mean	Std dev	Mean	Std dev
EPItu	5.0	4.1	5.2	5.2	7.1	1.2	5.8	0.6	5.8	0.8
EPItl	3.3	0.5	3.5	0.7	2.7	0.3	4.2	0.4	5.1	0.5
PDIsl	-42.9	1.0	-44.0	1.3	-46.3	1.0	-41.9	1.6	-30.6	1.1
PDIsr	39.6	1.0	38.9	1.1	38.1	1.6	38.5	2.3	30.3	0.9
PDItl	-4.8	1.0	-3.2	0.7	2.6	0.7	4.8	1.0	5.8	0.7
PDItr	-6.5	1.0	-5.4	1.1	4.9	1.0	6.0	1.3	3.1	0.7

Table 3. Angular measurements from Tan study (degrees) (Tan, Teo and Chua, 2004)

1.3 Analysis

To find correlations present in the anthropometrics of the vertebral bodies in the cervical spine, statistical analysis was completed on each vertebral segment from C3 to C7. Initially, investigation into the C3 vertebra was completed, starting with the linear measurements. As an example, the C3_EPWu was compared to all 24 other measured parameters of the C3 vertebra. This resulted in 14 linear measurements compared to 24 other measurement parameters for the C3 vertebra, resulting in a total of 336 comparisons.

From analysis of the C3 linear measurements it was found that there were 8 significant correlations present among all 336 comparisons. These results are shown in Table 4. The dependent variables are listed first with the regressor/independent showing second. The first case illustrates that the C3_PDWr is the dependent variable and C3_VBHp is the regressor or independent variable. From analysis of the area measurements of the C3 vertebra, only one significant correlation was present among 120 comparisons (Table 5). Finally when comparing the angular measurements of the C3 vertebra, it was found that there were 2 significant correlations among a total of 144 comparisons (Table 6).

The examination of the other vertebral segments, from C4 to C7, was accomplished in a similar fashion. Analysis of the C4 vertebra resulted in extensively more significant relationships than were found in C3 with a total of 23 significant correlations.

Comparisons of the linear measurements of the C4 vertebra yielded 12 strong relationships, and these results are shown in Table 7. From investigation into the area measurements of the C4 vertebra, it was found that there were five comparisons of anthropometrics that had a considerable link among 120 comparisons (shown in Table 8). Finally when comparing the angular measurements of the C4 vertebra to the other 24 measurements (include all three forms of linear, area, and angular), there were 6 strong relationships found from the 144 total comparisons. All of the significant correlations of the angular measurements can be found in Table 9.

In completing the investigation into the C5 vertebra, it was again found to have increasingly more relationships, with a total of 40 strong correlations. The comparisons of the linear measurements of the C5 vertebra to the rest of the anthropometric measurements resulted in the most relationships; these are displayed in Tables 10 and 11. Of these comparisons there were 21 relationships found in the C5 vertebral body anthropometrics. With the investigation into the area measurements of the C5 vertebra, it was found that there were 10 significant correlations from a total of 120 comparisons completed. Finally in investigating

	ANOVA	Parameter Estimates		Y-intercept	Slope
	P	P (y-intercept)	P (regressor/ independent)		
PDWr vs. VBHp	0.0424	<0.0001	0.0424	8.34113	-0.3523
	Significant	Significant	Not Significant		
PDWr vs. SCW	0.0166	0.0043	0.0166	2.39346	0.10409
	Significant	Significant	Significant		
SCD vs. PDHl	0.0085	<0.0001	0.0085	12.88191	-0.3927
	Significant	Significant	Significant		
TPW vs. EPItl	0.0324	<0.0001	0.0324	40.20551	0.3705
	Significant	Significant	Not Significant		
TPW vs. PDIsr	0.0068	<0.0001	0.0068	50.21888	-0.22291
	Significant	Significant	Significant		
VBHa vs. EPDu	0.0062	<0.0001	0.0062	17.51686	-0.55302
	Significant	Significant	Significant		
VBHa vs. PDAl	0.0149	<0.0001	0.0149	11.47438	-0.05326
	Significant	Significant	Significant		
VBHp vs. PDWr	0.0024	<0.0001	0.0024	10.52205	0.15663
	Significant	Significant	Significant		

Table 4. C3 Linear measurements

	ANOVA	Parameter Estimates		Y-intercept	Slope
	P	P (y-intercept)	P (regressor/ independent)		
PDAl vs. EPAu	0.0061	0.0024	0.0061	14.53586	0.08468
	Significant	Significant	Significant		

Table 5. C3 Area measurements

	ANOVA	Parameter Estimates		Y-intercept	Slope
	P	P (y-intercept)	P (regressor/ independent)		
EPItl vs. EPWu	0.0124	0.0413	0.0124	15.04667	-0.85215
	Significant	Not Significant	Significant		
EPItu vs. EPDu	0.0419	0.0355	0.0419	33.19859	-2.07824
	Significant	Not Significant	Not Significant		

Table 6. C3 Angular measurements

	ANOVA	Parameter Estimates			
	P	P (y-intercept)	P (regressor/ independent)	Y-intercept	Slope
EPDl vs. SPL	0.0245 Significant	0.0006 Significant	0.0245 Significant	9.21517	0.19714
EPDu vs. EPWu	0.02 Significant	<0.0001 Significant	0.02 Significant	10.32639	0.24868
EPWl vs. PDItr	0.0326 Significant	<0.0001 Significant	0.0326 Not Significant	15.10987	0.01808
EPWu vs. PDWr	0.0248 Significant	<0.0001 Significant	0.0248 Significant	15.14845	-0.10029
PDHr vs. EPDl	0.0303 Significant	<0.0001 Significant	0.0303 Not Significant	8.23208	-0.10071
PDWl vs. EPDl	0.0076 Significant	0.0003 Significant	0.0076 Significant	2.66256	0.12866
SCD vs. EPItu	0.0207 Significant	<0.0001 Significant	0.0207 Significant	10.30332	0.01353
SCW vs. VBHa	0.0056 Significant	<0.0001 Significant	0.0056 Significant	13.58683	0.56164
SCW vs. TPW	0.0323 Significant	<0.0001 Significant	0.0323 Not Significant	23.82069	-0.11106
TPW vs. EPDu	0.0108 Significant	<0.0001 Significant	0.0108 Significant	68.19806	-1.95975
TPW vs. SCW	0.0323 Significant	<0.0001 Significant	0.0323 Not Significant	49.4967	-0.41322
VBHp vs. PDAl	0.0348 Significant	<0.0001 Significant	0.0348 Not Significant	9.47331	0.06488

Table 7. C4 Linear Measurements

the relationships present in the C5 vertebra angular measurements and the other anthropometric measurements, 9 significant correlations were found. The strong relationships that were present in the C5 vertebra's angular measurements are displayed in Table 13.

	ANOVA	Parameter Estimates			
	P	P (y-intercept)	P (regressor/ independent)	Y-intercept	Slope
EPAu vs. EPWu	0.0104	0.66909	0.0104	-34.27382	13.8368
	Significant	Not Significant	Significant		
EPAu vs. VBHa	0.0219	<0.0001	0.0219	204.59603	-3.58479
	Significant	Significant	Significant		
EPAu vs. PDWl	0.0288	<0.0001	0.0288	143.55496	5.63237
	Significant	Significant	Not Significant		
EPAu vs. PDItl	0.0046	<0.0001	0.0046	161.94009	-2.29035
	Significant	Significant	Significant		
SCA vs. SCW	0.0066	0.0093	0.0066	78.19896	4.24208
	Significant	Significant	Significant		

Table 8. C4 Area Measurements

	ANOVA	Parameter Estimates			
	P	P (y-intercept)	P (regressor/ independent)	Y-intercept	Slope
EPItl vs. EPDl	0.0191	0.3069	0.0191	-2.63366	0.40237
	Significant	Not Significant	Significant		
EPItl vs. VBHp	0.0064	0.0781	0.0064	-6.16074	0.8523
	Significant	Not Significant	Significant		
EPItl vs. SCW	0.0345	0.4489	0.0345	-1.91019	0.27823
	Significant	Not Significant	Not Significant		
PDIsl vs. SCW	0.0437	<0.0001	0.0437	-55.42757	0.59539
	Significant	Significant	Not Significant		
PDIsl vs. EPItl	0.0178	<0.0001	0.0178	-42.30098	-0.46784
	Significant	Significant	Significant		
PDIsr vs. VBHp	0.0076	<0.0001	0.0076	24.45626	1.27529
	Significant	Significant	Significant		

Table 9. C4 Angular Measurements

	ANOVA	Parameter Estimates		Y-intercept	Slope
	P	P (y-intercept)	P (regressor/ independent)		
EPDl vs SCW	0.0272	<0.0001	0.0272	11.78609	0.1654
	Significant	Significant	Not Significant		
EPDl vs SCA	0.0014	<0.0001	0.0014	17.20746	-0.01233
	Significant	Significant	Significant		
EPDu vs SCA	0.0415	<0.0001	0.0415	13.85796	0.0026
	Significant	Significant	Not Significant		
EPDu vs PDItl	0.0405	<0.0001	0.0405	14.22233	0.02818
	Significant	Significant	Not Significant		
EPWl vs PDWr	0.0394	<0.0001	0.0394	15.30981	0.11428
	Significant	Significant	Not Significant		
EPWu vs VBHa	0.0353	<0.0001	0.0353	15.79112	-0.09338
	Significant	Significant	Not Significant		
EPWu vs EPAu	0.0465	<0.0001	0.0465	14.35019	0.00291
	Significant	Significant	Not Significant		
PDHl vs PDWl	0.0207	0.0426	0.0207	2.90559	0.70698
	Significant	Not Significant	Significant		
PDHl vs PDIsr	0.0488	<0.0001	0.0488	7.36108	-0.02888
	Significant	Significant	Not Significant		
PDWl vs PDHl	0.0207	<0.0001	0.0207	4.23321	0.0756
	Significant	Significant	Significant		
PDWl vs PDAr	0.0189	<0.0001	0.0189	5.30502	-0.02109
	Significant	Significant	Significant		
PDWr vs EPWl	0.0394	0.722	0.0394	-1.01063	0.37265
	Significant	Not Significant	Not Significant		
PDWr vs EPItu	0.026	<0.0001	0.026	4.64358	0.03596
	Significant	Significant	Not Significant		
PDWr vs PDItl	0.0213	<0.0001	0.0213	4.7488	0.06132
	Significant	Significant	Significant		
SCD vs VBHp	0.033	0.4903	0.033	2.50118	0.69204
	Significant	Not Significant	Not Significant		
SCD vs EPAu	0.0167	<0.0001	0.0167	12.49697	-0.01164
	Significant	Significant	Significant		
SCW vs EPDl	0.0272	<0.0001	0.0272	15.81713	0.29501
	Significant	Significant	Not Significant		
SPL vs PDItr	0.0117	<0.0001	0.0117	34.72062	-0.23666
	Significant	Significant	Significant		
VBHa vs EPWu	0.0353	<0.0001	0.0353	16.67726	-0.47571
	Significant	Significant	Not Significant		

Table 10. C5 Linear Measurements (Part 1)

	ANOVA	Parameter Estimates		Y-intercept	Slope
	P	P (y-intercept)	P (regressor/ independent)		
VBHp vs SCD	0.033	<0.0001	0.033	10.60996	0.06583
	Significant	Significant	Not Significant		
VBHp vs PDAl	0.0163	<0.0001	0.0163	10.72456	0.0205
	Significant	Significant	Significant		

Table 11. C5 Linear Measurements (Part 2)

	ANOVA	Parameter Estimates		Y-intercept	Slope
	P	P (y-intercept)	P (regressor/ independent)		
EPAl vs. EPDl	0.0126	<0.0001	0.0126	418.92547	-8.70812
	Significant	Significant	Significant		
EPAu vs. EPDl	0.0356	0.0001	0.0356	122.32733	4.3305
	Significant	Significant	Not Significant		
EPAu vs. SCD	0.0317	<0.0001	0.0317	21.37129	2.08213
	Significant	Significant	Not Significant		
EPAu vs. PDAl	0.0282	<0.0001	0.0282	156.25957	1.14813
	Significant	Significant	Not Significant		
PDAl vs. VBHp	0.0163	0.7504	0.0163	-4.13231	2.80491
	Significant	Not Significant	Significant		
PDAr vs. PDWl	0.0189	<0.0001	0.0189	40.73714	-2.60613
	Significant	Significant	Significant		
PDAr vs. PDIsr	0.0178	<0.0001	0.0178	23.56205	0.12563
	Significant	Significant	Significant		
PDAr vs. PDItr	0.0356	<0.0001	0.0356	29.53373	-0.21079
	Significant	Significant	Not Significant		
SCA vs. EPDu	0.0415	0.5788	0.0415	-61.91478	16.06134
	Significant	Not Significant	Not Significant		
SCA vs. EPDl	0.0014	<0.0001	0.0014	290.35634	-8.10387
	Significant	Significant	Significant		

Table 12. C5 Area Measurements

	ANOVA	Parameter Estimates		Y-intercept	Slope
	P	P (y-intercept)	P (regressor/ independent)		
EPItu vs. PDWr	0.026	0.8775	0.026	0.46263	1.37892
	Significant	Not Significant	Not Significant		
PDIsl vs. PDItl	0.0488	<0.0001	0.0488	-45.50392	-0.24512
	Significant	Significant	Not Significant		
PDIsr vs. PDHl	0.0488	<0.0001	0.0488	47.54716	-1.35216
	Significant	Significant	Not Significant		
PDIsr vs. PDAr	0.0178	<0.0001	0.0178	26.44105	0.44528
	Significant	Significant	Significant		
PDItl vs. EPDu	0.0405	0.0701	0.0405	-18.8499	1.4951
	Significant	Not Significant	Not Significant		
PDItl vs. PDWr	0.0213	0.3461	0.0213	-1.71369	0.86345
	Significant	Not Significant	Significant		
PDItl vs. PDIsl	0.0488	0.1931	0.0488	-4.82456	-0.15925
	Significant	Not Significant	Not Significant		
PDItr vs. SPL	0.0117	0.0001	0.0117	13.95113	-0.26654
	Significant	Significant	Significant		
PDItr vs. PDAr	0.0356	0.0002	0.0356	10.99851	-0.21019
	Significant	Significant	Not Significant		

Table 13. C5 Angular Measurements

In the analysis of the C6 vertebra 22 strong relationships, less than what was seen in the C5 and C4 vertebra but more than what was seen in the C3 vertebra. Investigation of the C6 linear measurements and comparisons between the other anthropometric measurements discovered 15 significant comparisons out of a total of 336 comparisons completed. These results are shown in Table 14. Exploration into the relationships present in the C6 vertebra area measurements in comparison to the other anthropometrics, showed that there were two significant correlations present (shown in Table 15). Finally analysis of the C6 vertebra and the angular measurements comparisons to the other anthropometrics, found there to be 5 strong relationships from a total of 144 comparisons made (Table 16).

In the analysis of the C7 vertebra there were 34 significant relationships found. Thus finding that the C7 vertebra has more correlations present than all the other vertebra's except for C5. Investigation of the C7 linear measurements and comparing them with the other anthropometrics discovered 18 comparisons with strong relationships from 336 comparisons completed. The result of this is displayed in Tables 17 and 18. Exploration into the relationships present in the C7 vertebra area measurements divulged that there were five significant correlations present (shown in Table 19). Finally analysis of the C7's angular measurements found 11 strong relationships out of 144 comparisons made (Table 20).

1.4 Discussion

Through investigation into correlations that may be present within the anthropometric data of each vertebra, there were a total of 130 significant relationships discovered:

- 11 in the C3 vertebra
- 23 in the C4 vertebra
- 40 in the C5 vertebra
- 22 in the C6 vertebra
- 34 in the C7 vertebra.

Some of these relationships were physiologically reconcilable, in particular for the C3 vertebral segment the upper endplate transverse inclination and the upper endplate depth (EPItu & EPDu). From looking at Figure 1 it can be seen how the EPItu would possibly increase in the same way as the EPDu increases based on a person's stature.

As for the C4 vertebral segment the correlations that make the most sense are the upper endplate area vs. the upper endplate width (EPAu vs. EPWu), the upper endplate depth vs. the upper endplate width (EPDu vs. EPWu), and the lower endplate transverse inclination vs. the lower endplate depth (EPItl vs. EPDl). In the study completed by Panjabi they found that modeling the area of the endplates, spinal canal, and pedicles as ellipses was "justified" (Liu, Clark and Krieger, 1986). So when looking at the case of the EPAu and the EPWu, this relationship can be explained by the area of an ellipse. Since the area of an ellipse is Area= πab where a and b are depicted in Figure 2 as the radius. In the same aspect since a radius of an ellipse is the diameter divided by 2 ($a=\frac{a'}{2} or\, b=\frac{b'}{2}$) then the area can also equate to Area= $\pi\left(\frac{a'}{2}\right)\left(\frac{b'}{2}\right)$, where a′ and b′ are depicted in Figure 2 as the diameters. In this case EPWu would be b′ and the area would be EPAu. So as the diameter EPWu increases so does the area EPAu.

As for the relationship found between the EPDu and the EPWu, the same argument may be placed that the depth of the end plate could be seen as the diameter as well, as shown here:

$$Area = \pi\left(\frac{a'}{2}\right)\left(\frac{b'}{2}\right)$$

$$EPAu = \pi\left(\frac{EPDu}{2}\right)\left(\frac{EPWu}{2}\right)$$

In the case for the relationship between the EPItl and the EPDl, the same statement as stated for the C3 vertebra in the case of the EPItu and the EPDu can be stated.

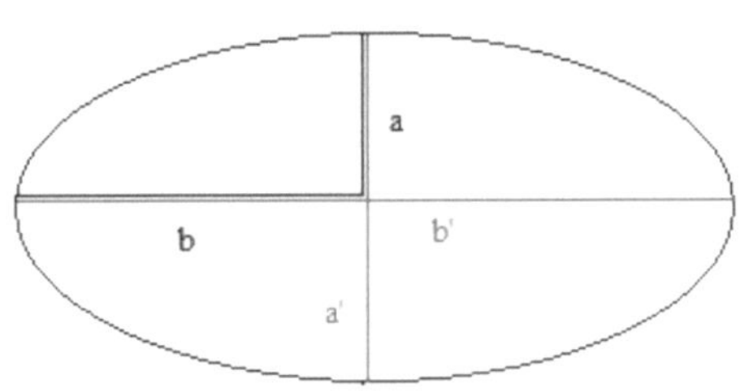

Fig. 2. Diagram of an ellipse to describe the area of an ellipse, where a and b are the radius and a′ and b′ are the diameters

	ANOVA	Parameter Estimates		Y-intercept	Slope
	P	P (y-intercept)	P (regressor/ independent)		
EPDl vs. EPWl	0.03	0.0158	0.03	8.2642	0.37956
	Significant	Significant	Not Significant		
EPDu vs. PDWl	0.0048	<0.0001	0.0048	13.16703	0.26669
	Significant	Significant	Significant		
EPWu vs. PDWr	0.0155	<0.0001	0.0155	15.80078	-0.0001436
	Significant	Significant	Significant		
PDHl vs. EPDu	0.0425	0.3398	0.0425	1.91753	0.28098
	Significant	Not Significant	Not Significant		
PDHr vs. EPDu	0.0181	<0.0001	0.0181	7.42367	-0.09719
	Significant	Significant	Significant		
PDHr vs. PDItl	0.0166	<0.0001	0.0166	5.88497	0.02407
	Significant	Significant	Significant		
SCD vs. EPDl	0.0198	<0.0001	0.0198	7.52521	0.1789
	Significant	Significant	Significant		
SCD vs. EPAl	0.0127	<0.0001	0.0127	7.53602	0.00885
	Significant	Significant	Significant		
SPL vs. PDItr	0.0405	<0.0001	0.0405	41.94107	-0.2777
	Significant	Significant	Not Significant		
TPW vs. PDItr	0.0321	<0.0001	0.0321	47.35028	0.16667
	Significant	Significant	Not Significant		
VBHa vs. EPDl	0.0179	<0.0001	0.0179	7.27923	0.19901
	Significant	Significant	Significant		
VBHa vs. VBHp	0.0283	0.0025	0.0283	6.0541	0.38446
	Significant	Significant	Not Significant		
VBHa vs. SCW	0.0451	<0.0001	0.0451	7.66983	0.13244
	Significant	Significant	Not Significant		
VBHa vs. EPAu	0.0046	<0.0001	0.0046	12.01819	-0.00769
	Significant	Significant	Significant		
VBHp vs. VBHa	0.0283	<0.0001	0.0283	10.00304	0.12525
	Significant	Significant	Not Significant		

Table 14. C6 Linear Measurements

For the C5 vertebral segment, the associations found that were physiologically reconcilable were in:

- The lower endplate area vs. the lower endplate depth (EPAl vs. EPDl)
- The upper endplate area vs. the lower endplate depth (EPAu vs. EPDl)
- The pedicle height on the left side vs. the pedicle width on the left side (PDHl vs. PDWl)
- The pedicle sagittal inclination on the left side vs. the pedicle transverse inclination on the left side (PDIsl vs. PDItl)
- The pedicle transverse inclination on the left side vs. the pedicle sagittal inclination on the left side (PDItl vs. PDIsl)
- The pedicle width on the left side vs. the pedicle height on the left side (PDWl vs. PDHl)

As for the correlations in the EPAl vs. EPDl, EPAu vs. EPDl, PDHl vs. PDWl, and PDWl vs. PDHl these can be explained in the same aspect as the relationships found in the C4 vertebra; with comparison of the area of an ellipse and the diameter of an ellipse, along with the diameter to diameter comparison of a ellipse. In the cases of the relationships present in the sagittal inclination and the transverse inclination, if looking at Figure 1 it can be seen how as one increases the other may increase.

In the C6 vertebral segment the relationships that were the most physiologically reconcilable are the lower endplate depth and the lower endplate width (EPDl and EPWl), this type of relationship was explained previously with the examination into the C4 vertebra and relationship present in diameter to diameter comparison of an ellipse. As for the relationship found between the anterior vertebral body height and the posterior vertebral body height (VBHa and VBHp), again if looking at Figure 1 it can be seen that if the height increases in either the anterior or posterior location of the vertebral body that there should be an increase in the former as well.

	ANOVA	Parameter Estimates		Y-intercept	Slope
	P	P (y-intercept)	P (regressor/ independent)		
EPAl vs. TPW	0.0289	<0.0001	0.0289	407.91242	-1.90819
	Significant	Significant	Not Significant		
PDAl vs. EPDl	0.0168	<0.0001	0.0168	44.79661	-0.9836
	Significant	Significant	Significant		

Table 15. C6 Area Measurements

Unlike the other vertebras, the C7 vertebra had no obvious relationships that were physiologically reconcilable. As for the other relationships found that were not described they were not physiological reconcilable. But they will help in further research as discussed earlier since they were found to be statistically significant.

It is of interest to investigate the findings further. In particular any relationships that was present and also present in the opposite comparison. As an example if a link was found between upper endplate width vs. the lower endplate width (EPWu vs. EPWl) and also a link between the lower endplate width vs. the upper endplate width (EPWl vs. EPWu).

	ANOVA	Parameter Estimates		Y-intercept	Slope
	P	P (y-intercept)	P (regressor/ independent)		
PDIsr vs. PDItl	0.0332	<0.0001	0.0332	41.66266	-0.60909
	Significant	Significant	Not Significant		
PDItl vs. PDHr	0.0166	0.1125	0.0166	-9.36715	2.3752
	Significant	Not Significant	Significant		
PDItl vs. PDIsr	0.0332	<0.0001	0.0332	7.77712	-0.07463
	Significant	Significant	Not Significant		
PDItr vs. SPL	0.0405	<0.0001	0.0405	12.26138	-0.1517
	Significant	Significant	Not Significant		
PDItr vs. TPW	0.0321	0.2446	0.0321	-7.19097	0.27594
	Significant	Not Significant	Not Significant		

Table 16. C6 Angular Measurements

For the C3 vertebra there was only one case that this was seen in:

- The pedicle width on the right side vs. the posterior vertebral body height and the posterior vertebral body height vs. the pedicle width on the right side (PDWr vs. VBHp and VBHp vs. PDWr).
- In the C4 vertebra there was also one case of this same type of connection which was seen in:
- The spinal canal width and the transverse process width (SCW and TPW).
- The C5 vertebra had increasingly more connections of this type and included the following:
- The lower endplate depth and the spinal canal width (EPDl and SCW)
- The lower endplate depth and the spinal canal area (EPDl and SCA)
- The upper endplate depth and the spinal canal area (EPDu and SCA)
- The upper endplate depth and the pedicle transverse inclination (EPDu and PDItl)
- The lower endplate width and the pedicle width on the right side (EPWl and PDWr)
- The upper endplate width and the anterior vertebral body height (EPWu and VBHa)
- The pedicle height on the left side and the pedicle width on the left side (PDHl and PDWl)
- The pedicle height on the left side and the pedicle sagittal inclination on the right side (PDHl and PDIsr)
- The pedicle width on the left side and the pedicle area on the right side (PDWl and PDAr)
- The pedicle width on the right side and the upper endplate transverse inclination (PDWr and EPItu)
- The pedicle width on the right side and the pedicle transverse inclination on the left side (PDWr and PDItl)
- The spinal canal depth and the posterior vertebral body height (SCD and VBHp)

- The spinal canal depth and the upper endplate area (SCD and EPAu)
- The spinous process length and the pedicle transverse inclination on the right side (SPL and PDItr)
- The posterior vertebral body height and the pedicle area on the left side (VBHp and PDAl)
- The pedicle area on the right side and the pedicle sagittal inclination on the right side (PDAr and PDIsr)
- The pedicle sagittal inclination on the left side and the pedicle transverse inclination on the left side (PDIsl and PDItl).

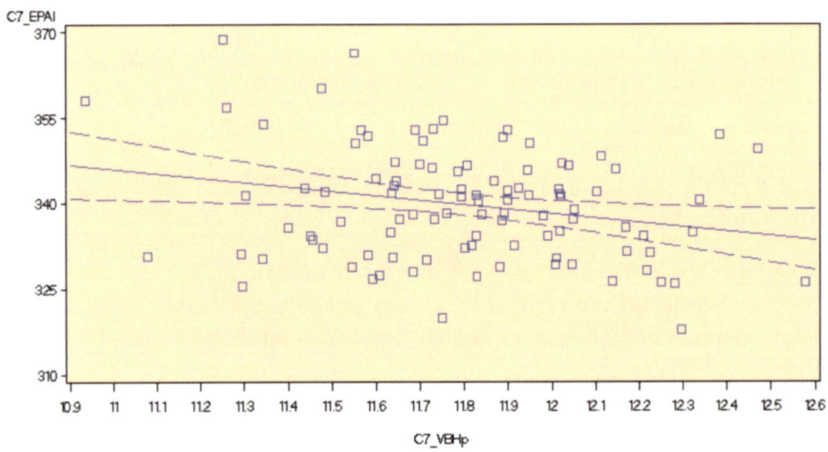

Fig. 3. Scatter plot of C7_EPAl vs. C7_VBHp

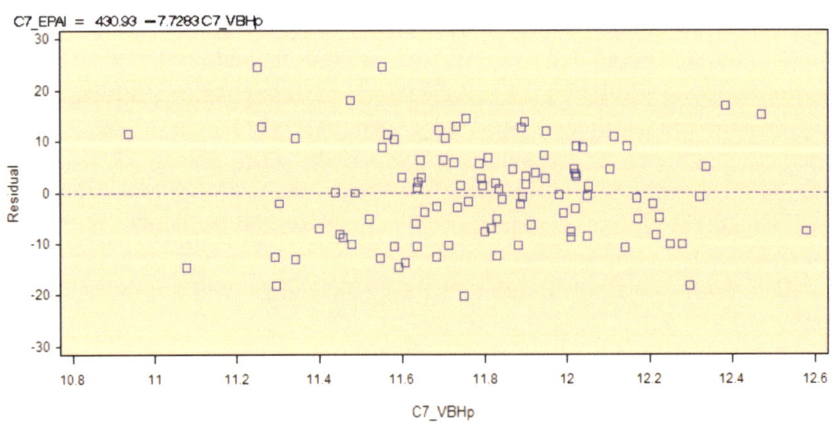

Fig. 4. Residual vs. C7_VBHp

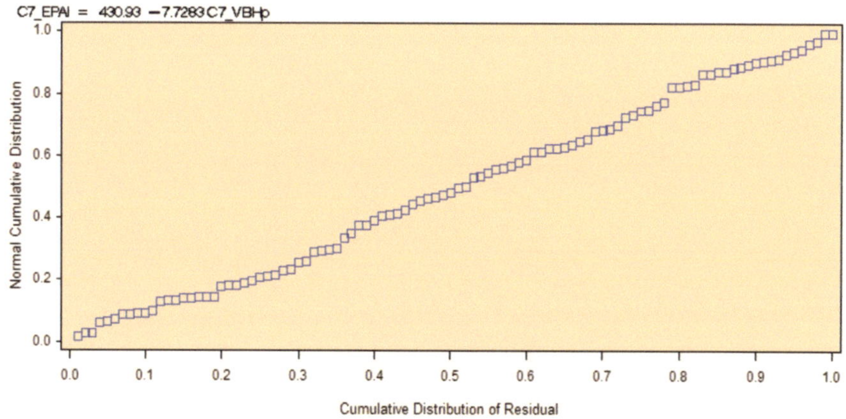

Fig. 5. Normal Cumulative Distribution vs. Cumulative Distribution of Residual

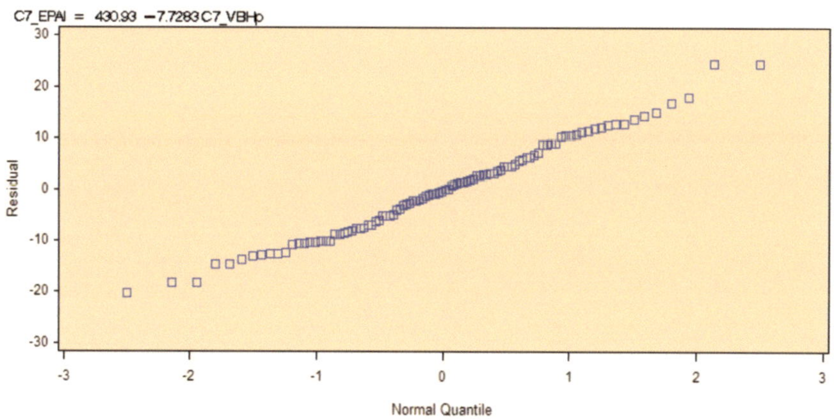

Fig. 6. Residual vs. Normal Quantile

In the C6 vertebra there were five connections of this type which were:

- The pedicle height on the right side and the pedicle transverse inclination on the left side (PDHr and PDItl)
- The spinous process length and the pedicle transverse inclination on the right side (SPL and PDItr)
- The transverse process width and the pedicle transverse inclination on the right side (TPW and PDItr)
- The anterior vertebral body height and the posterior vertebral body height (VBHa and VBHp)
- The pedicle sagittal inclination on the right side and the pedicle transverse inclination on the left side (PDIsr and PDItl).
- Finally for the C7 vertebra there were fifteen connections of this type which included:
- The lower endplate area and the upper endplate depth (EPAl and EPDu)
- The lower endplate area and the posterior vertebral body height (EPAl and VBHp)
- The upper endplate area and the lower endplate depth (EPAu and EPDl)

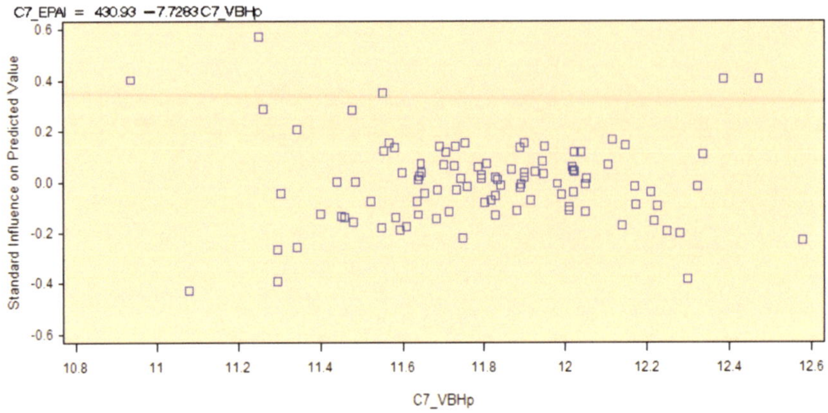

Fig. 7. Standard Influence on Predicted Value vs. C7_VBHP

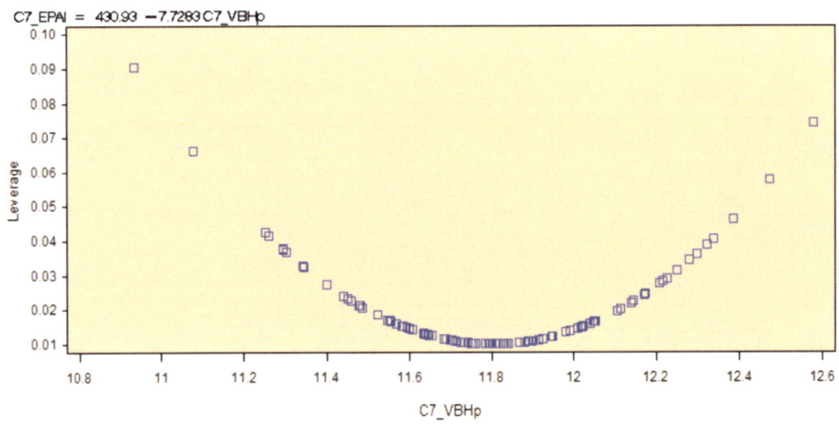

Fig. 8. Leverage vs. C7_VBHp

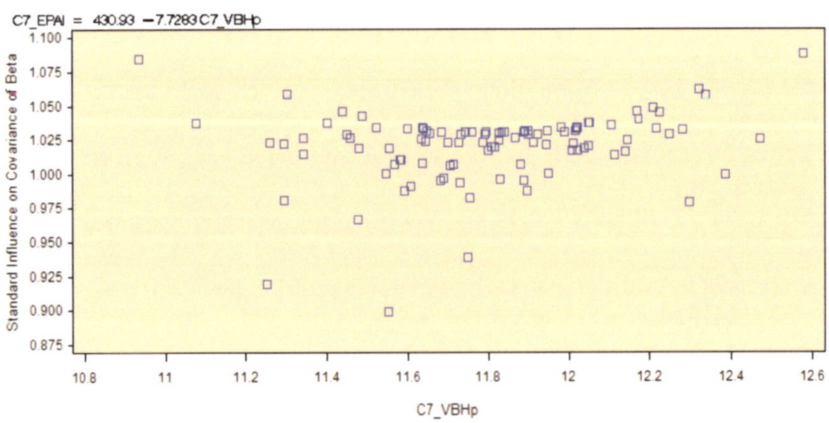

Fig. 9. Standard Influence on Covariance of Beta vs. C7_VBHp

- The lower endplate depth and the pedicle height on the right side (EPDl and PDHr)
- The upper endplate depth and the pedicle sagittal inclination on the left side (EPDu and PDIsl)
- The lower endplate transverse inclination and the spinal canal width (EPItl and SCW)
- The upper endplate transverse inclination and the posterior vertebral body height (EPItu and VBHp)
- The upper endplate transverse inclination and the pedicle area on the right side (EPItu and PDAr)
- The upper endplate width and the spinous process length (EPWu and SPL)
- The pedicle area on the left side and the pedicle sagittal inclination on the left side (PDAl and PDIsl)
- The pedicle height on the right side and the posterior vertebral body height (PDHr and VBHp)
- The pedicle height on the right side and the spinous process length (PDHr and SPL)
- The pedicle sagittal inclination on the left side and the anterior vertebral body height (PDIsl and VBHa)
- The pedicle sagittal inclination on the left side and the spinous process length (PDIsl and SPL)
- The pedicle transverse inclination on the right side and the spinal canal width (PDItr and SCW)

This type of relationship is important to note because it was thought that if there was a significance found in one comparison that the same type of relationship would be seen when doing the reciprocal comparison, but this was not always seen. When this type of relationship isn't seen it could be the result of the random number generation and the normal distribution of these numbers.

	ANOVA	Parameter Estimates		Y-intercept	Slope
	P	P (y-intercept)	P (regressor/ independent)		
EPDl vs. PDHr	0.0145	<0.0001	0.0145	19.19577	-0.59208
	Significant	Significant	Significant		
EPDl vs. EPAu	0.0227	<0.0001	0.0227	13.92929	0.00755
	Significant	Significant	Significant		
EPDu vs. EPAl	0.0499	<0.0001	0.0499	16.35341	-0.00366
	Significant	Significant	Not Significant		
EPDu vs. PDIsl	0.0149	<0.0001	0.0149	16.47022	0.04455
	Significant	Significant	Significant		
EPWu vs. SPL	0.0062	<0.0001	0.0062	20.24696	-0.02664
	Significant	Significant	Significant		
PDHr vs. EPDl	0.0145	<0.0001	0.0145	7.65744	-0.10034
	Significant	Significant	Significant		

Table 17. C7 Linear Measurements (Part 1)

	ANOVA	Parameter Estimates		Y-intercept	Slope
	P	P (y-intercept)	P (regressor/ independent)		
PDHr vs. VBHp	0.0153	<0.0001	0.0153	7.09761	-0.08508
	Significant	Significant	Significant		
PDHr vs. SPL	0.0454	<0.0001	0.0454	5.07412	0.01077
	Significant	Significant	Not Significant		
PDWr vs. PDIsr	0.0478	0.0002	0.0478	11.73842	-0.20069
	Significant	Significant	Not Significant		
SCW vs. EPItl	0.0143	<0.0001	0.0143	20.8603	-0.23411
	Significant	Significant	Significant		
SCW vs. PDItr	0.0218	<0.0001	0.0218	20.1399	-0.14708
	Significant	Significant	Significant		
SPL vs. EPWu	0.0062	<0.0001	0.0062	99.42223	-2.77358
	Significant	Significant	Significant		
SPL vs. PDHr	0.0454	<0.0001	0.0454	35.48384	1.84364
	Significant	Significant	Not Significant		
SPL vs. PDIsl	0.0208	<0.0001	0.0208	39.68142	-0.23022
	Significant	Significant	Significant		
VBHa vs. PDIsl	0.0274	<0.0001	0.0274	12.7251	0.04937
	Significant	Significant	Not Significant		
VBHp vs. PDHr	0.0153	<0.0001	0.0153	15.99891	-0.68864
	Significant	Significant	Significant		
VBHp vs. EPAl	0.0152	<0.0001	0.0152	14.38187	-0.00759
	Significant	Significant	Significant		
VBHp vs. EPItu	0.0064	<0.0001	0.0064	11.04194	0.10321
	Significant	Significant	Significant		

Table 18. C7 Linear Measurements (Part 2)

It is also of interest to see which correlations also existed in other vertebras, and if any existed throughout C3 to C7. There were eight correlations that were seen in more than just one vertebral segment. The link between the upper endplate width and the pedicle width on the right side (EPWu and PDWr) was seen in both the C4 vertebra and the C6 vertebra. As for the connection between the posterior vertebral body height and the pedicle area on the left side (VBHp and PDAl), this was seen in both the C4 vertebra and the C5 vertebra. The relationship between the spinous process length and the pedicle transverse inclination on the right side (SPL and PDItr) was seen in the C5 and C6 vertebra. The connection between the pedicle height on the right side and the lower endplate depth (PDHr and EPDl) was seen in both the C4 and C7 vertebra. The relationship between the upper endplate area and the lower endplate depth (EPAu and EPDl) was seen in both the C7 and C5 vertebra. The link between the lower endplate transverse inclination and the spinal canal width (EPItl and SCW) was seen in both C7 and C4. Also the connection between the upper endplate transverse inclination and the pedicle width on the right side (EPItu and PDWr) was seen in both the C5 and C7 vertebra. Finally the correlation between the pedicle transverse inclination on the right side and the spinous process length (PDItr and SPL) was seen in both the C5 and C6 vertebral segments. It would be of interest to investigate why this may be, and why there isn't a relationship present that was found in all 5 segments.

An example of some plots that resulted from the linear regression analysis of the vertebral segments through SAS® are shown in Figures 3-9. These figures are included to illustrate an example involving the lower endplate area and the posterior vertebral body height for the C7 vertebra (EPAl and VBHp).

In the study completed by Tan it was found that the endplate width and depth (EPW, EPD), and vertebral body height (VBH) are moderately constant from C3 to C5 and then increase to C7. The increase is more drastic in the endplate width than the end plate depth and vertebral body height, with the endplate depth and the vertebral body height having both a similar increasing trend. The posterior vertebral body height (VBHp), lower endplate width (EPWl), and the lower endplate depth (EPDl) are larger than their complementary measurements of anterior vertebral body height (VBHa), upper endplate width (EPWu) and the upper endplate depth (EPDu). As for the spinal canal, the width and depth (SCW and SCD) both are fairly constant from C3 to C6. The spinal canal width decreases sharply through to C7, and the spinal canal depth increases progressively into C7. Both the process lengths of the spinous and transverse (SPL and TPW) increase with a similar trend. The values of the left and right pedicle height (PDHl and PDHr) are comparatively similar, the same goes for the left and right pedicle width (PDWl and PDWr). The pedicle height (PDH) decreases gradually from C3 to C6 and then increases to C7. The pedicle width (PDW) increases throughout the cervical spine. The area of the lower and upper endplates (EPAl and EPAu) increases throughout the cervical spine, and the upper is at all times larger than the lower endplate area. The spinal canal area (SCA) increases from C3 to C5 and then decreases at C6 just to increase again to C7. Both the left and right pedicle area (PDAl and PDAr) are fairly constant from C3 to C5, but then the left increases to C7 while the right decreases to C7. The endplate inclinations for both the upper and lower regions (EPItu and EPitl) are angled toward the head with a steady inclination. The lower endplate transverse inclination is always smaller than the upper endplate transverse inclination. The pedicle sagittal inclination (PDIs) is fairly constant at about 40° from C3 to C6, but then at C6 the pedicles start to congregate towards each other. The pedicle transverse inclination (PDIt) is angled towards the back from C3 to C4 but then angle towards the head after C4 (Nissan and Gilad, 1984).

In comparing the endplate width and depth, the posterior vertebral body height, and the upper endplate area (EPW, EPD, VBHp, and EPAu) to Panjabi's study of the cervical spine of Caucasian subjects, Tan's measurements are smaller "by an average of 10.3%, 15.2%, 4.0%, and 8.3% respectively" than that of Panjabi's measurements. The lower endplate area (EPAl) are larger in Tan's study by 16.3% than Panjabi's. The trends of the vertebral body measurements are similar in both Panjabi's and Tan's study. The mean difference for the spinal canal width (SCW) is about "-19.8% and -31.8% for the SCD." The trend of the spinal canal width, depth, and area (SCW, SCD, and SCA) were different in the two studies. The pedicle height and width (PDH and PDW) had a mean difference of -16.1% and -25.7% respectively when using the average of the left and right measurements. Spinal implants have been developed based on measurements from Caucasian specimens, and as can be seen the difference in the pedicle width shows that the design of these implants would not be a good fit for the population studied by Tan. For example a 5-mm transpedicle screw would not be able to fit because the pedicle width is not wide enough for it in the Chinese Singaporeans. The spinous process length (SPL) is smaller in Tan's study by about 5.5% and the transverse process width (TPW) is smaller by about 15.6% than Panjabi's study. But the spinous process length is slightly larger from C5 to C7 than Panjabi's study (Nissan and Gilad, 1984; Tan, Teo and Chua, 2004).

In the study completed by Panjabi it was found that their results generally agreed with previous studies completed by Liu, Nissan, and Francis. The front to back endplate dimensions were generally within 2 mm of measurements completed by Nissan and Francis, and also followed the same tendency from C2 to C7. The measurements from Liu were smaller than Panjabi's which could be a result of how measurements were found. The endplate area (EPA) also differed from Liu which is believed to be because Panjabi's area did not include the uncovertebral facet area while Liu's more than likely did. Panjabi's study saw that there was a widening in the spinal canal from C5 to C6 and then decrease at C7, while Francis' stated that the canal was relatively small throughout. The vertebral body height (VBH) were smaller by 2 to 2.5 mm than Nissan and Francis' study which is believed to be again because of the different measurement techniques (Francis, 1955; Liu, Clark and Krieger, 1986; Panjabi et al., 1991; Nissan and Gilad, 1984) .

	ANOVA	Parameter Estimates		Y-intercept	Slope
	P	P (y-intercept)	P (regressor/ independent)		
EPAl vs. EPDu	0.0499	<0.0001	0.0499	499.16617	-10.55392
	Significant	Significant	Not Significant		
EPAl vs. VBHp	0.0152	<0.0001	0.0152	430.9265	-7.72835
	Significant	Significant	Significant		
EPAu vs. EPDl	0.0227	0.0166	0.0227	112.68776	6.86404
	Significant	Significant	Significant		
PDAl vs. PDIsl	0.0197	<0.0001	0.0197	53.33933	0.63139
	Significant	Significant	Significant		
PDAr vs. EPItu	0.0162	<0.0001	0.0162	28.57023	0.46784
	Significant	Significant	Significant		

Table 19. C7 Area Measurements

	ANOVA	Parameter Estimates		Y-intercept	Slope
	P	P (y-intercept)	P (regressor/ independent)		
EPItl vs. SCW	0.0143	<0.0001	0.0143	10.06297	-0.25521
	Significant	Significant	Significant		
EPItl vs. EPAu	0.026	<0.0001	0.026	7.8214	-0.01266
	Significant	Significant	Not Significant		
EPItu vs. VBHp	0.0064	0.7368	0.0064	-1.01525	0.71055
	Significant	Not Significant	Significant		
EPItu vs. PDWr	0.0423	<0.0001	0.0423	6.34333	0.18166
	Significant	Significant	Not Significant		
EPItu vs. PDAr	0.0162	0.0357	0.0162	3.43319	0.12299
	Significant	Not Significant	Significant		
PDIsl vs. EPDu	0.0149	<0.0001	0.0149	-50.57442	1.32445
	Significant	Significant	Significant		
PDIsl vs. VBHa	0.0274	<0.0001	0.0274	-41.61664	0.98543
	Significant	Significant	Not Significant		
PDIsl vs. SPL	0.0208	<0.0001	0.0208	-19.73933	-0.2317
	Significant	Significant	Significant		
PDIsl vs. PDAl	0.0197	<0.0001	0.0197	-33.49084	0.08599
	Significant	Significant	Significant		
PDItr vs. SCW	0.0218	0.0011	0.0218	10.15645	-0.35731
	Significant	Significant	Significant		
PDItr vs. EPAu	0.042	0.7155	0.042	-0.67451	0.01729
	Significant	Not Significant	Not Significant		

Table 20. C7 Angular Measurements

1.5 Conclusion

The analysis of the cervical spines vertebral segments anthropometrics resulted in 600 total comparisons being completed in each vertebral body from C3-C7, resulting in 3000 comparisons in total being done. From this it was found that there were 11 relationships in the C3 vertebra, 23 in the C4 vertebra, 40 in the C5 vertebra, 22 in the C6 vertebra, and 34 in the C7 vertebra which is a total of 130 relationships found from C3 to C7. From this analysis it was found that only about $4\frac{1}{3}\%$ of the 3000 comparisons were significant. There were only 8 comparisons that were significant in more than one vertebral segment.

The relationships found between the dimensional anatomy of the vertebrae of the cervical spine will assist in accurate modeling of the spine as well as for device development helping to eliminate possible failure of devices due to improper fit within the region of the spine. This anthropometric data can also enable a better understanding of disease occurrence in certain alignments of the spine, and susceptibility of specific race, gender, or age groups. The results of this research will also allow for a better understanding of the functionality of the cervical spine and its susceptibility to failure.

2.1 Cervical spine finite element modeling methods review

The cervical spine is one of the most important physiologic systems in the human body. The cervical spine offers primary stability to the head neck system along with protecting the spinal cord. The cervical spine features higher levels of motion and flexibility as compared to the other spine regions. The spines flexibility and motion does leave it susceptible to a higher rate of injury as compared to the other spine regions (Ng and Teo, 2001). It is therefore important to study the spine to gain a better understanding of the behavior spine. In-vivo studies of the cervical spine can provide information on the behavior of the spine under specific conditions. However, an in-vivo analysis of the spine cannot provide specific load response information at the vertebral and intervertebral disc level. In contrast, in-vitro analysis of the cervical spine can provide load displacement response at vertebral segments. In-vitro analyses of the cervical spine are limited to load displacement results; they cannot provide internal response characteristics such as stress and strain (Yoganandan et al., 1996; Panagiotopoulou, 2009). As such, there has been growing interest and application of finite element (FE) methods in the study of the cervical spine. Finite element models of the cervical spine have been used to study spine biomechanics, injuries, and response to medical interventions (Yoganandan, Kumaresan and Pintar, 2001; Pitzen et al., 2002; Pitzen, Matthis and Steudel, 2002)

Development of a finite element model of the spine involves several key areas of consideration. A finite element model of the spine must aim to accurately represent the anatomical features of the spine. Spinal vertebrae, intervertebral discs, ligaments, and their interrelation must all be carefully considered in the development of a model (Kallemeyn, Tadepalli and Shivanna, 2009; Bogduck and Yoganandan, 2001). The methods applied in constructing the finite element model play an important role in its ability to accurately represent the cervical spine. The finite element methods applied in analyzing a cervical spine model play are also of extreme import (Yoganandan, Kumaresan and Pintar, 2001; Yoganandan et al., 1996). In order to gain a better understanding of cervical spine finite element modeling and analysis, a review of the pertinent literature was performed.

2.2 Cervical Spine Analysis

Mathematical modeling approaches allow for both static and dynamic analysis of the cervical spine. Dynamic analyses of the spine often aim to characterize the response of the cervical region during an impact with the goal of better understanding vehicular injury scenarios such as whiplash. Dynamic models of the cervical spine often include the entire cervical spine, and the head. Vertebral bodies have been modeled as rigid bodies, with soft tissues such as spinal ligaments represented by linear springs (Esat, 2005; Brolin and Halldin, 2005; Brolin and Halldin, 2004; Stemper et al., 2006). This modeling approach somewhat limits the load response data that can be derived for specific vertebral bodies and intervertebral discs.

Static FE analyses focus on analysis of load response characteristics of cervical spine segments. In an effort to represent the load response as accurately as possible, static FE models are constructed with as much detail as possible. (Kallemeyn, Tadepalli and Shivanna, 2009; Panzer and Cronin, 2009; Goel and Clausen, 1998; Ha, 2006). In contrast to dynamic models, static models often focus on two to three vertebral bodies as opposed to the complete cervical spine. These functional spinal units (FSU) can provide important internal load and segment displacement data (Ng et al., 2003). Static analyses also allow for corroboration of FE results with in vitro study load displacement results. Static analyses have been used to analyze a variety of topics including spinal column biomechanics, soft tissue effects on behavior, soft and hard tissue injuries, and even prosthetic disc replacements (Zhang et al., 2006; Voo et al., 1997; Noailly et al., 2007; Ha, 2006; Galbusera et al., 2008).

As stated, static element analyses lend themselves well to validation of cervical spine finite element models. Validation of any finite element model is an extremely important process the confirms that the model and assumptions there in, adequately represent that actual physical spine. There have been in-vitro studies of the cervical spine and spine segments that can act as comparison and validation cases for finite element studies (Moroney et al., 1988; Panjabi et al., 2001; Richter et al., 2000). In order to use an in-vitro study as a comparison case, test conditions including loading and constraints must be equivalent. This does not however limit the loading cases applied to finite element studies to those already employed in-vitro. By verifying a study under known in-vitro conditions investigators can assume the response of the finite element model is valid for a certain range then continue to test different scenarios (Ng et al., 2003). The following summary table, Table 21, provides study types, load conditions and validation methods employed.

Author	Year	Study Type	Spine Levels	Loading	BC	Validation
Li et al. (Li and Lewis, 2010)	2010	Static Surgery	All Segment	0.33 - 2 Nm Flexion Extension Lateral Bending Axial Rotation 1 Nm + 73.6 Compression	Inferior Endplate Fully Fixed	Panjabi et al. 2001 Wheeldon et al. 2006
Kallemeyn et al. (Kallemeyn, Tadepalli and Shivanna, 2009)	2009	Static Biomechanics	2 Segment	1 Nm Flexion Extension Lateral Bending Axial Rotation + 73.6 N Compression 600 N Compression	Inferior Endplate Fully Fixed	(Moroney et al., 1988; Traynelis et al., 1993; Pintar et al., 1995)
Panzer et al. (Panzer and Cronin, 2009)		Static Biomechanics	2 Segment	0.3 – 3.5 Nm Flexion Extension Lateral Bending Axial Rotation	Inferior Endplate Fully Fixed	Goel et al. 1988 Voo et al. 1997 Maurel et al. 1997 Moroney et al. 1998

Author	Year	Study Type	Spine Levels	Loading	BC	Validation
Galbuseara et al. (Galbusera et al., 2008)	2008	Static Prosthesis	4 Segment	2.5 Nm Flexion Extension + 100 N Compression	Inferior Endplate Fully Fixed	In-vitro (Wheeldon et al., 2006)
Greaves et al. (Greaves, Gadala and Oxland, 2008)		Static Injury	3 Segment	Injury based deflection	Injury based	In-vivo Hung et al. 1979 Maiman et al. 1989
Wheeldon et al. (Wheeldon et al., 2008)		Static Biomechanics	4 Segment	0 - 2 Nm Flexion Extension Axial Rotation	Inferior Endplate Fully Fixed	Gilad & Nissan 1986 Panjabi et al. 1991
Teo et al. (Teo et al., 2007)		Static Mesh Generation	7 Segment	N/A	Inferior Endplate Fully Fixed	N/A
Ha (Ha, 2006)	2006	Static Prosthesis	4 Segment	1 Nm Flexion Extension Lateral Bending Axial Rotation	Inferior Endplate Fully Fixed	Moroney et al. 1991 Pelker et al. 1987 Goel et al. 1998 Teo & Ng et al. 2001
Zhang et al. (Zhang et al., 2006)		Static Biomechanics	8 Segment	1 Nm Flexion Extension Lateral Bending Axial Rotation 50 N Compression	Inferior Endplate Fully Fixed	Goel et al. 1984 Moroney et al. 1988 Goel & Clausen 1998 Panjabi et al. 2001
Haghpanahi & Mapar (Haghpanahi, 2006)		Static Biomechanics	5 Segment	1.8 Nm Flexion Extension	Inferior Endplate Fully Fixed	Lopez-Espinea (FEA) 2004 Goel et al. Voo et al. Maurel et al. Moroney et al.
Esat et al. (Esat, 2005)	2005	Dynamic Biomechanics	3 Segment	1.6 Nm Flexion Extension 73.6 N Compression	Inferior Endplate Fully Fixed	Shea et al. 1991
Brolin et al. (Brolin and Halldin, 2004)	2004	Static Biomechanics	2 Segment	1.5, 10 Nm Flexion Extension Lateral Bending Axial Rotation 1500 N Tension	Inferior Endplate Fully Fixed	Panjabi et al. 1991 Panjabi et al. 1991 Van et al. 2000 Goel et al. 1990

Author	Year	Study Type	Spine Levels	Loading	BC	Validation
Ng et al. (Ng et al., 2003)	2003	Static Injury	3 Segment	1.8 Nm Flexion Extension Lateral Bending Axial Rotation 73.6 N Compression	Inferior Endplate Fully Fixed	Shea et al. 1991 Moroney et al. 1988 Pelker et al. 1991 Maurel et al. 1997 Goel et al. 1998
Bozkus et al. (Bozkus et al., 2001)	2001	Static Injury	1 Segment	200 - 1200 N Compression	Inferior Endplate Fully Fixed	Cadaver Study
Teo et al. (Teo and Ng, 2001)		Static Biomechanics	3 Segment	1 mm Axial Displacement	Inferior Endplate Fully Fixed	Shea et al. 1991 Yoganandan et al. 1996 (FEA)
Graham et al. (Graham et al., 2000)	2000	Static Injury	1 Segment	1279, 1736 N Compression	Inferior Endplate Fully Fixed	Doherty et al 1993
Kumaresan et al. (Kumaresan et al., 2000)		Static Biomechanics	3 Segment	0.5 Nm Flexion Extension 200 N Compression	Inferior Endplate Fully Fixed	FEA Kumaresan et al. 1997
Zheng et al. (Zheng, Young-Hing and Watson, 2000)		Static Surgery	5 Segment	196 N Compression	Injury Case Dependent	
Kumaresan et al. (Kumaresan et al., 1999)	1999	Static Biomechanics	3 Segment	0.5 - 1.8 Nm Flexion Extension Lateral Bending Axial Rotation	Inferior Endplate Fully Fixed	Cadaver Study Pintar et al. 1995
Kumaresan et al. (Kumaresan, Yoganandan and Pintar, 1999)		Static Biomechanics	3 Segment	1.8 Nm Flexion Extension Lateral Bending Axial Rotation 125 - 800 N Compression	Inferior Endplate Fully Fixed	Moroney et al. 1988
Goel et al. (Goel and Clausen, 1998)	1998	Static Biomechanics	2 Segment	1.8 Nm Flexion Extension Lateral Bending Axial Rotation 73.5 N Compression	Inferior Endplate Fully Fixed	Moroney et al. 1988 Clausen et al. 1996 Goel et al. 1988 Teo et al. (FEA) 1994

Author	Year	Study Type	Spine Levels	Loading	BC	Validation
Kumaresan et al. (Kumaresan et al., 1998)		Static Biomechanics	2 Segment	Flexion Extension Lateral Bending Compression	Inferior Endplate Fully Fixed	N/A
Maurel et al. (Maurel, Lavaste and Skalli, 1997)	1997	Static Biomechanics	5 Segment	0 - 1.6 Nm Flexion Extension Lateral Bending Axial Rotation 6 N Compression	Inferior Endplate Fully Fixed	Cressend 1992 Panjabi et al. 1986 Wen 1993 Wen et al. 1993 Moroney et al. 1984, 1998
Voo et al. (Voo et al., 1997)		Static Surgery	3 Segment	1.8 Nm Flexion Extension Lateral Bending Axial Rotation	Inferior Endplate Fully Fixed	Liu et al. 1982 Moroney et al. 1988
Yoganandan et al. (Yoganandan et al., 1996)	1996	Static Biomechanics	3 Segment	1 mm Compression	Inferior Endplate Fully Fixed	Shea et al. 1991
Bozic et al. 1994 (Bozic et al., 1994)	1994	Static Injury	1 Segment	3400 N Compression	Inferior Endplate Fixed by Spring	

Table 21. Cervical Spine Finite Element Modeling Summary Table

The study by Esat et al. (Esat, 2005) combines both static and dynamic analysis methods. The investigators aimed to simulate the response of the head and neck system under frontal and rear impact scenarios. A multi-body dynamic head and neck computational model was developed and validated using human volunteer experimental data. The investigators take the analysis further by developing a finite element model of the cervical spine and intervertebral discs. The finite element model was used to study the response of the intervertebral discs to the dynamic load cases (Esat, 2005). The study illustrates the flexibility of employing the finite element method in the analysis of the cervical spine. A study by Sung Kyu Ha employed a finite element model of the cervical spine to study the effects of spinal fusion and the implantation of a prosthetic disc on spine behavior (Ha, 2006). Spinal fusion was modeled by applying a graft with material properties of the cortical bone between adjacent vertebral segments. The disc prosthesis was modeled by replacing the entire intervertebral disc with an elastomer core. Efforts were made to select an elastomer core with similar properties to that of the intervertebral disc. The analysis results showed that spinal fusion led to a 50 - 70% reduction in range of motion for the fused spinal segment. The introduction of a prosthetic disc did not change the range of motion seen in the motion segment (Ha, 2006). Using a validated finite element model of the cervical spine, the study was able to help predict the effect of two interventions that are often employed in spinal injury cases.

2.3 Hard tissue modeling

As stated, the accuracy of an FE model at representing the cervical spine anatomy is of extreme importance. There are two prominent modeling methods in the development of cervical spine vertebral body models. Multi axis digitizers can be used to map points along the vertebral bodies. The data set of points can then be used to create a model via a computer aided drafting package. This approach can be applied to the development of two dimensional (2D) and three dimensional (3D) models (Zhang et al., 2006; Esat, 2005; Haghpanahi, 2006; Panzer and Cronin, 2009). Haghpanahi et al. used the data point approach to create a parameterized 2D model of the C3 - C7 vertebral model. Intervertebral discs were modeled in relation to adjacent vertebral pairs (Haghpanahi, 2006).

Digitizing the surface geometry of cervical spine segments is somewhat limited by the number of points plotted. A look at the vertebral segment by Haghpanahi shows that surfaces are somewhat linear. The vertebral endplates and posterior elements are represented by straight line segments which do not convey the actual curvature and undulations of the vertebra. An alternative hard tissue modeling approach is to use computed tomography (CT) scan data. The process involves digitizing CT scans and using the data to create a vertebral model. In a study by Yoganandan et al., investigators used NIH-Image and an edge detection algorithm they developed to process the CT scans of the spine. The data extracted from NIH-Image provided edge locations for the vertebral bodies which were used to create wire frames of each vertebral body (Yoganandan et al., 1997). A decade later, a study by Sung Kyu Ha used the Amira image processing software to digitize CT scans, with 3D models and meshes generated in RapidForm and Ansys respectively (Ha, 2006). Though the two methods both yielded anatomically correct vertebral models, the process employed by Ha involved much less manual tasks and offered a higher level of refinement.

Regardless of the methods employed to develop the 3D model of the vertebral bodies, for the purposes of finite element analysis, a finite element mesh of the part must be developed. Element selection is of paramount importance in developing any finite element mesh. Element selection is dependent on several factors including, the type of analysis to be performed, and the geometry of the body to be meshed to name a few. Cervical spine vertebral bodies can be adequately meshed with 4 noded solid tetrahedral elements; however 8 noded hexahedral elements are preferred (Bozkus, 2001; Teo et al., 2007). Vertebral bodies are made up of two bone regions, the cancellous core and cortical shell. The cortical shell can be modeled as separate region of distinct thickness. The region can be modeled with a separate set of solid or shell elements (Yoganandan, Kumaresan and Pintar, 2001). The final hard tissue areas that must be considered during modeling are the vertebral body facet joints. The facet joints play an important role in stabilizing and constraining the motion of adjacent vertebral bodies. There are a myriad of modeling methods employed in approximating facet joints and their behavior. A summary of mesh methods employed in vertebral body modeling is provided in Table 22.

Author	Year	Source	Cancellous	Cortical	Facet Joints
Yuan et al. (Li and Lewis, 2010)	2010	CT	4 node tetrahedral	3 node shell element	

Author	Year	Source	Cancellous	Cortical	Facet Joints
Kallemeyn et al. (Kallemeyn, Tadepalli and Shivanna, 2009)	2009	CT	8 node hexahedral	8 node hexahedral	Pressure over closure relationship
Panzer et al. (Panzer and Cronin, 2009)		CAD	3D hexahedral	2D quadrilateral	Squeeze film bearing relationship
Galbuseara et al. (Galbusera et al., 2008)	2008	CT	8 node hexahedral	8 node hexahedral	Frictionless surface-based contact
Greaves et al. (Greaves, Gadala and Oxland, 2008)		CT	8 node brick	8 node brick	
Wheeldon et al. (Wheeldon et al., 2008)		CT	Solid	Solid	Solid / fluid hydraulic incompressibl e
Teo et al. (Teo et al., 2007)		CT	Hexahedral Tetrahedral	Hexahedral Tetrahedral	
Ha (Ha, 2006)	2006	CT	20 node brick	8 node shell	Non-linear contact element
Zhang et al. (Zhang et al., 2006)		CAD	8 node brick	8 node brick	Surface to surface contact
Haghpanahi & Mapar (Haghpanahi, 2006)		CAD	solid	solid	
Esat et al. (Esat, 2005)		CAD	8 node brick	8 node brick	
Brolin et al. (Brolin and Halldin, 2004)	2004	CT	8 node brick	4 node shell	Sliding contact with friction
Ng et al. (Ng et al., 2003)	2003	CAD	8 node solid	8 node solid	Nonlinear contact
Bozkus et al. (Bozkus et al., 2001)	2001	CT	Solid / 4 node tetrahedral		
Teo et al. (Teo and Ng, 2001)		CAD		8 node solid	
Graham et al. (Graham et al., 2000)	2000	CT	tetrahedral	Tetrahedral thin shell	
Kumaresan et al. (Kumaresan et al., 2000)		CT	8 node brick	8 node brick	8 node, fluid, membrane elements
Zheng et al. (Zheng, Young-Hing and Watson, 2000)		CT	10 node tetrahedral	10 node tetrahedral	
Kumaresan et al. (Kumaresan et al., 1999)	1999	CT	8 node brick	8 node brick	8 node, fluid, membrane elements

Author	Year	Source	Cancellous	Cortical	Facet Joints
Kumaresan et al. (Kumaresan, Yoganandan and Pintar, 1999)		CT	8 node brick	8 node brick	8 node, fluid, membrane elements
Goel et al. (Goel and Clausen, 1998)	1998	CT	8 node brick	8 node brick	
Kumaresan et al. (Kumaresan et al., 1998)		CT	8 node brick	8 node brick	8 node, fluid, membrane elements
Maurel et al. (Maurel, Lavaste and Skalli, 1997)	1997	CT	8 node	8 node	Gap element
Voo et al. (Voo et al., 1997)		CT	8 node solid	thin shell	
Yoganandan et al.	1996	CT	8 node solid	thins shell	
Bozic et al. 1994 (Bozic et al., 1994)	1994	CT	8 node solid	8 node solid	

Table 22. Cervical Spine Vertebral Modeling Methods

2.4 Intervertebral disc modeling

Intervertebral discs (IVD) are extremely important to the behavior of the spine. Intervertebral discs act as dampers responding to compressive forces within the spine (Yoganandan, Kumaresan and Pintar, 2001). Discs are made up of two distinct regions, the outer annulus fibrosus ring, and an inner nucleus pulposus core (Ha, 2006). Both regions are largely fluid based. The annulus fibrosus is made up of collagen fibers embedded in an extracellular matrix composed of water and elastin fibers. Collagen fibers are arranged as a structure of rings throughout the annulus region. Fibers are oriented between 25° and 45° with respect to the horizontal plane. Collagen fibers provide primary stiffness to the annulus region (Ambard and Cherblanc, 2009; Noailly, Lacoix and Planell, 2005). Discs interact with adjacent vertebral bodies via the cartilaginous endplates.

Considerations must be made to accurately model IVD behavior. Modeling IVD must be approached in a different manner than the vertebral bodies as CT scans do not provide soft tissue data. Cryomicrotomy images can be used as an alternative to fill in the missing soft tissue data (Yoganandan, Kumaresan and Pintar, 2001; Voo et al., 1997). An alternative to employing cryomicrotomy is to model intervertebral discs in reference to their interaction with related solid bodies (Yoganandan, Kumaresan and Pintar, 2001). An advantage of IVD modeling is their relative simple geometry in comparison with vertebral bodies. An IVD can be modeled with a CAD package as a cylindrical disc (Meakin and Huskins, 2001). For finite element analysis purposes the intervertebral disc annulus is often modeled as a fiber reinforced composite. Solid brick elements will be reinforced by a fiber or rebar element matrix of alternating angular orientation. The reinforcing fibers often employ a nonlinear response behavior unique to that of the solid annuls elements they are suspended within. The nucleus has been modeled as an incompressible fluid (Eberlein, Holzapfel and Froelich, 2004). This approach can involve modeling the nucleus with specific incompressible fluid elements. Though ideal, this approach presents a level of complexity that cannot be attained in all studies. The alternative involves applying general modulus and poison's ratio to the nucleus region (Ha, 2006) Table 23. summarizes finite element modeling approaches employed for the IVD.

Author	Year	Disc Components	Elements
Li et al. (Li and Lewis, 2010)	2010	Annulus fibrosus Nucleus pulposus	8 node brick 4 node tetrahedral
Kallemeyn et al. (Kallemeyn, Tadepalli and Shivanna, 2009)	2009	Annulus fibrosus Nucleus pulposus	8 node tetrahedral Hydrostatic fluid
Panzer et al. (Panzer and Cronin, 2009)		Annulus fibrosus Nucleus pulposus	Hexahedral element Incompressible element
Galbuseara et al. (Galbusera et al., 2008)	2008	Annulus fibrosus Nucleus pulposus	Hexahedral element Tension only truss
Wheeldon et al. (Wheeldon et al., 2008)		Annulus fibrosus Nucleus pulposus	Solid element Rebar element Incompressible fluid
Palomar et al. (Palomar, Calvo and Doblare, 2008)		Annulus fibrosus Nucleus pulposus	Solid element Linear tetrahedral Incompressible fluid
Schmidt et al. (Schmidt, 2007)	2007	Annulus fibrosus Nucleus pulposus	8 node solid element 3D spring element Incompressible Hyper elastic
Ha (Ha, 2006)	2006	Annulus fibrosus Nucleus pulposus	20 node solid element Tension only spar
Zhang et al. (Zhang et al., 2006)		Annulus fibrosus Nucleus pulposus	8 node brick
Eberlin et al. (Eberlein, Holzapfel and Froelich, 2004)	2004	Annulus fibrosus Nucleus pulposus	8 & 20 node hexahedral Incompressible fluid
Meakin et al. (Meakin and Huskins, 2001)	2001	Annulus fibrosus Nucleus pulposus	Solid element Fluid element
Kumaresan et al. (Kumaresan et al., 2000)	2000	Annulus fibrosus Nucleus pulposus	8 node solid Tension only rebar 3D fluid element
Kumaresan et al. (Kumaresan et al., 1999)	1999	Annulus fibrosus Nucleus pulposus	8nnode solid Rebar element ncompressible fluid
Maurel et al. (Maurel, Lavaste and Skalli, 1997)	1997	Annulus fibrosus Nucleus pulposus	8 node element Cable element
Voo et al. (Voo et al., 1997)		Uniform disc	8 node element
Yoganandan et al. (Yoganandan et al., 1996)	1996	Uniform disc	8 node element
Bozic et al. 1994 (Bozic et al., 1994)	1994	Uniform disc	Springs element

Table 23. Cervical Spine Intervertebral Disc Modeling Methods

The IVD disc modeling summary table illustrates an acceptance of modeling the two distinct regions, the nucleus pulposus and intervertebral disc. As stated, the approaches employed do vary. In modeling the annulus fibrosus, the inclusion or exclusion of the fiber reinforcing matrix is a key modeling point. A study by Palomar (Palomar, Calvo and Doblare, 2008) illustrates the level of detail that can be employed in modeling the annulus fibrosus fiber matrix. The authors used in-vitro data sourced from a specific analysis of the tensile behavior of multiple layers of annulus under very slow strain (Ebara et al., 1996). The data was used to adjust material properties of a strain energy function developed for annulus fibers (Holzapfel, 2000). The mathematical model was then implemented via a UMAT user subroutine in the Abaqus finite element software package (Palomar, Calvo and Doblare, 2008). It is clear that this approach focused on developing a realistic intervertebral disc model. The model allowed for greater understanding of internal stress response of the intervertebral discs.

2.6 Cervical spine ligament modeling

Ligaments are the supportive connective structures of the spine. Ligaments of the spine include the ligamentum flavum (LF), interspinous ligament (ISL), capsular ligament (CL) and intertransverse (ITL) ligaments. This set of ligaments function to support individual vertebra. The anterior longitudinal (ALL), posterior longitudinal (PLL), and the supraspinous ligament (SSL) act as supports for series of vertebra (Yoganandan, Kumaresan and Pintar, 2001). Spinal ligaments are often modeled based on knowledge of their anatomical makeup, locations, and relation to vertebra and intervertebral discs as they are not represented in CT images. There is data available providing ligament cross sectional area, length, and mechanical behavior. For finite element purposes, ligaments are most often represented as non linear tension only entities. Spring, cable, truss, and tension only elements have all been employed in the modeling of ligaments (Yoganandan, Kumaresan and Pintar, 2001). A summary of some ligament modeling techniques applied is provided in Table 24.

Author	**Year**	**Ligaments**	**Behavior**	**Elements**
Li et al. (Li and Lewis, 2010)	2010	ALL, PLL, CL, LF, ISL, TL, APL	Nonlinear	Tension-only spar
Kallemeyn et al. (Kallemeyn, Tadepalli and Shivanna, 2009)	2009	ALL, PLL, CL, LF, ISL	Nonlinear	2 node truss
Panzer et al. (Panzer and Cronin, 2009)		ALL, PLL, CL, LF, ISL	Nonlinear	1D tension only
Galbuseara et al. (Galbusera et al., 2008)	2008	ALL, PLL, CL, LF, ISL	Nonlinear	Spring element
Greaves et al. (Greaves, Gadala and Oxland, 2008)		ALL, PLL, CL, LF, ISL	Nonlinear	2 node link
Palomar et al. (Palomar, Calvo and Doblare, 2008)		ALL, PLL, YL, ISL, ITL	Nonlinear	Tension only truss
Wheeldon et al. (Wheeldon et al., 2008)		ALL, PLL, LF, CL, ISL	Nonlinear	Spring element
Schmidt et al. (Schmidt, 2007)	2007	ALL, PLL, CL, LF, ISL, SSL	Force deflection curve	Spring element
Ha (Ha, 2006)	2006	ALL, PLL, LF, ISL, CL	Nonlinear	Tension only spar
Zhang et al. (Zhang et al., 2006)		ALL, PLL, SSL, ISl, LF, CL, AL, TL, NL, APL	Linear	2 node link
Brolin et al. (Brolin and Halldin, 2004)	2004	ALL, PLL, TL, LF, CL, ISL	Force deflection curve	Tension only spring
Eberlin et al. (Eberlein, Holzapfel and Froelich, 2004)		ALL, PLL, TL, LF, CL, ISL	Nonlinear	Membrane element

Author	Year	Ligaments	Behavior	Elements
Kumaresan et al. (Kumaresan et al., 2000)	2000	ALL, PLL, CL, LF, ISL	Nonlinear	Tension only element
Kumaresan et al. (Kumaresan, Yoganandan and Pintar, 1999)	1999	ALL, PLL, CL, LF, ISL	Nonlinear	Tension only element
Maurel et al. (Maurel, Lavaste and Skalli, 1997)	1997	ALL, PLL, CL, Lf, ISl, SSL	Nonlinear	Tension only cable element
Voo et al. (Voo et al., 1997)		ALL, PLL, CL, LF, ISL	Linear	2 node uniaxial
Yoganandan et al. 1996 (Yoganandan et al., 1996)	1996	N/A	N/A	N/A
Bozic et al. (Bozic et al., 1994)	1994	N/A	N/A	N/A

Table 24. Cervical Spine Ligament Modeling Methods

The summary table clearly illustrates that despite the difficulties of visualizing spinal ligaments for modeling purposes; they are still included in most cervical spine finite element models. It is also evident that the majority of investigators aim to capture the nonlinear behavior of cervical spine ligaments. The degree to which ligament nonlinearity has been captured does vary amongst studies. The use of finite elements with nonlinear characteristics has been applied and deemed adequate (Ha, 2006). Non linearity can be further implemented by employing strain dependent modulus of elasticity values to the finite element model ligaments. Strain dependent moduli of elasticity are often sourced from in vitro experimentation of cervical spine segments (Kallemeyn, Tadepalli and Shivanna, 2009; Yoganandan, Kumaresan and Pintar, 2000). Strain dependent moduli of elasticity invariably add complexity to any mathematical analysis procedure. Additionally strain limits are vary greatly depending on the in-vitro data sourced and are subject to variability and questions of applicability to the current study. Despite the shortfalls it is clear from a review of the literature that investigators are continually developing and applying sophisticated modeling techniques to spinal ligaments.

2.7 Discussion

The review of cervical spine modeling techniques has illustrated the FEA can be a powerful tool in the study of cervical spine behavior, injury, and treatment. There have been studies the focus on finite element models of the as tools in the design of spine prostheses (Galbusera et al., 2008; Ha, 2006; Meakin and Huskins, 2001). Ha et al. developed a multi segment model of the cervical spine and continued to analyze its behavior with and without an elastomer-type prosthetic disc. The study aimed to design the prosthetic disc that would most closely reflect the behavior of the spinal unit with a disc present. The study found that a disc with a modulus of 5.9 MPa would maintain biomechanical behavior of the complete spine. The authors even note that the modulus value found could be achievable using polyurethane. Determining a modulus value numerically provides a good basis for which to start designing an IVD prosthesis that maintains biomechanical function (Ha, 2006).

Finite element analysis models have even begun to be applied to juvenile spinal models including juvenile anatomical features such as joint plates (Wheeldon et al., 2008; Sairyo et al., 2006; Sairyo et al., 2006). Models have also continued to better represent the spine not only in geometry but in behavior. Studies have been undertaken to develop accurate material and behavioral models based on extensive concurrent in-vitro testing (Yoganandan, Kumaresan and Pintar, 2001; Eberlein, Holzapfel and Froelich, 2004).

In studying fracture in the atlas Teo et al. developed a single entity model of the atlas. Material property data from the literature were employed and a series of load to failure FE analyses were performed. The study predicted stress distributions along the atlas and maximal loads. Though this data cannot be directly applied in a clinical sense, it can be qualitatively applied to future in-vitro or mathematical modeling (Teo and Ng, 2001). Models have progressed from going without ligaments (Yoganandan et al., 1996; Bozic et al., 1994), to focusing specifically on calibrating ligament material properties (Brolin and Halldin, 2004). Similar advances can be seen in IVD disc modeling which has gone from a simple spacer of uniform material properties (Bozic et al., 1994) to highly advanced nonlinear constitutive models of IVD behavior (Palomar, Calvo and Doblare, 2008). The variation in study types and analysis methods underscore the effectiveness, and flexibility of applying FE methods to the study of the cervical spine.

3.1 Finite element analysis of superior C3 cervical vertebra endplate and cancellous core under static loads

Subsidence is a failure mechanism that can occur after implantation of an intra vertebral implant device. Subsidence is clinically defined as the loss of postoperative intervertebral disc height and has been shown to occur in as many as 77% of patients after fusion surgeries (Choi and Sung, 2006). According to actuarial rates subsidence occurs at 63.4 and 70.7 percent at 12 and 16 weeks respectively (Choi and Sung, 2006). Occurrences of subsidence could be due to bone failure, which may be attributed to compressive stresses, or a failure of the implanted device specifically bone graft material (Jost et al., 1998). While a loss of height is common, measuring it may be contentious. Identifying the edge of the device proves difficult due to bone in-growth and the shadow of the apophyseal ring. Significant subsidence has been defined differently for the lumbar and cervical regions of the spine. Losses of disc height of 2mm in the lumbar spine and 3mm in the cervical spine have been considered relevant benchmarks (Choi and Sung, 2006; Van Jonbergen et al., 2005; Kulkarni et al., 2006). Another indication of subsidence is the change in lordic curve of the cervical spine. Changes in angle between the endplates, at the surgical level in the case of fusion, would indicate that the device is sinking into the vertebral bodies. Angle changes have been measured at a lordic increase of 1.6 degrees postoperatively to a follow up lordic decrease of 2.5 degrees (Kulkarni et al., 2006). The reduction in angle indicates that either the anterior or posterior part of the implanted device had subsided into the vertebral body. This failure is also a localized failure that is initiated by high contact forces generated by implanted disc devices.

Understanding the endplate morphology and biomechanics is crucial to the future success rates of implanted devices. The previous section of this chapter talked extensively about the type of data collected and measured. Several studies have been aimed at determining the thickness, strength and density of the vertebral endplates of the cervical spine by directly measuring cadaver specimens. The thickest regions are in the posterior region of the superior endplate and the anterior region of the inferior endplate with the central region being the thinnest area (Panjabi et al., 2001; Pitzen, 2004; Edwards et al., 2001). Mechanically the thicker regions of the endplate are stronger than the thinner areas (Grant et al, 2001; Oxland, 2003). Oxland showed that the thinner, middle lumbar region had a mean failure load between approximately 60-100 N, and increased toward the endplate's periphery, thicker regions, to a load of approximately 175 N (Grant et al, 2001). Locations of thicker endplate bone are indicative of other factors that affect the biomechanical quality of the

endplate. Density scans of the endplate, as measured by peripheral quantitative computed tomography (pQCT) scans, reveal that the endplate bone is denser in thicker regions (Ordway et al., 2007). Results show that an increase in bone density from 150 to 375 mg/mm^3 equates to an approximate stiffness increase from 100 to 200 N/mm. These same regions, which have a greater density and are thicker, also have an increased mineral deposition than thinner regions of the cervical endplates (Muller-Gerbl et al., 2008; Panzer et al., 2009). The increased mineral deposits were located in areas of the endplate that typically have the highest indentation test results and therefore higher failure limits (Grant et al, 2001; Oxland, 2003; Muller-Gerbl et al., 2008; Panzer et al., 2009).
Causes of subsidence can be modeled using finite element models. As previously discussed finite element modeling allows the investigation of several parameters like stress strain and deformations of irregularly shaped objects. The complex anatomical features previously discussed can be modeled with finite element methods to create a theoretical model that can be validated using experimental methods. This should increase the reproducibility of the model in many different scenarios to analyze different aspects of the cervical spine, in this case specifically the vertebral bodies. Frequently theoretical vertebral geometry is constructed from anthropometric data (Polikeit et al., 2003; Denoziere and Ku, 2006). The anthropometric data is typically compiled from measurements taken on a large sample group of cadavers. Theoretical models usually assume geometric properties of parameters that are difficult to measure directly and cost effectively, for example cortical shell thickness. Experimental models built from CT's also have material property limitations but are well suited for replicating anthropometric geometry for a single user. In both cases some assumptions need to be made concerning shell thicknesses. Several studies simplify the cortical shell and endplates as a shell with constant or only a slight variation in the endplate. The goal of the following study is to determine if an endplate thickness of a half-millimeter is an adequate approximation for the vertebral endplate by comparing endplate stresses.

3.2 Methods

A 3-dimensional linear elastic model of the C3 vertebrae was constructed from CT images of a 25-year old female that consisted of the vertebrae's bony structure. MIMICS 13.0 (Materialise, Ann Arbor, Michigan, USA) was used to convert the CT images to a 3-D model. The 3D model was smoothed and meshed using 3-Matic (Materialise, Ann Arbor, Michigan, USA). From 3-Matic an orphan mesh was imported into Abaqus 6.9 (Simulia, Providence, Rhode Island, USA) finite element design suite for post-processing. This experiment considers the thickness of the superior vertebral endplate. The superior endplate was modeled in four different ways, labeled Model 1 through Model 4. The first model, Model 1, used half-millimeter thick shell elements as an approximation for the superior endplate. Model 2 assumes the endplate has been completely removed. The removal was modeled by the actual removal of the shell elements exposing the volume elements of the core. Model 3 had a superior endplate that is divided into three regions (Panjabi et al., 2001). Model 4 had a superior endplate divided into seven regions (Pitzen et al., 2004). The thickness and region distributions are presented in Figure 10.
The finite element model was constructed with 60697 tetrahedral elements and 13651 nodes. The cortical shell was created with 4552 offset shell elements, less for the model with the removed endplate. The shells of the inferior endplate and the radial cortical shell were set to a half-millimeter thickness and used the same offset shell method. Figure 10 shows how the endplates were sectioned. The cartilaginous endplate was not considered in this analysis

because it is often removed during surgery and does not contribute significantly to the stiffness of the endplates (Polikeit et al., 2003(2)).
Assigned material properties have been previously well documented in literature and are presented in Table 25.

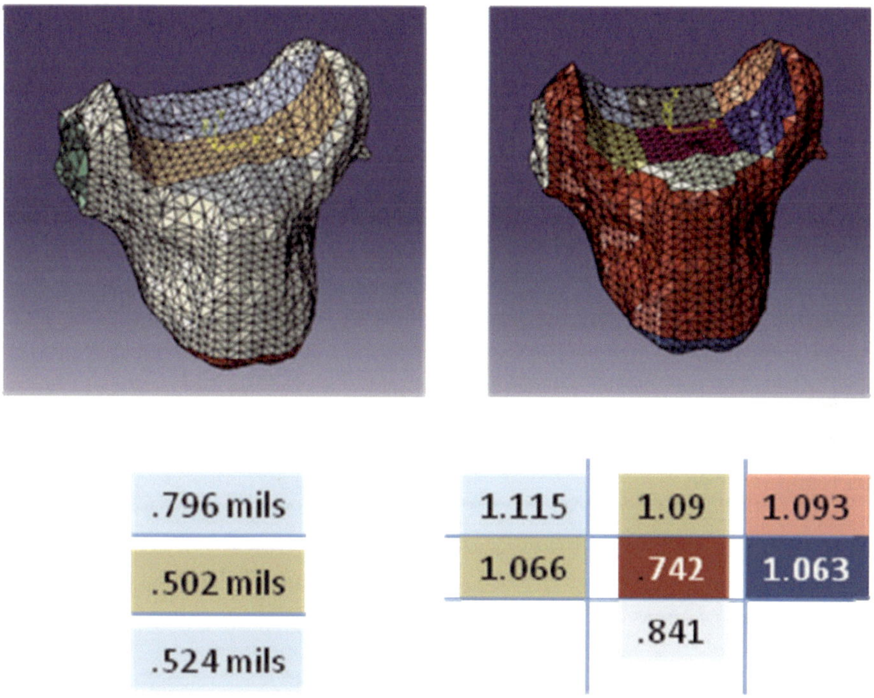

Fig 10. Finite element models of the C3 vertebrae. Left image is Model 3 and the right is Model 4. Below each model is the thickness of the endplate in each region. The thickness regions of the endplate correspond to the colored regions below

	Modulus of Elasticity (MPa)	**Poisson's Ratio**
Cortical Shell[1]	10,000	.3
Cancellous Core[2]	Ezz = 344, G1,2 = 63	.11
	Eyy = 144, G1,3 = 53	.17
	Exx = 100, G2,3 = 45	.23
Superior Endplate[3,4]	1,000	.3
Inferior Endplate[3,4]	1,000	.3
Posterior Elements[3,4]	3,500	.25

1 - Li and Lewis, 2010
2 - Rohlmann et al., 2006
3 - Polikeit[1] et al., 2003
4 - Polikeit[2] et al., 2003

Table 25. List of material properties applied to the finite element model. This list was compiled from a large group of finite element studies

Material properties were considered to be homogenous. This is not physiologically accurate. The assumption was made that on the macro level the irregularities would be evenly distributed throughout the material sections and represented by the assigned values. The properties were made continuous from point to point and assigned in a hierarchical structure, which separates different bone categories, i.e. cortical and cancellous, into different material groups. This is clinically relevant since the material property definitions simulate bone's various material distributions and can be adapted to replicate disease or injury. The entire vertebra was broken down into posterior elements, cancellous core, radial cortical shell and the superior and inferior endplates. All elements were assigned linear elastic element types. The cancellous core of the vertebral body was assumed to be anisotropic. The axial direction is the strongest due to the difference in cortical bone structure and alignment in the axial direction along lines of stress (Panjabi and White, 1990; Boos and Aebi, 2008).

The models were statically loaded with an axial force of 1000 N and flexion and extension moment of 7.5 Nmm. To avoid the concentration of stress from point loads a pressure distribution was applied to the superior endplate. In this scenario, a higher stress peak develops in the same direction as an applied moment. For example a flexion moment would have a resultant distributed load with a compressive stress peak in the anterior region of the vertebral body. The boundary conditions consisted of fixing the inferior endplate in translation and rotation.

3.3 Results

The results show that the endplate stresses are all approximately the same in magnitude and location. The values of stress calculated in this analytical model are presented in the Table 26 and Figure 11.

	Endplate Flexion (MPa)	Endplate Extension (Mpa)	Percent Diff, Model 1 vs. Model 3,4	Core Stress Flexion (Mpa)	Core Stress Extension (Mpa)
Model 1	24.6	25.57	N/A	17.1	34.5
Model 2	N/A	N/A	N/A	74.8	38.2
Model 3	20.7	15.7	17.2, 47.8	13.12	8.5
Model 4	19.5	19.5	22.5, 26.9	20.5	30.14

Table 26. Max stress values in MPa in the core and the endplate from flexion, extension and axial loading

The von Mises stresses range from a minimum of 15.7 MPa, Model 3 in extension, to a maximum of 25.57 MPa, Model 1 in extension. These values are consistent with other studies listed in Table 27. The endplate stresses are also well under the failure stress for cortical bone. The cancellous core stresses are less consistent. A stress range of 8.5 MPa, Model 3 in extension, to 34.5 MPa, Model 1 in extension, was recorded in cases with endplates present. These values are greater than that of the listed failure stress for cancellous bone of 4 MPa, but are in line with some of the previously modeled vertebra in Table 27. In the models with the removed endplate, core stresses reach a maximum of 74.8 MPa, which is much greater than the 4 MPa failure limit.

The von Mises stresses were also analyzed at various depths of the vertebral core. This was done to examine how the stress propagated through the cancellous core. Measurements were taken in 4 spots in the axial plane and at 4 different depths in the sagittal or coronal plane for a total of 16 measurements. The locations of the stress chosen in the axial plane were measured where the stress should have been highest in the cases of flexion and extension. The first set of measurements was taken directly beneath the vertebral endplate. The second set was taken at approximately 1/3 of the height of the vertebral body beneath the superior endplate. The third set was measured at approximately 1/3 of the height of the vertebral body above the inferior endplate. Partial results are presented in the Figure 12 with the complete set of figures in the appendix.

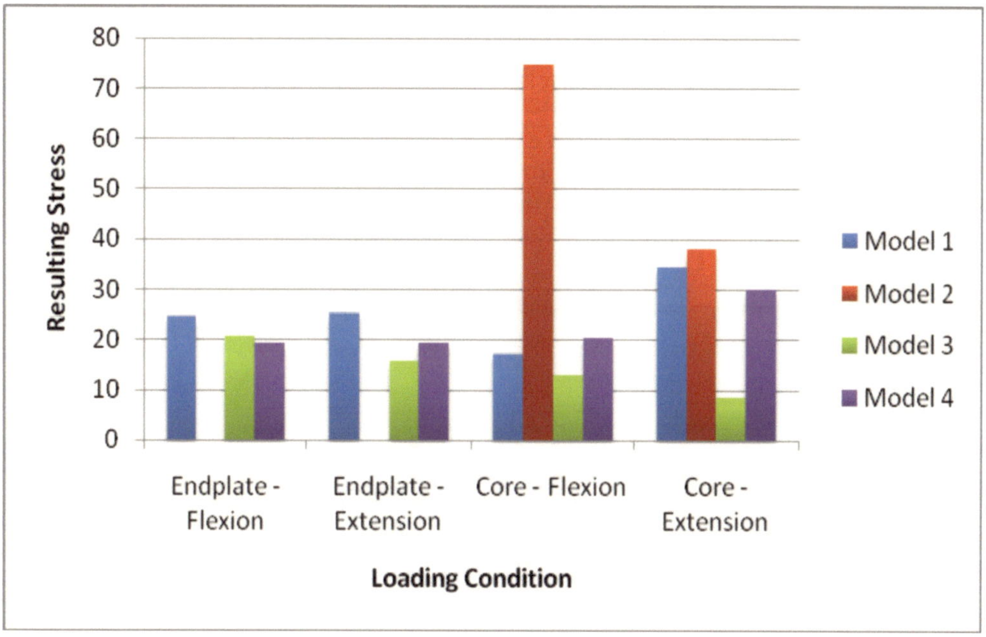

Fig. 11. Stress comparisons between models focusing on endplates and cancellous cores

Researcher	Study Topic	Loads	Max Endplate Stress (MPa)	Max Core Stress (MPa)	Study Level
Galbusera, 2008	Anterior Cervical Fusion	100 N Axial, 2.5 Nmm Bending	2.80	N/A	C5-C6
Denoziere, 2006	Fusion/Mobile Disc	720-1300 N Axial, 11.45 Nmm Axial rotation	90	3.5	L3-L4
Polikeit, 2003	Fusion	1000 N Axial, 12 Nm m Bending	Stress values recorded as percentage increases		L2-L3

Langrana, 2006	Curvature	N/A	40	N/A	L4-L5
Zhang, 2010	Bone Filling Material	400 N Axial, 7.5/3.75 Nmm flex, ext	9.503	.584	L1-L2
Zander, 2002	Bone Graft Location with Fixators	250 N Axial, 7.5 Nmm flex, ext, lat bend	25	N/A	L2-L5
Dai, 1998	Osteoporosis	1200 N Axial, 30 Nmm flex, ext	5.17	24.03	Lumbar
Adams, 2003	Fusion	1310 N Axial	25	N/A	L5

Table 27. The first listed researcher and the emphasis of the study are in columns one and two. The loading condition is in column three and the stress results in the core and endplate are in columns four and five. The level of the spinal column modeled is in column six

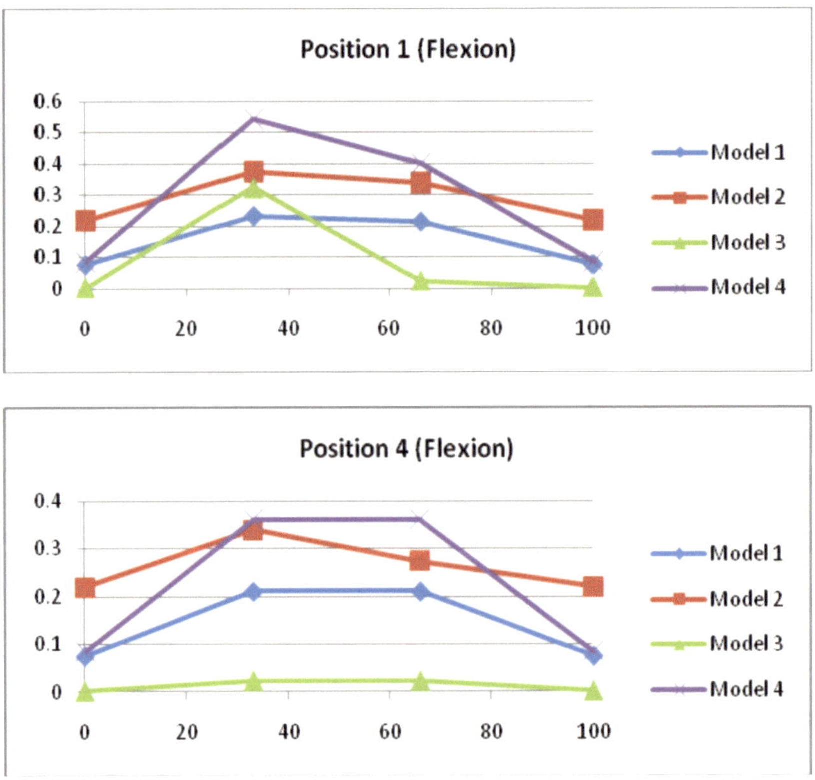

Fig. 12. Each figure represents the cancellous core stress through the height of the vertebral core. The entire chart is present in the appendix. Position 1 is posterior/right, Position 4 is anterior/left. The X-axis represents the percentage of height from top down. The Y-axis is the von Mises stress at each level

The stress results from this test were compared to studies conducted examining the stress in the endplate and vertebral body, and the loads used to obtain these stresses. These results are presented in Table 27. Direct comparisons are difficult because of the wide range of loading conditions, vertebral levels, and different study conditions i.e. fusion, curvature and bone grafts. The results of this study are however within these investigated ranges, which suggest the model is representative of the C3 cervical level.

The differences in reported von Mises stresses can be attributed to different loading conditions and boundary conditions among other things. Few studies go into detail about exactly how loads are applied to finite element models or how the models are bounded. Both factors can have large effects on the outcomes of stress maximums. Research has shown that a stress of 4 MPa is the failure limit for trabecular bone and 131-224 for cortical bone (Linde, 1994; Nigg and Herzog, 1994). These limits can be assumed as a benchmark for the onset of bone failure in the endplates and cancellous core.

The stresses developed in this study indicate that a half-millimeter approximation for the vertebral endplate is adequate. The half-millimeter approximation in Model 1 has a maximum/minimum percent difference from the anthropometric models of 47.8% and 17.2% respectively (percent differences presented in Table 26). The stress generated in Model 1 is also greater than the other models lending to a conservative design if these values are used for mechanical design considerations. The ability to model the endplate with a constant thickness saves time ultimately making the analysis more efficient.

While the endplate stresses were well under its failure limit of 133 MPa the maximum cancellous core stresses in the Model 2 (removed endplate) were much greater than its failure stress of 2 MPa. For example under flexion the core experienced a maximum stress of 74.8 MPa, which is approximately 35 times its failure limit. Subchondral failure was not investigated in this study so its contribution to failure cannot be addresses at this time.

Von Mises values were also recorded through the height of the vertebral core to examine stress propagation. For all cases the 2 MPa core failure stress was not reached except in the case of Model 1, extension, in the posterior right region of the vertebral body where the stress reached 3.98 MPa. This stress is slightly under the upper failure limit of 4 MPa. Table 4 shows average values for stress in each level of the vertebral body under each loading condition: flexion or extension, while also ignoring Model 2 since it does not have an endplate. Table 5 charts stresses associated with flexion and extension in either the posterior or anterior areas of the vertebral core.

Height (as percentage from bottom)	Average von Mises Stress in Flexion (MPa)	Average von Mises Stress in Extension (MPa)
100%	.129	.900 (.620)
66%	.180	.561
33%	.229	.510
0%	.054	.351

Table 28. Average stress propagation through the vertebral body in flexion and extension. The number in parentheses is not considering the highest possibly outlying stress value

Position	Flexion	Extension
1,2	.191	.691
3,4	.105	.467

Table 29. Average stress in the posterior or anterior areas of the vertebral body in flexion and extension. The number in parentheses is not considering the highest possibly outlying stress value

A general trend, in Figure 12, can be seen that the stress is increases towards the center of the vertebral core. In the upper endplate under extension the trend does not hold even if the highest stressed element is not considered. It's likely that there is some load sharing between the endplate and the vertebral core that redistributes load away from the core at the top and bottom near the endplates. The middle the vertebral body seems sufficiently removed from the endplates thus the higher reported stresses. Table 28 also indicates that the posterior of the vertebral body is stressed higher than the anterior portion under both flexion and extension.

3.4 Discussion

Removal of the cortical endplate has a significant effect on the cancellous core stress. Ideally the endplate should be left intact as much as possible. From the evidence above the minimum cancellous core stress was 38.2 MPa. This stress is almost 10 times that of failure limit for cancellous bone using a upper failure limit.

This investigation only analyzes a pressure load that is evenly distributed on the vertebral body. Unless a cage or artificial disc fits perfectly in the disc space with continuous contact, stresses will greatly increase at areas of contact (Adam et al., 2003). Curvature is particularly important in the cervical spine. Unlike the lumbar region that has large flat endplates the cervical spine has a large curvature in the coronal plane that comes from the uncanate processes (Bogduk and Mercer, 2006; Langrana et al., 2006).

The previous analysis shows that a half-millimeter endplate approximation can be used to adequately represent the cortical endplate experimentally. When compared to morphologically complex models the resulting half-millimeter endplate stress was 25.57 MPa and core stresses were 34.5 MPa similar to stresses in other research. It was found that the vertebral body can be modeled analytically without experimentation and can use simplified modeling parameters to save time and cost. Further understanding of regional stress characteristics will be valuable for the design of implantable devices.

4. Appendix

Appendix. Each graph is a specific position in the axial plane of the vertebral body. Position 1 is posterior/right, Position 2 is posterior/left, Position 3 is anterior/right, Position 4 is anterior/left. The X-axis on each chart is the height position in the vertebral body with 100 percent being just under the superior endplate. The Y-Axis is the resulting von Mises stress in MPa.

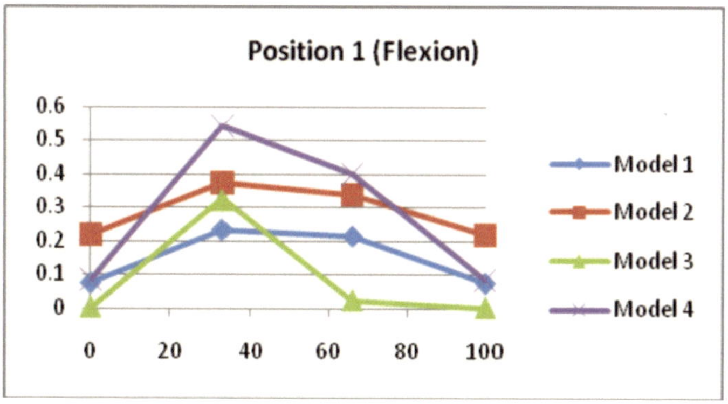
Position 1 (Flexion)
0.6
0.5
0.4
0.3
0.2
0.1
0
0
20
40
60
80
100
Model 1
Model 2
Model 3
Model 4

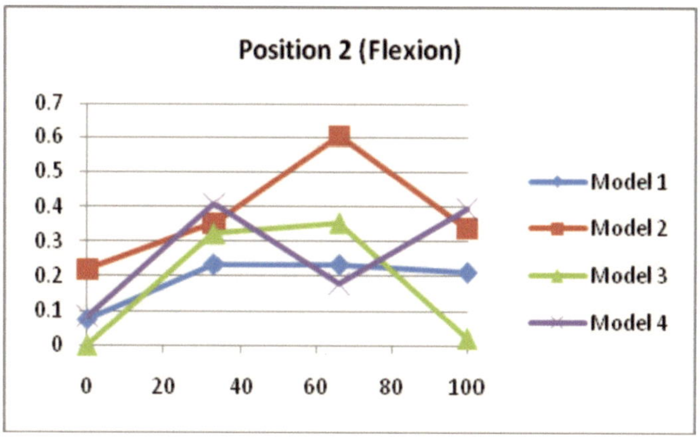
Position 2 (Flexion)
0.7
0.6
0.5
0.4
0.3
0.2
0.1
0
0
20
40
60
80
100
Model 1
Model 2
Model 3
Model 4

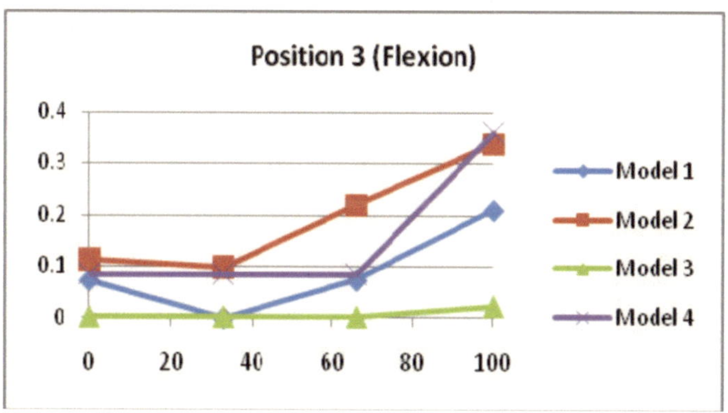
Position 3 (Flexion)
0.4
0.3
0.2
0.1
0
0
20
40
60
80
100
Model 1
Model 2
Model 3
Model 4

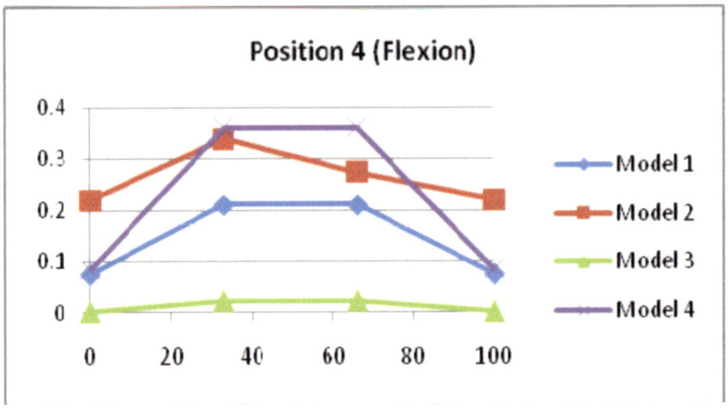
Position 4 (Flexion)
0.4
0.3
0.2
0.1
0
0
20
40
60
80
100
Model 1
Model 2
Model 3
Model 4

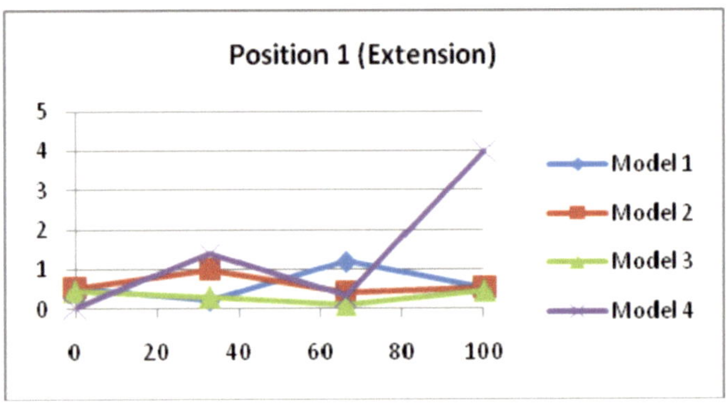
Position 1 (Extension)
5
4
3
2
1
0
0
20
40
60
80
100
Model 1
Model 2
Model 3
Model 4

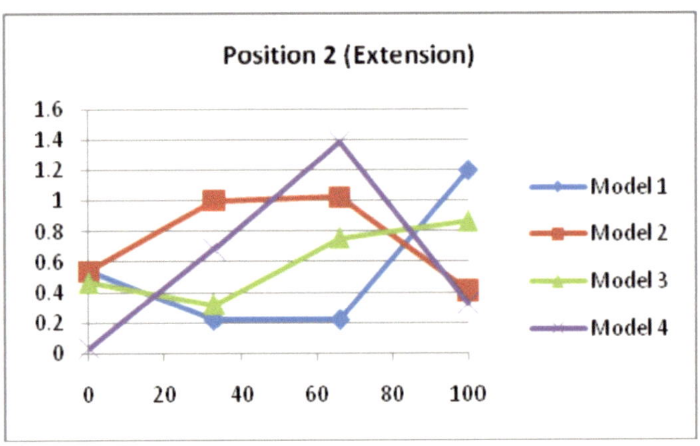
Position 2 (Extension)
1.6
1.4
1.2
1
0.8
0.6
0.4
0.2
0
0
20
40
60
80
100
Model 1
Model 2
Model 3
Model 4

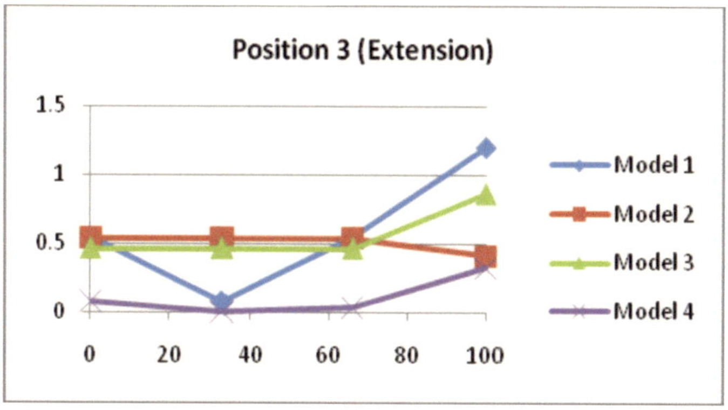

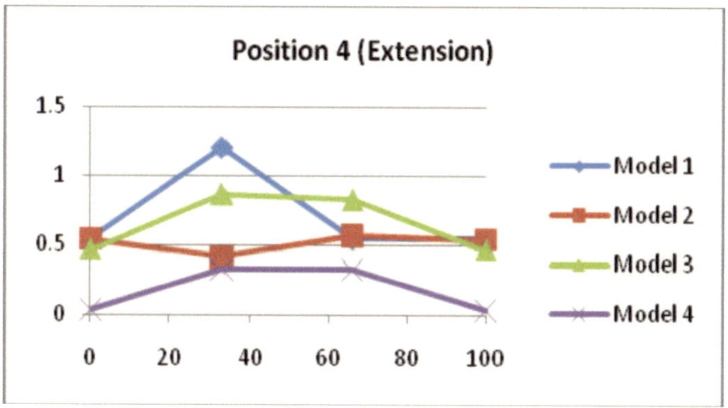

5. Acknowledgments

The authors would like to thank Miami Valley Hospital (Dayton, OH) for support on this project. Specifically Dr. David Udin in the Clinical Research Office, Scott Calvin manager of the Miami Valley Imaging Group, and Matt Binkley fourth year medical student, for assistance in collecting CT images.

6. References

Ambard, D. and Cherblanc, F. (2009) 'Mechanical Behavior of Annulus Fibrosus: A Microstructural Model of Fibers Reorientation', *Annals of Biomedical Engineering*.

Adam, C. Pearcy, M. McCombe, P. 2003. Stress analysis of interbody fusion - Finite element modeling of inetervertebral implant and vertebral body. Clinical Biomechanics 18, 265-272

Bogduck, N. and Yoganandan, N. (2001) 'Biomechanics of the cervical spine Part 3: minor injuries', *Clinical Biomechanics*, pp. 267 - 75.

Boos, N., Aebi, M., 2008 Spinal Disorders: Fundamentals of Diagnosis and Treatment. Springer, Zurich, pp. 41-62

Bozic, K.J., Keyak, J.H., Skinner, H.B., Beuff, H.U. and Bradford, D.S. (1994) 'Three-dimensional finite element modeling of a cervical vertebra: An investigation of burst fracture mechanism', *Journal of Spinal Disorders*, pp. 102 - 10.

Bozkus, H.A.K.M.H.M.U.E.B.A.S. (2001) 'Finite element model of the Jefferson fracture: comparison with a cadaver model', *Eur Spine J*, pp. 257-263.

Bozkus, H., Karakas, A., Hanci, M., Uzan, M., Bozdag, E. and Sarioglu, A. (2001) 'Finite element model of the Jefferson fracture: comparison with a cadaver model', *European Spine Journal*, pp. 257-263.

Brolin, K. and Halldin, P. (2004) 'Development of a finite element model of the upper cervical spine and a parameter study of ligament characteristics', *Spine*.

Brolin, K. and Halldin, P. (2004) 'Development of a Finite Element Model of the Upper Cervical Spine and a Parameter Study of Ligament Characteristics', *Spine*, pp. 376-385.

Brolin, K. and Halldin, P. (2005) 'The effect of muscle activation on neck response', *Traffic Injury Prevention*.

Camacho, D.L.A., Nightingale, R.W., Robinette, J.J., Vanguri, S.K., Coates, D.J. and Myers, B.S. (1997) 'Experimetnal flexibility measurements for the development of a computational head-neck model validated for near-vertex head impact', *Proceedings from the 41st Stapp car Crash Conference*, pp. 473 - 486.

Choi, J. Sung, K., 2006. Subsidence after anterior lumbar interbody fusion using paired stand-alone rectangular cages. European Spine Journal 15, 16-22

Dai, L., 1998. The relationship between vertebral body deformity and disc degeneration in lumbar spine of the senile. European Spine Journal 7, 40-44

Denoziere, G. Ku, D., 2006. Biomechanical comparison between fusion of two vertebrae and implantation of an artificial disc. Journal of Biomechanics 39, 766-775

Doherty, B., Heggeness, M. and Esses, S. (1993) 'A biomechanical study of odontoid fractures and fracture fixation', *Spine*, pp. 178 - 84.

Douglas C. Montgomery, a.E.A.P. (1982) *Introduction to Linear Regression Analysis*, New York: John Wiley & Sons, Inc.

Ebara, S., Iatridis, J.C., Setton, L.A., Foster, R.J., Mow, V.C. and Weidenbaum, M. (1996) 'Tensile properties of nondegenerate human lumbar annulus fibrosus', *Spine*, pp. 452 - 61.

Eberlein, R., Holzapfel, G.A. and Froelich, M. (2004) 'Multi-segment FEA of the human lumbar spine including the heterogeneity of the annulus fibrosus', *Computational Mechanics*.

Edwards, W.T. Ferrarra, L.A. Yuan, H.A., 2001. The effect of the vertebral cortex in the thoracolumbar spine. Spine 26, 218-225

Esat, V.L.D.A.M. (2005) 'Combined Multi-Body Dynamic and FE Models of Human Head and Neck', IUTAM Proceedings on Impact Biomechanics, 91-100.

Francis, C.C. (1955) 'Dimensions of the cervical vertebrae', *Anat Rec*, vol. 122, pp. 603-609.

Francis, C.C. (1955) 'Dimensions of the cervical vertebrae', *Anat Rec*, vol. 122, pp. 603-609.

Galbusera, F., Bellini, C.M., Raimondi, M.T., Fornari, M. and Assietti, R. (2008) ' Cervical spine biomechanics following implantation of a disc prosthesis', *Medical Engineering and Physics*, pp. 1127-1133.

Gamst, G., Meyers, L.S. and Guarino, A.J. (2008) *Analysis of Variance Designs: A conceptual and Computational Approach with SPSS and SAS*, New York: Cambridge University Press.

Gilad, I. and Nissan, M. (1986) 'A study of vertebra and disc geometic relations of the human cervical and lumbar spine', *Spine*, pp. 154 - 7.

Glenn Gamst, L.S.M.A.J.G. (2008) *Analysis of Variance Designs: A conceptual and Computational Approach with SPSS and SAS*, New York: Cambridge University Press.

Goel, V.K., Clark, C.R., Harris, K.G. and Schulte, K.R. (1988) 'Kinematics of the cervical spine: effects of multiple total laminectomy and facet wiring', *Jouranl of Orhopaedic Research*, pp. 611 - 19.

Goel, V.K. and Clausen, J.D. (1998) 'Prediction of Load Sharing Among Spinal Components of a C5-C6 Motion Segment Using the Finite Element Approach', *Spine*, pp. 684-691.

Graham, R., Oberlander, E., Stewart, J. and Griffiths, D. (2000) 'Validation and use of a finite element model of C-2 for determination of stress and fracture patterns of anterior odontoid loads', *Journal of Neurosurgery*.

Grant, J. Oxland, T. Dvorak, M., 2001. Mapping the structural properties of the lumbosacral vertebral endplates. Spine 8, 889-896

Greaves, C.Y., Gadala, M.S. and Oxland, T.R. (2008) 'A three-dimensional finite element model of the cervical spine with spinal cord: an investigation of three injury mechanisms', *Annals of Biomedical Engineering*, pp. 396-405.

Ha, S.K. (2006) 'Finite element modeling of multi-level cervical spinal segments (C3-C6) and biomechanical analysis of an elastomer-type prosthetic disc', *Medical Engineering & Physics*, pp. 534-541.

Haghpanahi, M.M.R. (2006) 'Development of a Parametric Finite Element Model of Lower Cervical Spine in Sagital Plane', Proceedings of the 28th IEEE EMBS Annual International Conference, New York City, USA, 1739-1741.

Holzapfel, G.A. (2000) *Nonlinear Solid Mechanics*, New York: Wiley.

Jost, B. et. al., 1998. Compressive strength of interbody cages in the lumbar spine: the eddect of cage shape, posterior instrumentation and bone density. European Spine Journal 7, 132-144

Kallemeyn, N.A., Tadepalli, S.C. and Shivanna, K.H. (2009) ' An interactive multiblock approach to meshing the spine', *Computer Methods and Programs in Biomedicine*, pp. 227-235.

KenethS. Saladin PhD, a.L.M.M.S.N. (2004) *Anatomy & Physiology: The Unity of Form and Function*, Third Ed. edition, New York: McGraw-Hill.

Kulkarni, A. D'Orth Hee, H. Wong, H., 2007. Solis cage (PEEK) for anterior cervical fusion: preliminary radiological results with emphasis on fusion and subsidence. The Spine Journal 7, 205-209

Kumaresan, S., Yoganandan, N. and Pintar, F. (1999) 'Finite Element Analysis of the Cervical Spine: A Material Property Sensitivity Study', *Clinical Biomechanics*, pp. 41-53.

Kumaresan, S., Yoganandan, N., Pintar, F. and Maiman, D. (1999) 'Finite element modeling of the cervical spine: role of intervertebral disc under axial and eccentric loads', *Medical Engineering & Physics*, pp. 689-700.

Kumaresan, S., Yoganandan, N., Pintar, F. and Maiman, D. (1999) 'Finite element modeling of the lower cervical spine: role of intervertebral disc under axial and eccentric loads', *Medical Engineering & Physics*, pp. 689-700.

Kumaresan, S., Yoganandan, N., Pintar, F. and Maiman, D. (1999) 'Finite element modeling of the lower cervical spine: role of intervertebral disc under axial and eccentric loads', *Medical Engineering & Physics*, pp. 689-700.

Kumaresan, S., Yoganandan, N., Pintar, F., Maiman, D. and Kuppa, S. (2000) 'Biomechanical study of pediatric human cervical spine: a finite element approach', *Journal of Biomechanical Engineering*, pp. 60-71.

Kumaresan, S., Yoganandan, N., Voo, L. and Pintar, F. (1998) 'Finite Element Analysis of the Human Lower Cervical Spine', *Journal of Biomechanics*, pp. 87-92.

Langrana, N. Kale, S. Edwards, T. Lee, C. Kopacz, K., 2006. Measurement and analyses of the effects of adjacent

Li, Y. and Lewis, G. (2010) 'Influence of surgical treatment for disc degeneration diseases at C5-C6 on changes in some biomechanical parameters of the cervical spine', *Medical Engineering & Physics*, pp. 593 - 503.

Linde, F., 1994. Elastic and viscoelastic properties of trabecular bone by a compression testing approach. Danish Medical Bulletin 41, 119–138.

Liu YK, C.C.K.K. (1986) 'Quantitative geometry of young human male cervical vertebrae', *Mechanism of Head and Spine Trauma*, pp. 417-431.

Liu, Y.K., Clark, C.R. and Krieger, K.W. (1986) 'Quantitative geometry of young human male cervical vertebrae', *Mechanism of Head and Spine Trauma*, pp. 417-431.

Manohar M. Panjabi, P.J.D.M.V.G.P.T.O.M.a.K.T.M. (1991) 'Cervical Human Vertebrae: Quantitative Three-Dimensional Anatomy of the MIddle and Lower Regions', *Spine*, vol. 16, no. 8, pp. 861-869.

Maurel, N., Lavaste, F. and Skalli, W. (1997) 'A three-dimensional parameterized finite element model of the lower cervical spine. Study of the influence of the posterior articular facets', *Journal of Biomechanics*, pp. 921-931.

Meakin, J. and Huskins, D.W.L. (2001) 'Replacing the nucleus pulposus of the intervertebral disk: prediction of suitable properties of a replacement material using finite element analysis', *Journal of Materials Science*.

Montgomery, D.C. and Peck, E.A. (1982) *Introduction to Linear Regression Analysis*, New York: John Wiley & Sons, Inc.

Moroney, S.P., Schultz, A.B., Miller, J.A.A. and Andersson, G.B.J. (1988) 'Load-displacement properties of lower cervical spine motion segments', *Journal of Biomechanics*, vol. 21, pp. 769-779.

Muller-Gerbl, M. Weiber, S. Linsenmeier, U., 2008. The distribution of mineral density in the cervical vertebral endplates. European Spine Journal 17, 432-438

Nigg, B.M., Herzog, W.H., 1994. Biomechanics of the musculo- skeletal system (Eds.), John Wiley & Sons, Chichester.

Ng, H.-W. and Teo, E.-C. (2001) 'Nonlinear Finite-Element Analysis of the Lower Cervical Spine (C4-C6) Under Axial Loading', *Journal of Spinal Disorders*, pp. 201-10.

Ng, H.-W., Teo, E.-C., Lee, K.-K. and Qiu, T.-X. (2003) 'Finite element analysis of cervical spinal instability under physiologic loading', *Journal of Spinal Disorders & Techniques*, pp. 55 - 63.

Nightingale, R.W., Chancey, V.C., Otaviano, D., Luck, J.F., Tran, L., Prange, M. and Myers, B.S. (n.d) 'Flexion and extension structural properties and strenghts for male cervical spine segments', *Journal of Biomechanics*, pp. 534 - 42.

Nissan M, G.I. (1984) 'The cervical and lumbar vertebrae: An anthropometric model', *Eng Med*, vol. 13, no. 3, pp. 111-114.

Nissan, M. and Gilad, I. (1984) 'The cervical and lumbar vertebrae: An anthropometric model', *Eng Med*, vol. 13, no. 3, pp. 111-114.

Noailly, J., Lacoix, D. and Planell, J.A. (2005) 'Finite element study of a novel intervertebral disc substitute', *Spine*.

Noailly, J., Wilke, H.-J., Planell, J.A. and Lacroix, D. (2007) 'How does the geometry affect the internal biomechanics of a lumbar spine bi-segment finite element model? Consequences on validation process', *Journal of Biomechanics*, pp. 2414-2425.

Ordway, N. Lu, Y. Zhang, X. Cheng, C. Fang. H, Fayyazi, A., 2007. Correlation of the cervical endplate strength with CT measured subchondral bone density. European Spine Journal 16, 2104-2109

Oxland, T.R., 2003. Effects of endplate removal on the structural properties of the lower lumbar vertebral bodies. Spine 8, 771-777

Palomar, A., Calvo, B. and Doblare, M. (2008) 'An accurate finite element model of the cervical spine under quasi-static loading', *Journal of Biomechanics*, pp. 523-531.

Panagiotopoulou, O. (2009) 'Finite element analysis (FEA): applying an engineering method to functional morphology in anthropology and human biology', *Annals of Human Biology*, pp. 609 - 23.

Panjabi, M., White. A., 1990. Clinical Biomechanics of the Spine. Lippencott, Philidelphia

Panjabi, M. Chen, M.C. Wang, J.L., 2001. The cortical architecture of the human cervical vertebral bodies. Spine 22, 2478-2484

Panjabi, M.M., Crisco, J.J., Vasavada, A., Oda, T., Cholewicki, J., Nibu, K. and Shin, E. (2001) 'Mechanical properties of the human cervical spine as shown by three-dimensional load-displacement curves', *Spine*, pp. 2692 - 2700.

Panjabi, M., Duranceau, J., Goel, V., Oxland, T. and Takata, K. (1991) 'Cervical human vertebrae. Quanatative three-dimensional anatomy of the middle and lower regions', *Spine*, pp. 861 - 9.

Panjabi, M.M., Duranceau, J., Goel, V., Oxland, T. and Takata, K. (1991) 'Cervical Human Vertebrae: Quantitative Three-Dimensional Anatomy of the MIddle and Lower Regions', *Spine*, vol. 16, no. 8, pp. 861-869.

Panzer, M.B. and Cronin, D.S. (2009) 'C4-C5 segment finite element model development, validation, and load-sharing investigation', *Journal of Biomechanics*, pp. 480-490.

Pintar, F., Yoganandan, N., Pesigan, M., Reinartz, J., Sances, A. and Cusick, J. (1995) 'Cervical vertebral strain measurements under axial and eccentric loading', *Journal of Biomechanical Engineering*, pp. 474 - 8.

Pintar, F.A., Yoganandan, N., Pesigan, M., Reinartz, J., Sances, A. and Cusick, J.F. (1995) 'Cervical vertebral strain measurements under axial and eccentricl loading', *Journal of Biomechanical Engineering*, pp. 474 - 8.

Pitzen, T., Geisler, F., Matthis, D., Muller-Storz, H., Barbier, D., Steudel, W.-I. and Feldges, A. (2002) 'A finite element model for predicting the biomechanical behaviour of the human lumbar spine.', *Control Engineering Practice*, pp. 83-90.

Pitzen, T., Geisler, F.H., Matthis, D., Muller-Storz, H., Pedersen, K. and Steudel, W.-I. (2001) 'The influence of cancellous bone density on load sharing in human lumbar spine: a comparison between an intact and a surgically altered motion segment', *European Spine Journal*, pp. 23-29.

Pitzen, T., Matthis, D. and Steudel, W.-I. (2002) 'Posterior Element Injury and Cervical Spine Flexibility Following Anterior Cervical Fusion and Plating', *European Journal of Trauma*, pp. 24-30.

Pitzen, T. et. al., 2004. Variation of endplate thickness. European Spine Journal 13, 235-240

Polikeit[1], A. et. al., 2003. Factors influencing stresses in the lumbar spine after the insertion of intervertebral cages: Finite element analysis. European Spine Journal 12, 413-420

Polikeit[2], A. et. al., 2003. The importance of the endplate for interbody cages in the lumbar spine. European Spine Journal 12, 556-561

Richter, M., Wilke, H.J., Kluger, P., Claes, L. and Puhl, W. (2000) 'Load-displacement properties of the normal and injured lower cervical spine in vitro', *European Spine Journal*, pp. 104 - 8.

Rohlmann, A. Zander, T. Bergmann, G., 2006. Effects of fusion-bone stiffness on the mechanical behavior of the lumbar spine after vertebral body replacement. Clinical Biomechanics 21, 221-227

S.H. Tan, E.C.T.a.H.C.C. (2004) 'Quantitative three-dimensional anatomy of cervical, thoracic and lumbar vertebrae of Chinese Singaporeans', *European Spine Journal*, vol. 13, pp. 137-146.

Sairyo, K., Goel, V.K., Masuda, A., Vishnubhotla, S., Faizan, A., Biyani, A., Ebraheim, N. and Yonekura, D.e.a. (2006) ' Three-dimensional finite element analysis of the pediatric lumbar spine. Part 1: pathomechanism of apophyseal bony ring fracture', *European Spine Journal*, pp. 923-929.

Sairyo, K., Goel, V.K., Masuda, A., Vishnubhotla, S., Faizan, A., Biyani, A., Ebraheim, N. and Yonekura, D.e.a. (2006) 'Three-dimensional finite element analysis of the pediatric lumbar spine. Part II: biomechanical change as the initiating factor for pediatric ishmic spondylolisthesis after growth plate', *European Spine Journal*, pp. 930-935.

Sairyo, K., Goel, V.K., Masuda, A., Vishnubhotla, S., Faizan, A., Biyani, A., Ebraheim, N. and Yonekura, D.e.a. (206) ' Three-dimensional finite element analysis of the pediatric lumbar spine. Part II: biomechanical change as the initiating factor for pediatric ishmic spondylolisthesis after growth plate', *European Spine Journal*, pp. 930-935.

Saladin, K.S. and Miller, L. (2004) *Anatomy & Physiology: The Unity of Form and Function*, Third Ed. edition, New York: McGraw-Hill.

Schmidt, H.H.F.D.J.K.Z.C.L.W.H. (2007) 'Application of a calibration method provides more realistic results for a finite element model of a lumbar spinal segment', *Clinical Biomechanics*, pp. 377-384.

Shirazi-Adl, A. (2006) ' Analysis of large compression loads on lumbar spine in flexion and in torsion using a novel wrapping element.', *Journal of Biomechanics*, pp. 267-275.

Stemper, B.D., Yoganandan, N., Pintar, F.A. and Rao, R.D. (2006) 'Anterior longitudinal ligament injuries in whiplash may lead to cervical instability', *Medical Engineering & Physics*.

Tan, S.H., Teo, E.C. and Chua, H.C. (2004) 'Quantitative three-dimensional anatomy of cervical, thoracic and lumbar vertebrae of Chinese Singaporeans', *European Spine Journal*, vol. 13, pp. 137-146.

Teo, J.C.M., Chui, C.K., Wang, Z.L., Ong, S.H., Yan, C.H., Wang, S.C., Wong, H.K. and Teoh, S.H. (2007) 'Heterogenous meshing and biomechanical modeling of human spine.', *Medical Engineering and Physics*, pp. 277-290.

Teo, E.C. and Ng, H.W. (2001) 'Evaluation of the role of ligaments, facets and disc nucleus in lower cervical spine under compression and sagittal moments using finite element method', *Medical Engineering & Physics*, pp. 155 - 64.

Teo, E.C. and Ng, H.W. (2001) 'First cervical vertebra (atlas) fracture mechanism studies using finite element method', *Journal of Biomechanics*, pp. 13-21.

Traynelis, P.A., Donaher, R.M., Roach, R.M. and Goel, v.K. (1993) 'Biomechanical comparison of anterior caspar plate and three-level posterior fixation techniques in human cadaveric model', *Journal of Neurosurgery*, pp. 96 - 103.

Van Jonbrgen, H.P. Spruit, M. Anderson, P. Pavlov, P., 2005. Anterior cervical interbody fusion with a titanium box cage: early radiological assessment of fusion and subsidence. The Spine Journal 5, 645-649

Voo, L.M., Kumaresan, S., Yoganandan, N., Pintar, F. and Cusick, J.F. (1997) 'Finite element analysis of cervical facetectomy', *Spine*.

Wheeldon, J.A., Pintar, F.A., Knowles, S. and Yoganandan, N. (2006) 'Experimental flexion/extension data corridors for validation of finite element models of the young, normal cervical spine', *Journal of Biomechanics*, pp. 375 - 80.

Wheeldon, J.A., Stempter, B.D., Yoganandan, N. and Pintar, F.A. (2008) 'Validation of finite element model of the young normal lower cervical spine', *Annals of Biomedical Engineering*, pp. 1458-1469.

Yoganandan, N., Kumaresan, S. and Pintar, F.A. (2000) 'Geometric and mechanical properties of human cervical spine ligaments', *Journal of Biomechanical Engineering*, pp. 623 - 29.

Yoganandan, N., Kumaresan, S. and Pintar, F.A. (2001) 'Biomechanics of the cervical spine part 2. Cervical spine soft tissue responses and biomechanical modeling', *Clinicals Biomechanics*, pp. 1-27.

Yoganandan, N., Kumaresan, S., Voo, L. and Pintar, F.A. (1996) 'Finite elemnt applications in human cervical spine modeling', *Spine*.

Yoganandan, N., Kumaresan, S., Voo, L. and Pintar, F.A. (1997) 'Finite Element Model of the Human Lower Cervical Spine: Parametric Analysis of the C4-C6 Unit', *Journal of Biomechanical Engineering*, pp. 81-92.

Yoganandan, N., Kumaresan, S., Voo, L., Pintar, F. and Larson, S. (1996) 'Finite Element Analysis of the C4-C6 Cervical Spine Unit', *Medical Engineering & Physics*, pp. 569-574.

Zander, T. Rohlmann, A. Klockner, C. Bergmann, G., 2002. Effect of bone graft characteristics on the mechanical behavior of the lumbar spine. Journal of Biomechanics 35, 491-497

Zhang, Q.H., Teo, E.C., Ng, H.W. and Lee, V.S. (2006) 'Finite element analysis of moment-rotation relationships for human cervical spine', *Journal of Biomechanics*.

Zheng, P.D..N., Young-Hing, M.D..K. and Watson, P.D..L.G. (2000) 'Morphological and biomechanical studies of pedicle screw fixation for the lower cervical spine', *Journal of Systems Integration*, pp. 55-66.

7

Biomechanics of the Temporomandibular Joint

Shirish M. Ingawalé[1] and Tarun Goswami[1,2]
[1]*Biomedical, Industrial and Human Factors Engineering, Wright State University, Dayton, OH*
[2]*Orthopaedic Surgery and Sports Medicine, Wright State University, Dayton, OH*
U.S.A.

1. Introduction

Temporomandibular joint (TMJ) connects the mandible or the lower jaw to the skull and regulates the movement of the jaw (see Figure 1). The TMJ is one of the most complex, delicate and highly used joints in a human body (Alomar et al., 2007). The most important functions of the TMJ are mastication and speech. Temporomandibular disorder (TMD) is a generic term used for any problem concerning the jaw joint. Injury to the jaw, the TMJ, or muscles of the head and neck can cause TMD. Other possible causes include grinding or clenching the teeth; dislocation of the disc; presence of osteoarthritis or rheumatoid arthritis in the TMJ; stress, which can cause a person to tighten facial and jaw muscles or clench the teeth; aging (Bakke et al., 2001; Detamore et al., 2007; Ingawalé and Goswami, 2009; Tanaka et al., 2000). The most common TMJ disorders are pain dysfunction syndrome, internal derangement, arthritis, and traumas (Breul et al., 1999; Chen et al., 1998). TMDs are seen most commonly in people between the ages of 20 and 40 years, and occur more often in women than in men (Detamore and Athanasiou, 2003; Detamore et al., 2007; Tanaka et al., 2008a). Some surveys have reported that 20-25% of the population exhibit one or more symptoms of TMD (Detamore et al., 2007; Ingawalé and Goswami, 2009).

With a large part of population suffering from TMDs, it is a problem that should be looked at more fully. Relations between muscle tensions, jaw motions, bite and joint force, and craniofacial morphology are not fully understood. A large fraction of TMD causes are currently unexplained. There is a great need of better understanding of the etiology of TMDs to develop methods to prevent, diagnose, and cure joint disorders (Beek et al., 2003; Ingawalé and Goswami, 2009). This chapter provides a state-of-the-art review of TMJ anatomy, disorders, and biomechanics; and briefly discusses our approach toward three-dimensional (3D) anatomical and finite element (FE) modeling to understand the interaction between structure and function of the TMJ.

2. TMJ anatomy and function

TMJ is a bi-condylar joint in which the condyles, the movable round upper ends of the mandible, function at the same time (see Figure 1). Between the condyle and the articular fossa is a disc made of fibrocartilage that acts as a cushion to absorb stress and allows the condyle to move easily when the mouth opens and closes (AAOMS, 2007; Ide et al., 1991).

The bony structures consist of the articular fossa; the articular eminence, which is an anterior protuberance continuous with the fossa; and the condylar process of the mandible that rests within the fossa. The articular surfaces of the condyle and the fossa are covered with cartilage (Ide et al., 1991). The disc divides the joint cavity into two compartments - superior and inferior (Ide et al., 1991; Tanaka et al., 2008b). The two compartments of the joint are filled with synovial fluid which provides lubrication and nutrition to the joint structures (Tanaka et al., 2008b). The disc distributes the joint stresses over broader area thereby reducing the chances of concentration of the contact stresses at one point in the joint. The presence of the disc in the joint capsule prevents the bone-on-bone contact and the possible higher wear of the condylar head and the articular fossa (Beek et al., 2001; Tanaka et al., 2008b). The bones are held together with ligaments. These ligaments completely surround the TMJ forming the joint capsule.

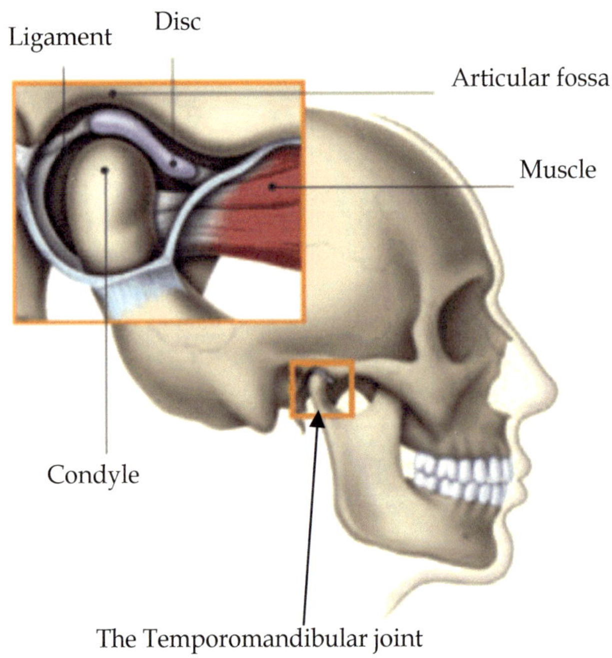

Source: American Association of Oral and Maxillofacial Surgeons (AAOMS, 2007).

Fig. 1. Anatomical structure of the temporomandibular joint (TMJ)

Strong muscles control the movement of the jaw and the TMJ. The temporalis muscle which attaches to the temporal bone elevates the mandible. The masseter muscle closes the mouth and is the main muscle used in mastication (see Figure 2) (Hylander, 1979). Movement is guided by the shape of the bones, muscles, ligaments, and occlusion of the teeth. The TMJ undergoes hinge and gliding motion (Alomar et al., 2007). The TMJ movements are very complex as the joint has three degrees of freedom, with each of the degrees of freedom associated with a separate axis of rotation. Rotation and anterior translation are the two primary movements. Posterior translation and mediolateral translation are the other two possible movements of TMJ (Dutton, 2004).

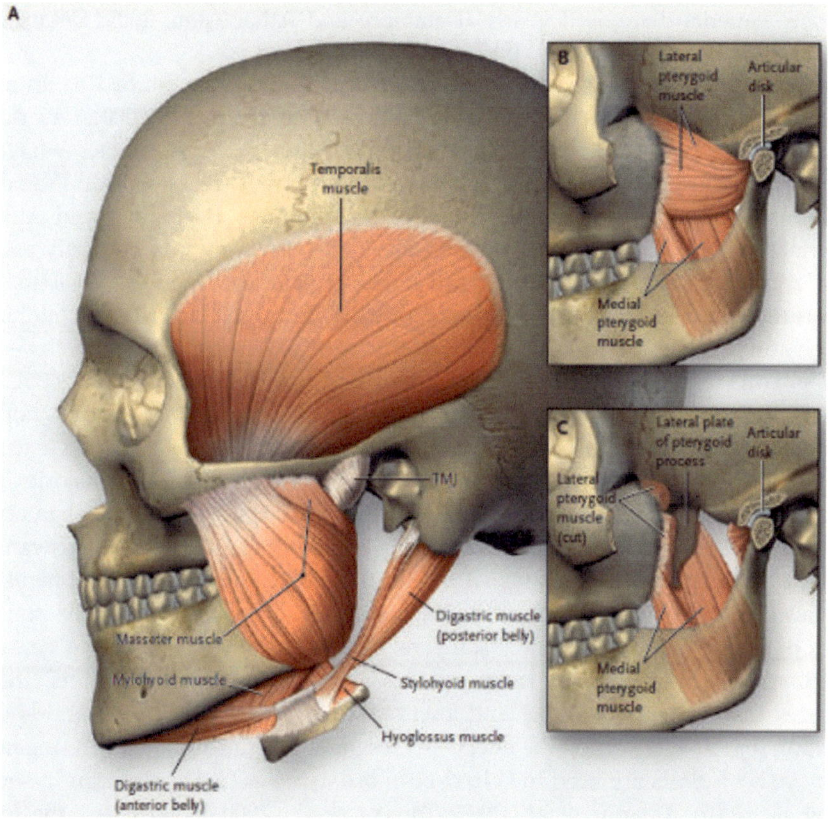

Source: Scrivani et al., 2008

Fig. 2. Normal anatomy of the jaw. The lateral view of the skull (Panel A) shows the normal position of the mandible in relation to the maxilla, the TMJ capsule, and the masticatory muscles - temporalis, masseter, mylohyoid, anterior and posterior digastrics, hyglossus, and stylohyoid. Also shown (Panels B and C) are the deep muscles associated with jaw function and the TMJ intra-articular disc.

3. TMJ disorders and treatment

Temporomandibular disorder (TMD) is a generic term used for any problem concerning the jaw joint. Injury to the jaw, temporomandibular joint, or muscles of the head and neck can cause TMD. Other possible causes include grinding or clenching the teeth, which puts a lot of pressure on the TMJ; dislocation of the disc; presence of osteoarthritis or rheumatoid arthritis in the TMJ; stress, which can cause a person to tighten facial and jaw muscles or clench the teeth; aging, etc (Bakke et al., 2001; Detamore et al., 2007; Ingawalé and Goswami, 2009; Tanaka et al., 2000). The most common TMJ disorders are pain dysfunction syndrome, internal derangement, arthritis, and traumas (Detamore and Athanasiou, 2003; Detamore et al., 2007). TMD is seen most commonly in people between the ages of 20 and 40 years, and occurs more often in women than in men (Detamore and Athanasiou, 2003; Detamore et al., 2007). Some surveys have reported that 20-25% of the population exhibit symptoms of TMD while it is estimated that 30 million Americans suffer from it, with approximately one million new patients diagnosed yearly (Detamore and Athanasiou, 2003; Detamore et al., 2007; Tanaka et al., 2008b; Wolford, 1997).

Disc displacement is the most common TMJ arthropathy and is defined as an abnormal relationship between the articular disc and condyle (Tanaka et al., 2000). As the disc is forced out of the correct position, there is often bone on bone contact which creates additional wear and tear on the joint, and often causes the TMD to worsen (Tanaka et al., 2000). Almost 70% of TMD patients have disc displacement (Detamore and Athanasiou, 2003). Different types of functional malocclusion have been shown to be partly responsible for signs and symptoms of TMD. The functional unilateral posterior cross-bite, habitual body posture during sleep, juvenile chronic arthritis - a chronic arthritis in childhood with an onset before the age of 16 years and a duration of more than three months - are also reported as TMD risk (Bakke et al., 2001; Hibi and Ueda, 2005; Pellizoni et al., 2006).

Treatments for the various TMJ disorders range from physical therapy and nonsurgical treatments to various surgical procedures. Usually the treatment begins with conservative, nonsurgical therapies first, with surgery left as the last option. The majority of TMD patients can be successfully treated by non-surgical therapies and surgical interventions may be required for only a small part of TMD population (Ingawalé and Goswami, 2009). The initial treatment does not always work and therefore more intense treatments such as joint replacement may be a future option (Ingawalé and Goswami, 2009). The non-surgical treatment options include medication; self-care; physical therapy, to keep the synovial joint lubricated and to maintain full range of the jaw motion; wearing splints, the plastic mouthpieces that fit over the upper and lower teeth to prevent the upper and lower teeth from coming together, lessening the effects of clenching or grinding the teeth (Ingawalé and Goswami, 2009). Splints are used to help control bruxism - a TMD risk factor in some cases (Glaros et al., 2007; Kalamir et al., 2007; Tanaka et al., 2000a). However, the long-term effectiveness of this therapy has been widely debated and remains controversial (Glaros et al., 2007; Kalamir et al., 2007). Surgery can play an important role in the management of TMDs. Conditions that are always treated surgically involve problems of overdevelopment or underdevelopment of the mandible resulting from alterations of condylar growth, mandibular ankylosis, and benign and malignant tumors of the TMJ (Laskin et al., 2006). The surgical treatments include arthrocentesis, arthroscopy, discectomy, and joint replacement. While more conservative treatments are preferred when possible, in severe cases or after multiple operations, the current end stage treatment is joint replacement (Tanaka et al., 2008b). However, before a joint replacement option is ever considered for a patient, all non-surgical, conservative treatment options must be exhausted; and all conservative surgical methodologies should be employed (Quinn, 199; Quinn, 2000).

4. Biomechanical behavior of the TMJ

Mandibular motions result in static and dynamic loading in the TMJ. During natural loading of the joint, combinations of compressive, tensile, and shear loading occur on the articulating surfaces (Tanaka et al., 2008b). The analysis of mandibular biomechanics helps us understand the interaction of form and function, mechanism of TMDs; and aids in the improvement of the design and the behavior of prosthetic devices, thus increasing their treatment efficiency (Hansdottir and Bakke, 2004; Ingawalé and Goswami, 2009; Korioth and Versluis, 1997)

4.1 In-vivo assessment

Very few studies which report in-vivo biomechanical assessment of the TMJ can be found in the literature. In contrast to some earlier studies which reported the TMJ to be a force-free joint, Hylander (1979) demonstrated that considerable forces were exerted on the TMJ during occlusion as well as mastication. In face of these contrary reports, Breul et al. (1999) showed that the TMJ was subjected to pressure forces during occlusion as well as during mastication and it was slightly eccentrically loaded in all positions of occlusion.

Korioth and Hannam (1994) indicated that the differential static loading of the human mandibular condyle during tooth clenching was task dependent and both the medial and lateral condylar thirds were heavily loaded. Huddleston Slater et al. (1999) suggested that when the condylar movement traces coincide during chewing, there is compression in the TMJ during the closing stroke. However, when the traces do not coincide, the TMJ is not or only slightly compressed during chewing. Naeije and Hofman (2003) used these observations to study the loading of the TMJ during chewing and chopping tasks. Their analysis showed that the distances traveled by the condylar kinematic centers were shorter on the ipsilateral side than on the contralateral. The kinematic centers of all contralateral joints showed a coincident movement pattern during chewing and chopping. The indication that the ipsilateral joint is less heavily loaded during chewing than the contralateral joint may explain why patients with joint pain occasionally report less pain while chewing on the painful side.

Hansdottir and Bakke (2004) evaluated the effect of TMJ arthralgia on mandibular mobility, chewing, and bite force in TMD patients (categorized as disc derangements, osteoarthritis, and inflammatory disorders) compared to healthy control subjects. The pressure pain threshold (PPT), maximum jaw opening, and bite force were significantly lower in the patients as compared to that in controls. The patients were also found to have longer duration of chewing cycles. The bite force and jaw opening in patients were significantly correlated with PPT. The most severe TMJ tenderness (i.e., lowest PPT) and the most impeded jaw function with respect to jaw opening and bite force were found to be more severe in the patients with inflammatory disorders than the patients with disc derangement or osteroarthritis (Hansdottir and Bakke, 2004).

4.2 In-vitro assessment – mechanical testing and finite element modeling

As the TMJ components are difficult to reach and as the applications of experimental devices inside the TMJ cause damage to its tissue, the direct methods are not used often. Indirect techniques utilized to evaluate mandibular biomechanics have had limited success due to their ability to evaluate only the surface stress of the model but not its mechanical properties (Ingawalé and Goswami, 2009). Mechanical testing and finite element modeling (FEM) have been progressively used by TMJ researchers.

Excessive shear strain can cause degradation of the TMJ articular cartilage and collagen damage eventually resulting in joint destruction (Tanaka et al., 2008). Tanaka et al. (2008) attempted to characterize the dynamic shear properties of the articular cartilage by studying shear response of cartilage of 10 porcine mandibular condyles using an automatic dynamic viscoelastometer. The results showed that the shear behavior of the condylar cartilage is dependent on the frequency and amplitude of applied shear strain suggesting a significant role of shear strain on the interstitial fluid flow within the cartilage. Beek et al. (2001) performed sinusoidal indentation experiments and reported that the dynamic mechanical behavior of disc was nonlinear and time-dependent. Beek et al. (2003) simulated these experiments using axisymmetric finite element model and showed that a poroelastic material model can describe the dynamic behavior of the TMJ disc. Tanaka et al. (2006) carried out a series of measurements of frictional coefficients on 10 porcine TMJs using a pendulum-type friction tester. The results showed that the presence of the disc reduces the friction in the TMJ by reducing the incongruity between the articular surfaces and by increasing synovial fluid lubrication. This study highlighted the importance of preserving the disc through alternatives to discectomy to treat internal derangement and osteoarthritis of the TMJ.

The finite element modeling (FEM) has been used widely in biomechanical studies due to its ability to simulate the geometry, forces, stresses and mechanical behavior of the TMJ components and implants during simulated function (Beek et al., 2001; Chen et al., 1998; Koolstra and van Eijden, 2005, 2006; Perez del Palomar and Doblare, 2006b, 2008; Reina et al., 2007; Tanaka et al., 2000). Chen et al. (1998) performed stress analysis of human TMJ using a two-dimensional FE model developed from magnetic resonance imaging (MRI). Due to convex nature of the condyle, the compressive stresses were dominant in the condylar region whereas the tensile stresses were dominant in the fossa-eminence complex owing to its concave nature. Beek et al. (2001) developed a 3D linear FE model and analyzed the biomechanical reactions in the mandible and in the TMJ during clenching under various restraint conditions. Nagahara et al. (1999) developed a 3D linear FE model and analyzed the biomechanical reactions in the mandible and in the TMJ during clenching under various restraint conditions. All these FE simulations considered symmetrical movements of mandible, and the models developed only considered one side of the joint. Hart et al. (1992) generated 3D FE models of a partially edentulated human mandible to calculate the mechanical response to simulated isometric biting and mastication loads. Vollmer et al. (2000) conducted experimental and finite element study of human mandible to investigate its complex biomechanical behavior. Tanaka et al. (2001, 2004) developed a 3D model to investigate the stress distribution in the TMJ during jaw opening, analyzing the differences in the stress distribution of the disc between subjects with and without internal derangement. Tanaka et al. (2008c) suggested, from the results of finite element model of the TMJ based on magnetic resonance images, that increase of the frictional coefficient between articular surfaces may be a major cause for the onset of disc displacement. Sellers and Crompton, (2004) used sensitivity analysis to validate the predictions of 3D FE simulations.

In 2005, Koolstra and van Eijden developed a combination of rigid-body model with a FE model of both discs and the articulating cartilaginous surfaces to simulate the opening movement of the jaw. Using the same model, Koolstra and van Eijden (2006) performed FEA to study the load-bearing and maintenance capacity of the TMJ. The results indicated that the construction of the TMJ permitted its cartilaginous structures to regulate their mechanical properties effectively by imbibitions, exudation and redistribution of fluid. Perez-Palomar and Doblare (2006a) used more realistic FE models of both TMJs and soft components to study clenching of mandible. Perez del Palomar and Doblare (2006b) developed a 3D FE model that included both discs ligaments and the three body contact between all elements of the joints, and analyzed biomechanical behavior of the soft components during a nonsymmetrical lateral excursion of the mandible to investigate possible consequences of bruxism. This study suggested that a continuous lateral movement of the jaw may lead to perforations in the lateral part of both discs, conforming to the indications by Tanaka et al. (2001; 2004). Later, in 2007, Perez del Palomar and Doblare suggested that unilateral internal derangement is a predisposing factor for alterations in the unaffected TMJ side. However, it would be necessary to perform an exhaustive analysis of bruxism with the inclusion of contact forces between upper and lower teeth during grinding.

Whiplash injury is considered as a significant TMD risk factor and has been proposed to produce internal derangements of the TMJ (Kasch et al., 2002; Perez del Palomar and Doblare, 2008). However, this topic is still subject to debate (Detamore et al., 2007). In 2008, Perez del Palomar and Doblare, published the results of finite element simulations of the dynamic response of TMJ in rear-end and frontal impacts to predict the internal forces and deformations of the joint tissues. The results, similar to suggested by Kasch et al. (2002), indicated that neither a rear-end impact at low-velocity nor a frontal impact would produce damage to the soft tissues of the joint suggesting that whiplash actions are not directly related with TMDs. However; since this study has its own limitations such as analysis of only one model, for low-velocity impacts, without any restrictions like contact with some component of the vehicle; there is a need for more reliable finite element simulations to obtain more accurate numerical results.

A theoretical model developed by Gallo et al. (2000) for estimating the mechanical work produced by mediolateral stress-field translation in the TMJ disc during jaw opening/closing suggested that long-term exposure of the TMJ disc to high work may result in fatigue failure of the disc. In 2001, Gross et al. proposed a predictive model of occlusal loading of the facial skeleton while May et al. (2001) developed a mathematical model of the TMJ to study the compressive loading during clenching. Effect of mandibular activity on mechanical work in the TMJ, which produces fatigue that may influence the pathomechanics of degenerative disease of the TMJ, was studied by Gallo et al. (2006). Nickel et al. (2002) validated numerical model predictions of TMJ eminence morphology and muscle forces, and demonstrated that the mechanics of the craniomandibular system are affected by the combined orthodontic and orthognathic surgical treatments. Using this validated numerical model to calculate ipsilateral and contralateral TMJ loads for a range of biting positions and angles, Iwasaki et al. (2009) demonstrated that TMJ loads during static biting are larger in subjects with TMJ disc displacement compared to subjects with normal disc position.

4.3 Post-surgery assessment

TMJ reconstruction using the partial or total TMJ prosthetics, in most cases, improves range of motion and mouth opening in the TMJ patients. However, loss of translational movements of the mandible on the operated side has been often observed, especially in anterior direction, owing to various factors like loss of pterygoid muscle function, scarring of the joint region and the muscles of mastication (Yoon et al., 2007). Komistek et al. (1998) assessed in-vivo kinematics and kinetics of the normal, partially replaced, and totally replaced TMJs. Less translation was reported in the implanted fossa and total TMJs than in the normal joints. The study suggests that total TMJ implants only rotate and do not translate; and the muscles do not apply similar forces at the joint when the subject has a total TMJ implant, compared to a subject who has a normal, healthy TMJ.

In the post TMJ replacement follow-up studies, Mercuri et al. (2008) obtained the measures of mandibular interincisal opening and lateral excursions. The assessment showed a 24% and a 30% improvement in mouth opening after 2 years and 10 years, respectively. On the other hand, at 2 years post-implantation there was a 14% decrease in left lateral excursion and a 25% decrease in right lateral excursion from the pre-implantation data. As the loss of lateral jaw movement is a great disadvantage to total TMJ prosthesis replacement, a future prosthesis must allow some lateral translation as well as the anterior movement of mandible on the operated side when the mouth is opened (van Loon et al., 1995). Yoon et al., (2007) followed a kinematic method that tracked the condylar as well as incisors path of the TMJ motion. An electromagnetic tracking device and accompanying software were used to record the kinematics of the mandible relative to temporal bone during opening-closing, protrusive, and lateral movements (Yoon et al., 2007). Mean linear distance (LD) of incisors during maximal mouth opening for the surgical patient group was 18% less than the normal subjects. Mean LD for mandibular right and left condyles was symmetrical in the normal group; however, in the surgical patient group, measurements for operated condyle and unoperated condyle were asymmetric and reduced as compared with normal subjects by 57% and 36%, respectively (Yoon et al., 2007). In protrusive movements, operated and unoperated condyles of surgical patients traveled less and significantly differently as compared with condyles of normal subjects, which moved almost identically. For the surgical patient group, the mean incisor LD away from the operated side and toward the operated side as compared with the normal group incisors were reduced by 67% and 32%, respectively (Yoon et al., 2007).

5. Anatomical modeling and finite element analysis

The TMJ and associated components of masticatory system represent a complicated combination of several muscles and a mandible supported by two interlinked joints. Relations between muscle tensions, jaw motions, bite and joint forces, and craniofacial morphology are not fully understood, and critical information is often difficult to obtain by conducting experiments on living humans (Langenbach et al., 2002; Pileicikiene et al., 2007). Hence the mechanical forces, their distribution and impact in the TMJ and its associated structures cannot be measured directly in a non-destructive way. Therefore, to study mechanical behavior of the TMJ and attached artificial devices - to better understand the form and function, and to improve the design and performance of the prosthetic devices –, it is necessary to create an anatomically viable representation of the mandible, the TMJ and its associated structures. The TMJ surgeons, clinicians, and patient community have collectively expressed great interest in understanding the forces associated with translation, chewing, clenching, etc. Anatomical 3D models can be used to determine the relationships between the masticatory forces and the performance of the natural and/or reconstructed TMJ. The patient-specific force models would be highly valuable for comparison of pre- and post-operative conditions, and also to obtain data from people with healthy TMJ as a baseline group (Detamore et al., 2007). Our research focuses on developing computerized 3D models from medical images of the mandible and TMJs of men and women of different age groups. FEA of these models can provide useful information about contact stresses that possibly contribute to dysfunction of the mandible and the TMJ. Patient-specific FEMs are expected to add another dimension to TMD diagnosis, which is currently based on clinical, radiographic and morphological evaluations (Singh and Detamore, 2009).

5.1 Modeling approaches

Determining the actual shape of the TMJ components through medical images greatly increases the accuracy of the model. We tried two approaches for 3D reconstruction of mandible and the TMJ from computed tomography (CT) images. In the first method, a software tool, MATLAB, was used for image processing. The MATLAB code was developed in such a way that it converts the original gray scale CT images into binary images thus separating the region of interest from rest of the data in the images (see Figure 3). The MATLAB code, then, finds the co-ordinates of the boundary pixels of the region of interest in each slice of the scan. These co-ordinates were imported into another software package, ANSYS, to plot contours corresponding to each CT slice and, subsequently, to develop a 3D model by connecting the consecutive contours to form closed areas and, subsequently, the closed volume mesh (see Figure 4). This modeling approach is very time consuming and involves a lot of manual tasks for image processing and further modeling. Accuracy of image processing is affected when the CT images have scatter due to dental implants. This requires making approximations about the actual shape of the object of interest.

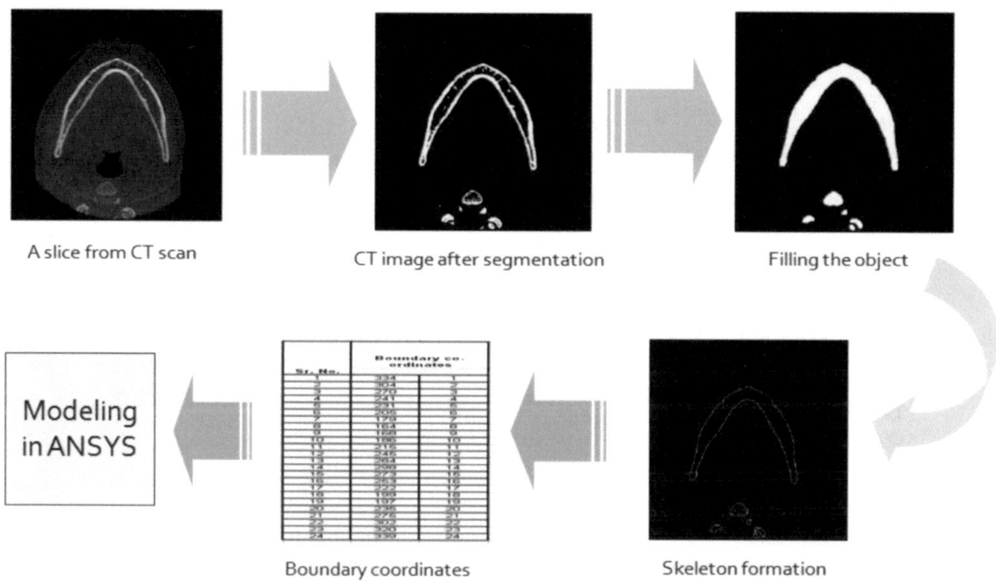

Fig. 3. Processing the CT images in MATLAB. Each gray-scale image (slice) in the scan is converted into a binary image after segmentation. After performing series of morphological operations to form the skeleton of the feature of interest (i.e., mandible in this example), the code returns the co-ordinates of the boundary pixels of the skeleton. These co-ordinates are then exported to ANSYS to create a 3D representation of the object of interest.

Due to the time consuming procedures and inaccuracies in the resultant models in the first approach, later we used a 3D modeling software Mimics® (Materialise, Ann Arbor, MI). Using Mimics one can translate CT or MRI data into complete 3D models for a variety of applications. Mimics® interactively reads CT/MRI data in the DICOM format. Once an area of interest is separated, it can be visualized in 3D. The segmentation task is made easier due to the ability to see the images in three different views: axial, sagittal, and coronal. We developed several subject-specific models of the mandible and TMJ. Improper segmentation of the medical images during reconstruction as well as less than optimal quality of medical images used for modeling hampers quality of the 3D models. Shorter the inter-slice distance in the medical images, better is the quality of resultant model. The inter-slice distance for the CT scans used to develop model 1 (see Figure 5) is 2 mm while that for the CT scans used for model 2 (see Figure 8) is 0.67mm.

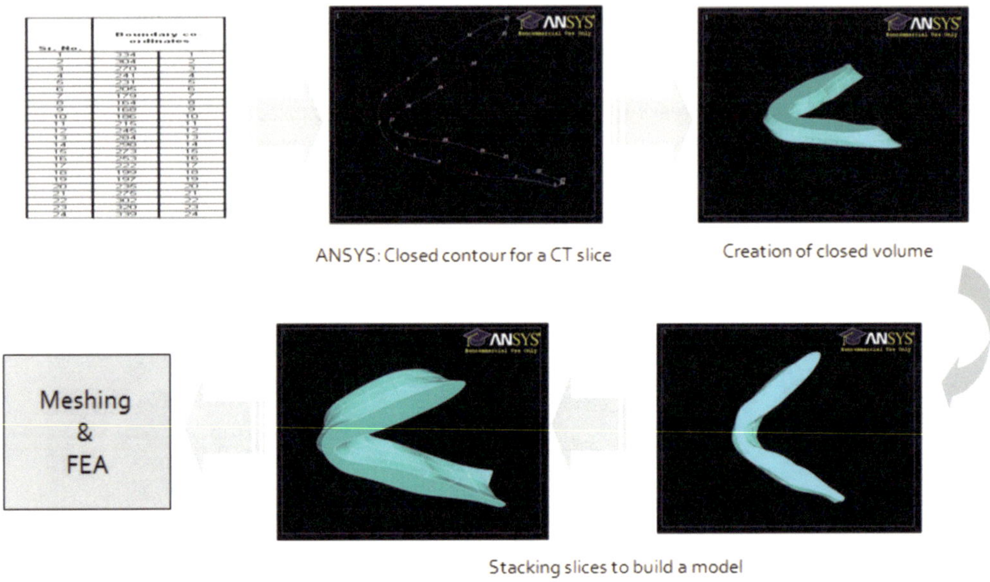

Fig. 4. The co-ordinates for each CT slice are imported in ANSYS (with the z-co-ordinate = the slice thickness) and plotted manually to form a contour that represents shape of the object in the CT slice. After plotting such contours for all slices, the consecutive slices are connected to form the solid model. Such a model can be meshed and used for FEA.

5.2 Model 1 - FEA

A subject-specific 3D model of mandible was developed in Mimics using CT data (see Figure 5). A surface mesh was formed from this solid model using 7074 triangular elements. More the number of elements, more exact is the FEA solution. However, the large number of elements means the model requires higher computing power and more time to run the FEA simulations. Therefore, we try to reduce the number of elements to an appropriate extent in such a way that the quality of the elements and the accuracy of the estimated FEA solution are not affected by the reduction in number of elements. In this process, it is made sure that the mesh has more elements in the areas of complex geometry.

Estimating the stresses occurring over the mandible and the TMJ during different bite patterns and bite forces can be useful in understanding the function of joint and the possible mechanism of TMDs. The 3D surface mesh of mandible was imported into ANSYS to investigate comparative stress development and distribution in the mandible as a result of bite forces during four different loading conditions: normal/balanced occlusion versus three parafunctional loading conditions - which are believed to contribute to the TMDs - unbalanced loading, teeth grinding (bruxism), and teeth clenching. Since von Mises failure criterion has been widely used for mechanical testing of the ductile materials and bone, we considered von Mises stress to assess stress profile of the mandible. Linear and isotropic material properties were assigned to the solid model. The Young's modulus of 15 GPa and Poisson's ratio of 0.3 were selected (Korioth and Versluis, 1997). The model was fixed at both the condylar heads. Ideally, for the condylar heads, some anterior-posterior and mediolateral displacement, and rotation should be allowed. The magnitudes for bite forces were selected based on the literature (Pizolato et al., 2007) and authors' judgment from discussions with clinicians.

In all loading conditions, maximum von Mises stress was observed at the condylar head, a component of the TMJ. FEA results showed the least maximum von Mises stress during balanced loading of the mandible. The maximum von Mises stress of increasing order were observed for unbalanced loading, teeth grinding, and clenching respectively (see Table 1 and Figure 6). Higher stresses were observed over the condylar region compared to the rest of the mandible. Overall, the results indicate two features: considerably more stress development at the condylar head; and relatively higher stress at the condylar head during unbalanced loading, bruxism, and clenching compared to the loading during balanced bite forces.

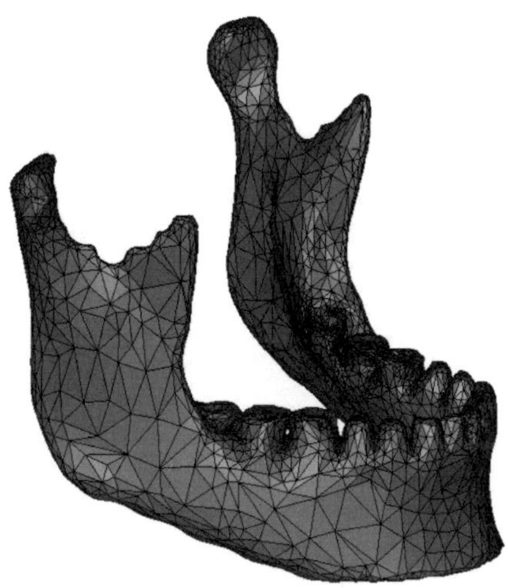

Fig. 5. 3D finite element surface mesh, with 7074 triangular elements, of a subject-specific mandibular model.

Loading Condition	Applied load (N)		Max. von Mises Stress (Pa)	Location of Max. von Mises stress
	Left side	Right side		
Balanced	400	400	0.884E+05	Right condylar head
Unbalanced	250	400	2.30E+05	Right condylar head
Teeth grinding	400(vertical), 300(transverse)	500(vertical), 300(transverse)	2.79E+05	Left condylar head
Clenching	600	600	8.96E+05	Right condylar head

Table 1. The maximum von Mises stress on the mandible for different loading conditions.

Two more FEA simulations were performed using the same 3D model with the same loading and boundary conditions; but Young's modulus of 10 GPa and 7 GPa. This was done to see if the bone quality has any effect on the stress development in mandible and, especially, the condylar head - a TMJ component. Both of these simulations resulted in the least and highest maximum stress on the condylar head during balanced loading and teeth clenching, respectively, in accordance with the first simulation (see Figure 7). However, in contradiction with the previous simulation, the new simulations showed lower maximum von Mises stress during teeth grinding than that during unbalanced loading.

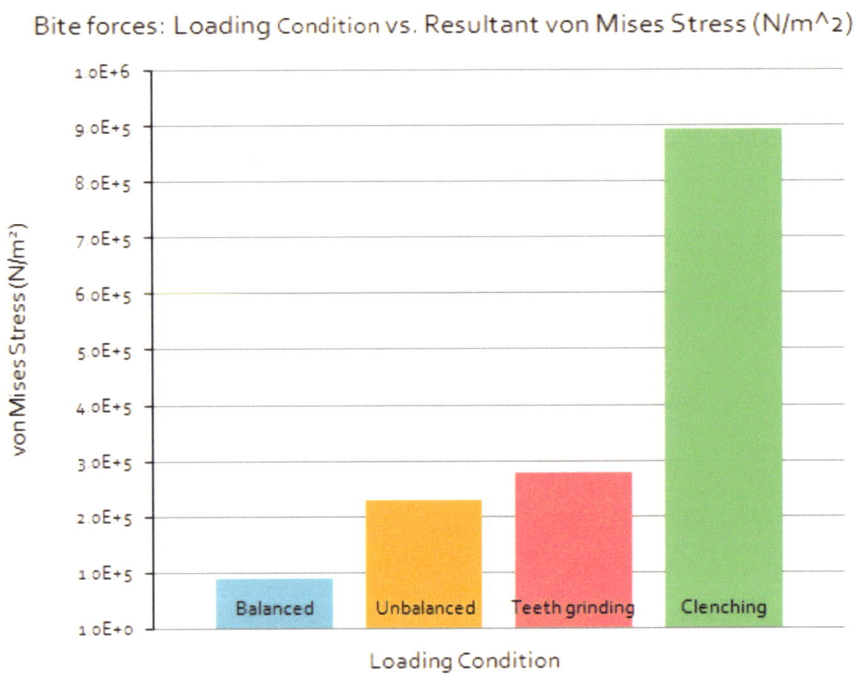

Fig. 6. Maximum von Mises stress developed over the mandibular 3D model during finite element simulation of teeth loading under four different bite conditions.

5.3 Model 2 - FEA

The second subject-specific anatomical 3D model of the mandible was developed in Mimics® from CT scan of a subject, aged 54 years, who reported moderate and intermittent pain in both TMJs. The CT images had ultra-high resolution with inter-slice thickness of 0.67mm. After importing the CT images in Mimics®; independent masks were created each for the cortical bone, cancellous bone, teeth, and articular fibrocartilage. After calculating 3D equivalent of the mandible, a volume mesh was generated using 37439 nodes and 23156 ten-node quadratic tetrahedral elements of type C3D10 (see Figures 8 and 9). Appropriate material properties were assigned to each component of the mandible using corresponding masks (see Table 2). The mandibular 3D volume mesh was, then, exported to a software package ABAQUS® (version 6.8) to perform comparative stress investigation in condylar cartilage under different loading conditions as in case of model-1.

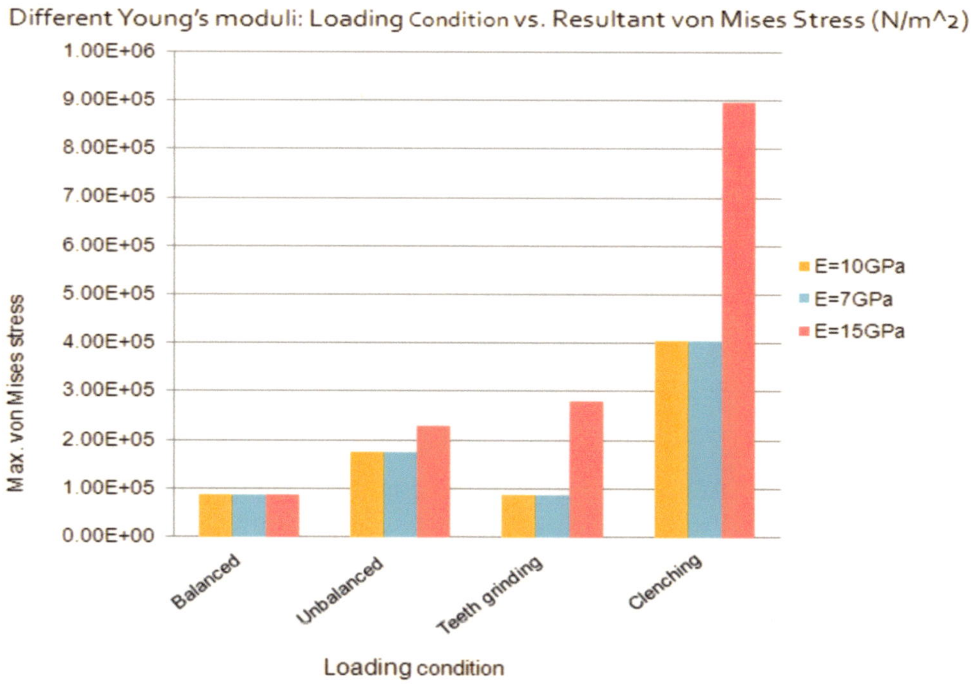

Fig. 7. Maximum von Mises stress developed over the mandibular 3D model during three FE simulations - for three values of Young's modulus (E) - under four loading conditions.

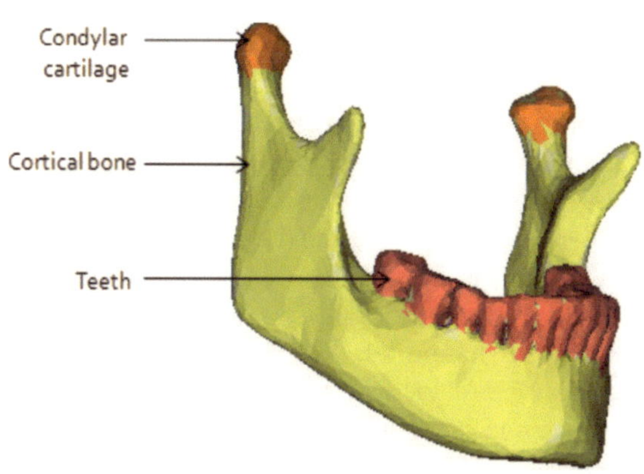

Fig. 8. Material properties were assigned to the 3D finite element volume mesh of the mandible using individual masks for each component. The cortical bone portion is indicated by yellow color, condylar cartilage by orange color, and teeth by red. As cancellous bone is covered by cortical bone, it is not visible in this figure.

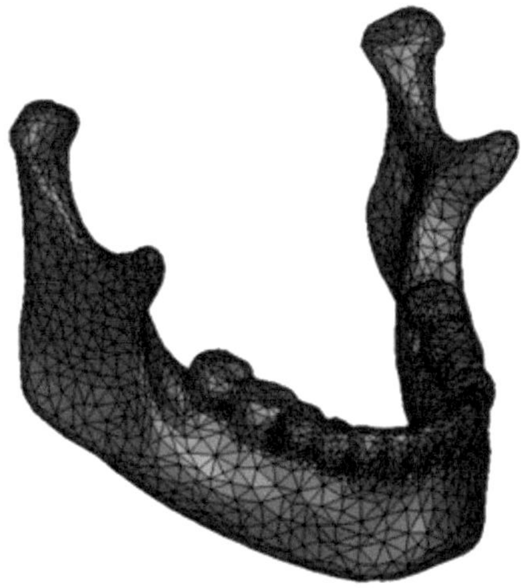

Fig. 9. Three-dimensional finite element volume mesh of the mandible. The volume mesh had 37439 nodes and 23156 ten-node quadratic tetrahedral elements (C3D10).

Part	Young's Modulus (MPa)[a, b]	Poisson's Ratio [a, b]
Cortical bone	1.47E+04	0.3
Cancellous bone	4.90E+02	0.3
Teeth	1.76E+04	0.25
Cartilage	6.1	0.49

Sources: [a]Ichim et al., 2006; [b]Reina et al., 2007

Table 2. Material properties assigned to different components of the mandibular FE model.

The mechanical behavior of the mandibular model was assumed to be linear-elastic, homogeneous, and isotropic. The model constraints were applied to imitate the in-vivo movements of the mandible as accurately as possible during each loading condition. Since the mastication forces are the result of the pressure in the teeth-food contact (Reina et al., 2007), the displacements were simply restrained at the nodes of the surface of the lower teeth that come in contact with the food or the upper teeth. During the balanced occlusive loading, both condyles were permitted translation of 10 mm in anterior-posterior direction and rotation of 11° along the medio-lateral axis. Same constraints were employed to simulate the unbalanced occlusive loading and bi-lateral molar clenching. During teeth grinding, the forces were applied on first and second molars and second premolar on right side only; and the right condyle was assumed free to move while the articular surface of the left condyle was constrained as during balanced loading.

The magnitudes of mandibular and TMJ loading reported in the literature differ significantly and there is currently no universally agreed upon value of TMJ loading (Ingawalé and Goswami, 2009). Conflicting views about type, magnitude, and orientation of masticatory forces used for FEMs were expressed by TMJ researchers at the TMJ

Bioengineering Conference, 2009 held at Boulder (CO, USA). Therefore, for this study, we selected the magnitudes of bite forces based on the literature and our discussions with dentists, and oral and maxillofacial surgeons (see Table 3). For balanced load simulation, 200N force was applied in vertically upward direction on the second molar on both sides of the mandible. During parafunctional activities, loading conditions are different from those under normal loading (Singh and Detamore, 2009). To simulate unbalanced loading, same location and orientation of force were used with 250N on the left second molar and 200N on the right second molar of the mandible. During teeth grinding, the bite forces - 350N vertically upward and 250N in medial direction - were applied on first and second molars and second premolar on only the right side of the mandible. Mandibular loading during clenching was simulated by applying 400N vertically upward bite force on all molars and premolars on both sides of the mandible.

Since material properties were assigned to the mandibular 3D mesh from independent masks for cortical bone, cancellous bone, teeth, and condylar cartilage; it was possible to investigate stress development in each of these components as well as the entire mandible. As the objective of this study is to study stress development in the articulating surfaces (condylar cartilage), we discuss the von Mises stresses in condylar cartilage hereon. Each loading condition was simulated thrice with the same model constraints, and location and magnitude of forces. These simulations are named Run1, Run2 and Run3. Applied bite forces and resultant maximum von Mises stresses in the condylar cartilage for all loading conditions are summarized in Table 3 (also see Figures 10 and 11).

Loading condition	Applied load (N)[a, b, c]		Max. von Mises Stress in condylar cartilage (*E+04 Pa)			Location of max. von Mises stress on condylar cartilage
	Left side	Right side	Run1	Run2	Run3	
Balanced	200	200	5.9	5.8	5.88	Right condyle
Unbalanced	250	150	5.97	5.97	5.9	Left condyle
Teeth grinding	---	350(vertically upward), 250 (medial)	7.23	7.2	7.21	Right condyle
Clenching	400	400	10.3	10.1	10.3	Right condyle

Table 3. Applied bite forces and resultant maximum von Mises stress in condylar cartilage
Source: [a]Abe et al., 2006; [b]Cosme et al., 2005; [c]Authors' discussions with Oral and Maxillofacial Surgeons.

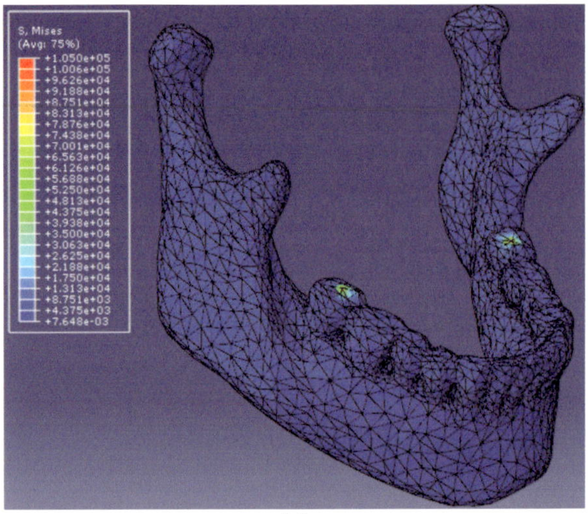
S, Mises
(Avg: 75%)
+1.050e+05
+1.006e+05
+9.626e+04
+9.188e+04
+8.751e+04
+8.313e+04
+7.876e+04
+7.438e+04
+7.001e+04
+6.563e+04
+6.126e+04
+5.688e+04
+5.250e+04
+4.813e+04
+4.375e+04
+3.938e+04
+3.500e+04
+3.063e+04
+2.625e+04
+2.188e+04
+1.750e+04
+1.313e+04
+8.751e+03
+4.375e+03
+7.648e-03

(a)

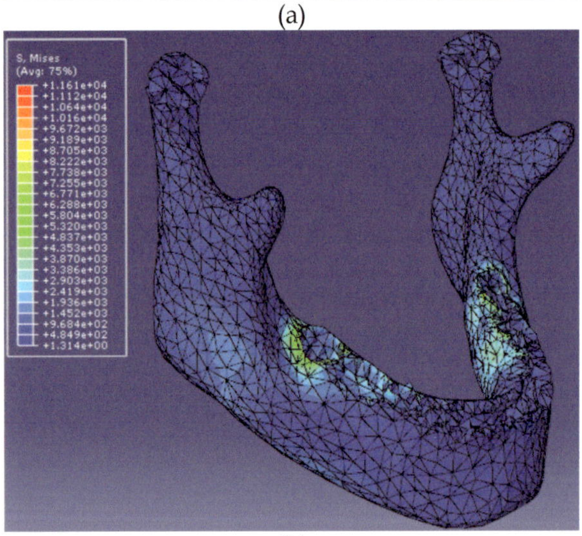
S, Mises
(Avg: 75%)
+1.161e+04
+1.112e+04
+1.064e+04
+1.016e+04
+9.672e+03
+9.189e+03
+8.705e+03
+8.222e+03
+7.738e+03
+7.255e+03
+6.771e+03
+6.288e+03
+5.804e+03
+5.320e+03
+4.837e+03
+4.353e+03
+3.870e+03
+3.386e+03
+2.903e+03
+2.419e+03
+1.936e+03
+1.452e+03
+9.684e+02
+4.849e+02
+1.314e+00

(b)

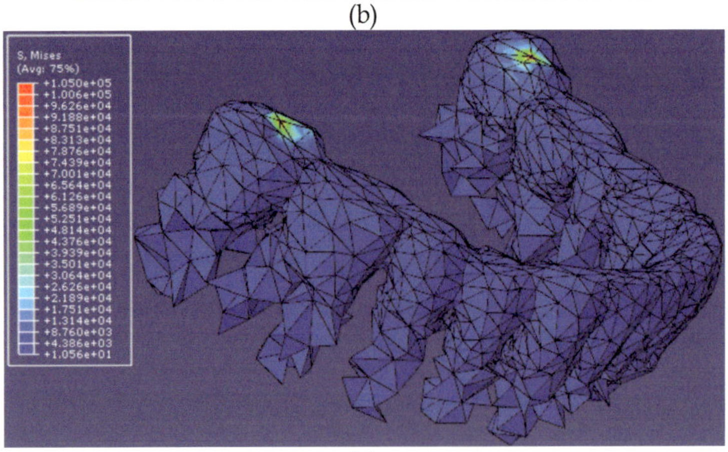
S, Mises
(Avg: 75%)
+1.050e+05
+1.006e+05
+9.626e+04
+9.188e+04
+8.751e+04
+8.313e+04
+7.876e+04
+7.439e+04
+7.001e+04
+6.564e+04
+6.126e+04
+5.689e+04
+5.251e+04
+4.814e+04
+4.376e+04
+3.939e+04
+3.501e+04
+3.064e+04
+2.626e+04
+2.189e+04
+1.751e+04
+1.314e+04
+8.760e+03
+4.386e+03
+1.056e+01

(c)

(d)

Fig. 10. von Mises stress [in (kg.mm/s^2); (1 kg.mm/s^2 = 1 kPa)] developed during balanced bilateral molar bite simulation in the entire mandible (a); and its components - cortical bone (b), teeth (c), and condylar articulating cartilage (d). (Note: The displayed sizes of components in panels c and d are not in proportion to each other and that of the components in other panels).

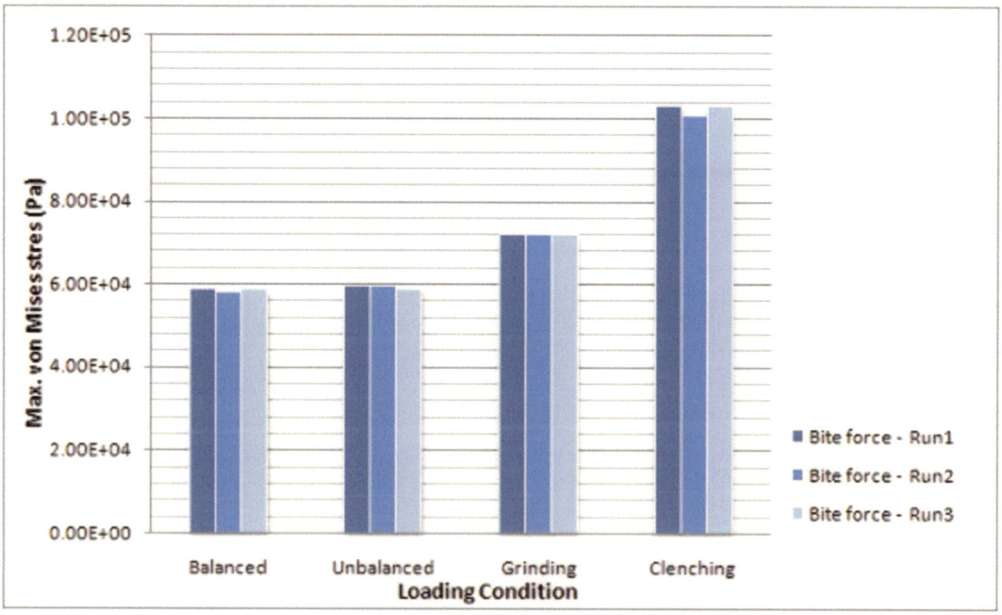

Fig. 11. A plot of maximum von Mises stress developed in the condylar articulating cartilage during four different occlusal static loading conditions - balanced molar bite, unbalanced molar bite, teeth grinding, and clenching - simulated thrice each. The FE simulations resulted in the highest mechanical stresses in the condylar cartilage during teeth clenching. Teeth grinding resulted in the mechanical stresses relatively less than during clenching, and higher than during unbalanced and balanced molar bites. The balanced loading produced the least stresses among all simulations.

The resultant stress data were analyzed using statistical analysis software JMP® (version 9). We employed the Tukey-Kramer HSD method to investigate the correlation between means of the peak von Mises stresses from three simulations/runs each of the four loading conditions under bite forces. From Tukey-Kramer HSD method, by comparing means of peak von Mises stresses for three runs/simulations of each loading condition, teeth grinding and clenching were found to result in significantly different (p-value <0.0001 at $\alpha = 0.05$) and higher von Mises stresses than balanced loading (see Figure 8). The von Mises stresses due to balanced and unbalanced loading were not significantly different from each other (at $\alpha = 0.05$, p-value = 0.4386).

Oneway Analysis of Bite Force: Max. von Mises Stress (*E+04 Pa) By Loading Condition

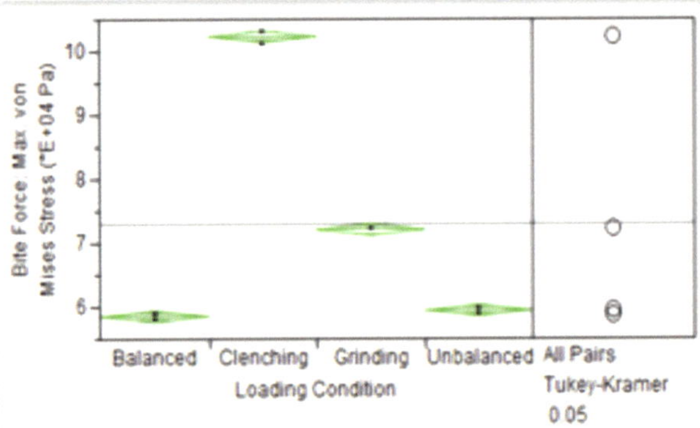

Means Comparisons

Comparisons for all pairs using Tukey-Kramer HSD

q*	Alpha
3.20234	0.05

Abs(Dif)-HSD

	Clenching	Grinding	Unbalanced	Balanced
Clenching	-0.1754	2.8446	4.1113	4.1979
Grinding	2.8446	-0.1754	1.0913	1.1779
Unbalanced	4.1113	1.0913	-0.1754	-0.0887
Balanced	4.1979	1.1779	-0.0887	-0.1754

Positive values show pairs of means that are significantly different.

Level		Mean
Clenching	A	10.233333
Grinding	B	7.213333
Unbalanced	C	5.946667
Balanced	C	5.860000

Levels not connected by same letter are significantly different.

Fig. 12. The Tukey-Kramer HSD statistical analysis by comparing means of maximum von Mises stresses for three runs of each loading condition revealed that teeth grinding and clenching resulted in statistically significantly different von Mises stresses than balanced loading. The von Mises stresses due to balanced and unbalanced loading were not significantly different from each other.

The resultant maximum von Mises stresses in the condylar cartilage during balanced loading and clenching lie in the range of those reported in the literature (Hu et al., 2003; Nagahara et al., 1999). However, since most of the studies have reported stress development in bones and disc of the TMJ, we could not find any reported values of stress in the condylar cartilage under unbalanced loading and teeth grinding conditions to compare our results with. Comparatively higher mechanical stresses during clenching and teeth grinding activities suggest that these activities may lead to and exacerbate the TMDs. This indication

of our study conforms to the attribution that teeth grinding and clenching (as a result of physical and/or psychological stress) may be the causative factors for TMDs.
Since we have applied the model constraints, material properties, and load values based on the literature, we consider the FEA results to be reliable and encouraging to advance our research efforts. We recognize that our FEA method has some limitations because we used simplified forces. We are developing subject-specific 3D models of the entire TMJ - including hard and soft tissues, and more refined FE mesh to perform biomechanical investigation under more realistic forces and model constraints. The proposed work promises to lead us to better understanding of the structural and functional aspects of natural and reconstructed TMJ. We also plan to validate the theoretical predictions of FEA through cadaver testing.

6. Summary

The TMJ literature underlines the importance of biomechanical analysis of the natural joint to better understand the structural and functional aspects; and of the reconstructed joint to assess the implant function and performance. Most of the methods reported in the literature have certain limitations due to the complex nature of the joint and also due to certain limitations of the techniques and software packages used for modeling and analysis. A more comprehensive biomechanical analysis of the natural and artificial TMJ is essential. The methodology used in this study for anatomical 3D reconstruction enables subject-specific modeling of complex structures and their constituent components. This feature can play a vital role in patient-specific anatomical modeling for diagnostic as well as therapeutic needs. Furthermore, such subject-specific anatomical models can be used to design custom prosthetic devices - which offer better fit, fixation, and efficiency - for a given anatomical structure. The FEA of such anatomical and prosthetic 3D models can be efficiently employed to better understand biomechanical behavior of the complex structures under investigation; and to improve the design, treatment efficiency, and durability of prosthetic devices. More comprehensive static and dynamic analyses of the mandible and TMJ coupled with experimental validation are necessary.

7. Acknowledgement

The authors would like to thank Dr. Deepak Krishnan (Assistant Professor, Oral and Maxillofacial Surgery, University of Cincinnati, Cincinnati, OH, USA) for sharing with us his clinical expertise and guiding our TMJ research.

8. References

Abe, M., Medina-Martinez, R. U., Itoh, K., Kohno, S., 2006. Temporomandibular joint loading generated during bilateral static bites at molars and premolars. Medical and Biological Engineering and Computing 44, 1017-1030.

Alomar, X., Medrano, J., Cabratosa, J., Clavero, J., Lorente, M., Serra, I., Monill, J., Salvador, A., 2007. Anatomy of the temporomandibular joint. Seminars in Ultrasound, CT, and MRI 28, 170-183.

American Association of Oral and Maxillofacial Surgeons (AAOMS), 2007. The temporomandibular joint (TMJ). Retrieved on 10/14/2007, from http://www.aaoms.org/tmj.php, 1.

Bakke, M., Zak, M., Jensen, B. L., Pedersen, F. K., Kreiborg, S., 2001. Orofacial pain, jaw function, and temporomandibular disorders in women with a history of juvenile chronic arthritis or persistent juvenile chronic arthritis Oral Surgery, Oral Medicine, Oral Pathology, Oral Radiology, and Endodontics 92, 406-414.

Beek, M., Aarnts, M. P., Koolstra, J. H., Feilzer, A. J., Van Eijden, T. M. G. J., 2001. Dynamic Properties of the Human Temporomandibular Joint Disc. Journal of Dental Research 80, 876-880.

Beek, M., Koolstra, J. H., van Eijden, T. M. G. J., 2003. Human temporomandibular joint disc cartilage as a poroelastic material Clinical Biomechanics 18, 69-76.

Beek, M., Koolstra, J. H., van Ruijven, L. J., van Eijden, T. M. G. J., 2001. Three-dimensional finite element analysis of the cartilaginous structures in the human temporomandibular joint. Journal of Dental Research 80, 1913-1918.

Breul, R., Mall, G., Landgraf, J., Scheck, R., 1999. Biomechanical analysis of stress distribution in the human temporomandibular-joint Annals of Anatomy 181, 55-60.

Chen, J., Akyuz, U., Xu, L., Pidaparti, R. M. V., 1998. Stress analysis of the human temporomandibular joint. Medical Engineering and Physics 20, 565-572.

Cosme, D. C., Baldisserotto, S. M., Canabarro Sde, A., Shinkai, R. S., 2005. Bruxism and voluntary maximal bite force in young dentate adults The International Journal of Prosthodontics 18, 328-332.

Detamore, M. S., Athanasiou, K. A., 2003. Structure and function of the temporomandibular joint disc: implications for tissue engineering Journal of Oral and Maxillofacial Surgery 61, 494-506.

Detamore, M. S., Athanasiou, K. A., Mao, J., 2007. A call to action for bioengineers and dental professionals: directives for the future of TMJ bioengineering Annals of Biomedical Engineering 35, 1301-1311.

Dutton, M., 2004. Orthopaedic Examination, Evaluation, & Intervention: A Pocket Handbook. McGraw-Hill, New York, pp548.

Gallo, L. M., Chiaravalloti, G., Iwasaki, L. R., Nickel, J. C., Palla, S., 2006. Mechanical work during stress-field translation in the human TMJ. Journal of Dental Research 85, 1006-1010.

Gallo, L. M., Nickel, J. C., Iwasaki, L. R., Palla, S., 2000. Stress-field translation in the healthy human temporomandibular joint. Journal of Dental Research 79, 1740-1746.

Glaros, A. G., Owais, Z., Lausten, L., 2007. Reduction in parafunctional activity: a potential mechanism for the effectiveness of splint therapy. Journal of Oral Rehabilitation 34, 97-104.

Gross, M. D., Arbel, G., Hershkovitz, I., 2001. Three-dimensional finite element analysis of the facial skeleton on simulated occlusal loading. Journal of Oral Rehabilitation 28, 684-694.

Hansdottir, R., Bakke, M., 2004. Joint tenderness, jaw opening, chewing velocity, and bite force in patients with temporomandibular joint pain and matched healthy control subjects. Journal of Orofacial Pain 18, 108-113.

Hart, R. T., Hennebel, V. V., Thongpreda, N., Van Buskirk, W. C., Anderson, R. C., 1992. Modeling the biomechanics of the mandible: a three-dimensional finite element study. Journal of Biomechanics 25, 261-286.

Hibi, H., Ueda, M., 2005. Body posture during sleep and disc displacement in the temporomandibular joint: a pilot study. Journal of Oral Rehabilitation 32, 85-89.

Hu, K., Qiguo, R., Fang, J., Mao, J. J., 2003. Effects of condylar fibrocartilage on the biomechanical loading of the human temporomandibular joint in a three-dimensional, nonlinear finite element model. 25, 107-113.

Huddleston Slater, J. J. R., Visscher, C. M., Lobbezoo, F., Naeije, M., 1999. The Intra-articular Distance within the TMJ during Free and Loaded Closing Movements. Journal of Dental Research 78, 1815-1820.

Hylander, W. L., 1979. An experimental analysis of temporomandibular joint reaction force in macaques. American Journal of Physical Anthropology 51, 433-456.

Ichim, I., Swain, M., Kieser, J. A., 2006. Mandibular biomechanics and development of the human chin. Journal of Dental Research 85, 638-642.

Ide, Y., Nakazawa, K., Garcia, L. T., 1991. Anatomical atlas of the temporomandibular joint. Quintessence Publ. Co., Chicago, pp. 116.

Ingawalé, S., Goswami, T., 2009. Temporomandibular joint: disorders, treatments, and biomechanics. Annals of Biomedical Engineering 37, 976-996.

Iwasaki, L. R., Crosby, M., Gonzalez, Y., McCall, W. D., Marx, D. B., Ohrbach, R., Nickel, J. C., 2009. Temporomandibular joint loads in subjects with and without disc displacement. Orthopedic Reviews 1, 90-93.

Kalamir, A., Pollard, H., Vitiello, A., Bonello, R., 2007. TMD and the problem of bruxism. A review. Journal of Bodywork and Movement Therapies 11, 183-193.

Kasch, H., Hjorth, T., Svensson, P., Nyhuus, L., Jensen, T. S., 2002. Temporomandibular disorders after whiplash injury: a controlled, prospective study Journal of Orofacial Pain 16, 118-128.

Komistek, R. D., Dennis, D. A., Mabe, J. A., Anderson, D. T., 1998. In vivo kinematics and kinetics of the normal and implanted TMJ. Journal of Biomechanics 31, 13.

Koolstra, J. H., van Eijden, T. M., 2006. Prediction of volumetric strain in the human temporomandibular joint cartilage during jaw movement. Journal of Anatomy 209, 369-380.

Koolstra, J. H., van Eijden, T. M., 2005. Combined finite-element and rigid-body analysis of human jaw joint dynamics Journal of Biomechanics 38, 2431-2439.

Korioth, T. W., Hannam, A. G., 1994. Mandibular forces during simulated tooth clenching Journal of Orofacial Pain 8, 178-189.

Korioth, T. W. P., Versluis, A., 1997. Modeling the mechanical behavior of the jaws and their related structures by finite element (FE) analysis. Critical Reviews in Oral Biology & Medicine 8, 90-104.

Langenbach, G. E. J., Zhang, F., Herring, S. W., Hannam, A. G., 2002. Modelling the masticatory biomechanics of a pig. Journal of Anatomy 201, 383-393.

Laskin, D. M., Greene, C. B., Hylander, W. L., 2006. Temporomandibular disorders an evidence-based approach to diagnosis and treatment. Quintessence Pub., Chicago, pp. 548.

May, B., Saha, S., Saltzan, M., 2001. A three-dimensional mathematical model of temporomandibular joint loading. Clinical Biomechanics 16, 489-495.

Mercuri, L. G., Ali, F. A., Woolson, R., 2008. Outcomes of total alloplastic replacement with periarticular autogenous fat grafting for management of reankylosis of the temporomandibular joint Journal of Oral and Maxillofacial Surgery 66, 1794-1803.

Naeije, M., Hofman, N., 2003. Biomechanics of the Human Temporomandibular Joint during Chewing. Journal of Dental Research 82, 528-531.

Nagahara, K., Murata, S., Nakamura, S., Tsuchiya, S., 1999. Displacement and stress distribution in the temporomandibular joint during clenching. The Angle Orthodontist 69, 372-379.

Nickel, J., Yao, P., Spalding, P. M., Iwasaki, L. R., 2002. Validated numerical modeling of the effects of combined orthodontic and orthognathic surgical treatment on TMJ loads and muscle forces. American Journal of Orthodontics and Dentofacial Orthopedics 121, 73-83.

Pellizoni, S. E., Salioni, M. A., Juliano, Y., Guimaraes, A. S., Alonso, L. G., 2006. Temporomandibular joint disc position and configuration in children with functional unilateral posterior crossbite: a magnetic resonance imaging evaluation American Journal of Orthodontics and Dentofacial Orthopedics 129, 785-793

Perez del Palomar, A., Doblare, M., 2008. Dynamic 3D FE modelling of the human temporomandibular joint during whiplash. 30, 700-709.

Perez del Palomar, A., Doblare, M., 2007. Influence of unilateral disc displacement on the stress response of the temporomandibular joint discs during opening and mastication. Journal of Anatomy 211, 453-463.

Perez del Palomar, A., Doblare, M., 2006a. The effect of collagen reinforcement in the behaviour of the temporomandibular joint disc. Journal of Biomechanics 39, 1075-1085.

Perez del Palomar, A., Doblare, M., 2006b. Finite element analysis of the temporomandibular joint during lateral excursions of the mandible. Journal of Biomechanics 39, 2153-2163.

Pileicikiene, G., Varpiotas, E., Surna, R., Surna, A., 2007. A three-dimensional model of the human masticatory system, including the mandible, the dentition and the temporomandibular joints. Stomatologija, Baltic Dental and Maxillofacial Journal 9, 27-32.

Pizolato, R. A., Gavião, M. B. D., Berretin-Felix, G., Sampaio, A. C. M., Trindade, A. S. J., 2007. Maximal bite force in young adults with temporomandibular disorders and bruxism. Brazilian Oral Research 21, 278-283.

Quinn, P. D., 2000. Lorenz Prosthesis. Oral and Maxillofacial Surgery Clinics of North America 12, 93-104.

Quinn, P. D., 199. Alloplastic Reconstruction of the temporomandibular joint. Selected Readings in Oral and Maxillofacial Surgery 7, 1-23.

Reina, J. M., Garcia-Aznar, J. M., Dominguez, J., Doblare, M., 2007. Numerical estimation of bone density and elastic constants distribution in a human mandible. Journal of Biomechanics 40, 828-836.

Scrivani, S. J., Keith, D. A., Kaban, L. B., 2008. Temporomandibular disorders The New England Journal of Medicine 359, 2693-2705.

Sellers, W., Crompton, R., 2004. Using sensitivity analysis to validate the predictions of a biomechanical model of bite forces. Annals of Anatomy 186, 89-95.

Singh, M., Detamore, M. S., 2009. Biomechanical properties of the mandibular condylar cartilage and their relevance to the TMJ disc. Journal of Biomechanics 42, 405-417.

Tanaka, E., Dalla-Bona, D. A., Iwabe, T., Kawai, N., Yamano, E., van Eijden, T., Tanaka, M., Miyauchi, M., Takata, T., Tanne, K., 2006. The Effect of Removal of the Disc on the Friction in the Temporomandibular Joint. Journal of Oral and Maxillofacial Surgery 64, 1221-1224.

Tanaka, E., Rodrigo, D. P., Miyawaki, Y., Lee, K., Yamaguchi, K., Tanne, K., 2000. Stress distribution in the temporomandibular joint affected by anterior disc displacement: a three-dimensional analytic approach with the finite-element method. Journal of Oral Rehabilitation 27, 754-759.

Tanaka, E., del Pozo, R., Tanaka, M., Asai, D., Hirose, M., Iwabe, T., Tanne, K., 2004. Three-dimensional finite element analysis of human temporomandibular joint with and without disc displacement during jaw opening. Medical Engineering & Physics 26, 503-511.

Tanaka, E., Detamore, M. S., Mercuri, L. G., 2008a. Degenerative disorders of the temporomandibular joint: etiology, diagnosis, and treatment. Journal of Dental Research 87, 296-307.

Tanaka, E., Detamore, M. S., Tanimoto, K., Kawai, N., 2008b. Lubrication of the temporomandibular joint Annals of Biomedical Engineering 36, 14-29.

Tanaka, E., Hirose, M., Koolstra, J. H., van Eijden, T. M., Iwabuchi, Y., Fujita, R., Tanaka, M., Tanne, K., 2008c. Modeling of the effect of friction in the temporomandibular joint on displacement of its disc during prolonged clenching Journal of Oral and Maxillofacial Surgery 66, 462-468.

Tanaka, E., Kikuchi, K., Sasaki, A., Tanne, K., 2000a. An adult case of TMJ osteoarthrosis treated with splint therapy and the subsequent orthodontic occlusal reconstruction: adaptive change of the condyle during the treatment American Journal of Orthodontics and Dentofacial Orthopedics 118, 566-571.

Tanaka, E., Rodrigo, D. P., Tanaka, M., Kawaguchi, A., Shibazaki, T., Tanne, K., 2001. Stress analysis in the TMJ during jaw opening by use of a three-dimensional finite element model based on magnetic resonance images International Journal of Oral and Maxillofacial Surgery 30, 421-430.

Tanaka, E., Rego, E. B., Iwabuchi, Y., Inubushi, T., Koolstra, J. H., van Eijden, T. M. G. J., Kawai, N., Kudo, Y., Takata, T., Tanne, K., 2008. Biomechanical response of condylar cartilage-on-bone to dynamic shear. Journal of Biomedical Materials Research 85A, 127-132.

van Loon, J. P., de Bont, L. G. M., Boering, G., 1995. Evaluation of temporomandibular joint prostheses Review of the literature from 1946 to 1994 and implications for future prosthesis designs. Journal of Oral and Maxillofacial Surgery 53, 984-996.

Vollmer, D., Meyer, U., Joos, U., Vegh, A., Piffko, J., 2000. Experimental and finite element study of a human mandible. 28, 91-96.

Wolford, L. M., 1997. Temporomandibular joint devices: Treatment factors and outcomes. Oral Surgery, Oral Medicine, Oral Pathology, Oral Radiology, and Endodontics 83, 143-149.

Yoon, H. J., Baltali, E., Zhao, K. D., Rebellato, J., Kademani, D., An, K. N., Keller, E. E., 2007. Kinematic study of the temporomandibular joint in normal subjects and patients following unilateral temporomandibular joint arthrotomy with metal fossa-eminence partial joint replacement Journal of Oral and Maxillofacial Surgery 65, 1569-1576.

Part 3

Nano Behavior

8

Design and Analysis of Key Components in the Nanoindentation and Scratch Test Device

Hongwei Zhao, Hu Huang, Jiabin Ji and Zhichao Ma
Jilin University
China

1. Introduction

In recent years, nanomechanics as an important branch of nanotechnology has been an effective method to study mechanical properties of structures and materials from micro- to nano-scale. It has been widely used to study mechanical behaviour and damage mechanism of nanotube, nanobelt as well as other nanostructures (Zhu & Espinosa, 2005; Han et al., 2007). Some researchers studied biomechanics properties of tissues and organs of human body, such as the red cell (Suresh, 2007; Lim et al., 2006) and bone (Tai et al., 2007; Hansma et al., 2006; Thurner, 2009; Zheng & Mak, 1996; Koff et al., 2010; Huja et al., 2010; Diez-Perez et al., 2010; Zhang et al., 2010) by means of experimental nanomechanics approaches, to reveal occurrence rules, prevention and control methods of some diseases.

Compared with tensile, torsion, bend and hardness tests etc. during which samples are usually in condition of simple stress state, nanoindentation and scratch test can obtain more parameters of materials including hardness, modulus, fracture toughness, creep property, fatigue, adhesion and so on (Doerner & Nix, 1986; Oliver & Pharr, 1992) because samples are in condition of complex stress state. So, it has been widely used in fields of materials science (Lucas & Oliver, 1999; Yang et al., 2007; Tao et al., 2010), nanotechnology, surface engineering (Jardret & Morel, 2003), semiconductor (Michler et al., 2005; Zhao et al., 2009), MEMS/NEMS (Abdel-Aal et al., 2005; Bhushan, 2007), biomedicine (Suresh, 2005), biomechanics (Bruet et al., 2008) and so on. In addition, it is a useful method to study multi-physical field coupled performance of materials (Bradby et al., 2003; Schuh et al., 2005; Nowak et al., 2009). So in past years, nanoindentation and scratch test gave a big boost to the development of related fields and also it was given huge attention in all over the world.

With further development of materials science and nanomechanics, more and more researchers tried to study principle of deformation and damage of materials (De Hosson et al., 2006; Rabe, 2006; Zhou & Komvopoulos, 2006). Nanoscale deformation: Seeing is believing (Hemker & Nix, 2008). So research on in situ nanoindentation and scratch test was proposed, through which process of deformation and damage of samples can be observed. However, because of large size and complex structure of existing commercial equipments (MTS NanoIndenter; Hysitron Incorporated; CSIRO.UMIS; Micro Materials Ltd.; CSM instruments), they can not be installed on the stage of SEM or TEM to realize in situ nanoindentation and scratch test. So, novel nanoindentation and scratch test devices are required. And this is advanced technology and up to now there is no mature product for in situ nanoindentation and scratch test.

In this chapter, we introduced principle of nanoindentation. A new method of indentation measurement through two displacement sensors and a displacement amplification structure was proposed. This measuring method was different from many commercial devices. Two key components of the proposed device including precise driving and precise measuring units were designed and analyzed. Hysteresis of two kinds of piezoelectric stacks were measured and analyzed. Here piezoelectric stacks and flexure hinges were used to realize precise positioning and precise loading and unloading of the diamond indenter. Flexure hinges with multi-structure forms were proposed and analyzed by finite element method. Based on the previous work, the prototype of nanoindentation and scratch device was designed and fabricated. Calibration experiments of sensors and displacement amplification structure were carried out and output performances of designed flexure hinges were measured and discussed. At last, nanoindentation experiments of optical glass were carried out. The relation curves between penetration load and depth were obtained, from which hardness of the glass was figured out. Nanoindentation morphology was obtained through high resolution optical microscope. Nanoindentation results indicated that the device presented in this chapter can realize the high precise nanoindentation test, but the testing resolution should be improved and the device also should be calibrated precisely. Though the accuracy was required to improve, this work was a bold attempt to combine the piezoelectric-driven mechanism and flexure hinge to realize precise motion in nanoindentation device. This transmission mode is very simple and can be used easily to realize miniaturization of the device and in situ nanoindentation test which is our future work.

2. Principle of nanoindentation test technology---the Oliver-Pharr method

Nanoindentation is a useful method to test the mechanical behavior and damage mechanism of materials from micro- to nano-scale. When a diamond indenter with sharp tip is penetrating into and then withdrawing from a sample surface, the load P and displacement h is continuously monitored by high resolution sensors. The load and displacement data is sent to processor during the indentation process and then converted to P-h curve which contains abundant information of material such as hardness, elastic modulus, yield stress and so on. Fig.1 is a typical P-h curve of nanoindentation. It mainly

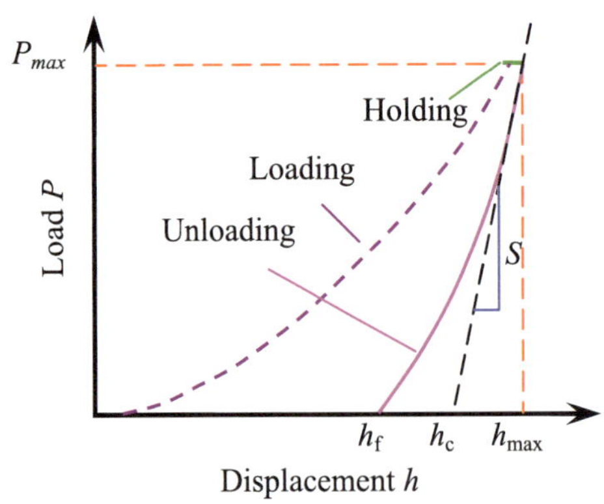

Fig. 1. A typical P-h curve of nanoindentation

consists of two portions, loading and unloading process. Sometimes, in order to study creep of materials, holding portion (holding for a certain time at maximum load) between loading and unloading is added. Fig.2 is cross-section of indentation and the related parameters. Up to now, the Oliver-Pharr method (Oliver & Pharr, 1992) based on results by Sneddon (Sneddon, 1965) is most commonly used to analyze the load and penetration depth for nanoindentation measurements.

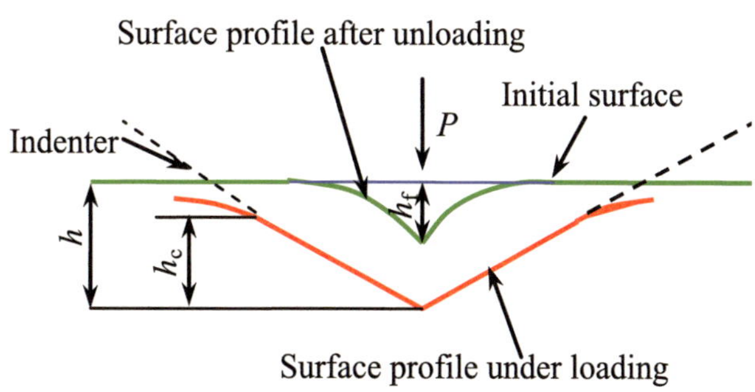

Fig. 2. Cross-section of indentation

According to Oliver-Pharr method, nanoindentation hardness is defined as the indentation load divided by the projected contact area of the indentation. In Fig.1, indentation hardness (H) can be obtained at the peak load given by

$$H = P_{max}/A_c \tag{1}$$

where P_{max} is the peak load and A_c is the projected contact area. And the projected contact area can be calculated from the relation as following

$$A_c = f(h_c) \tag{2}$$

where h_c is the contact depth which is given by

$$h_c = h_{max} - \varepsilon \frac{P_{max}}{S} \tag{3}$$

where ε is a constant and depends on the geometry of the indenter (ε=0.72 for cone indenter, ε=0.75 for paraboloid of revolution, and ε=1.00 for flat indenter) (Sneddon, 1965). h_{max} is the maximum penetration depth and S is the contact stiffness.

The contact stiffness S can be calculated from the slope of the initial portion of the unloading curve and $S=dP/dh$, which can be obtained by curve fitting of 25%-50% unloading data.

Based on relationships developed by Sneddon, the contact stiffness S can also be expressed by

$$S = 2\beta\sqrt{\frac{A}{\pi}}E_r \tag{4}$$

where β is a constant and depends on geometry of the indenter (β =1.034 for a Berkovich indenter, β =1.012 for a Vickers indenter and β =1.000 for a cylinder indenter).
Because both the sample and the indenter have elastic deformation during the indentation process, the reduced modulus E_r is defined by

$$\frac{1}{E_r} = \frac{1-\nu^2}{E} + \frac{1-\nu_i^2}{E_i} \tag{5}$$

where E and ν are the elastic modulus and Poisson's ratio for the sample; E_i and ν_i are the elastic modulus and Poisson's ratio for the indenter, respectively. For a diamond indenter, E_i =1141GPa, ν_i =0.07.
According to the Oliver-Pharr method mentioned above, the nanoindentation hardness, the contact stiffness and the elastic modulus of materials can be obtained.

3. Idea of device design

Fig.3 is the schematic diagram of indentation device which will be designed in this chapter. It mainly consists of two portions, precise driving unit including z-axis precise driving unit (4), x, y precise positioning platform (8) and precise measuring unit including capacitance displacement sensor (5), displacement amplification structure (9) and laser displacement sensor (10). Compared with most of commercial indentation equipments, the principle is different. Penetration load is not measured directly by a load sensor but it is obtained with the help of displacement amplification structure and laser displacement sensor. When the indenter is pushing into and withdrawing the sample located on the left of displacement amplification structure, the amplification structure will deform and the right will output enlarged displacement which is measured by high resolution laser displacement sensor. At the same time, displacement of the indenter is measured by the capacitance displacement sensor. Because deformation of displacement amplification structure is very small, it is elastic deformation.

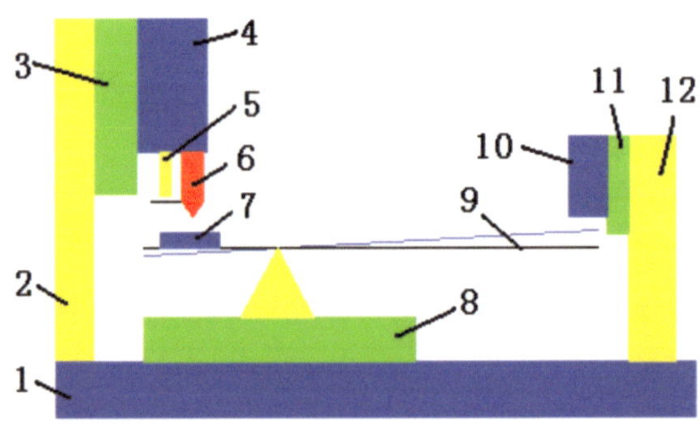

Fig. 3. Schematic diagram of the device which will be designed in this chapter 1 Base; 2(12) Supporting plates; 3(11) Macro-adjusting mechanism; 4 z-axis precise driving unit; 5 Capacitance displacement sensor; 6 Indenter; 7 Sample; 8 x, y precise positioning platform; 9 Displacement amplification structure; 10 Laser displacement sensor

Calculation model and related parameters are shown in Fig.4. P is penetration load and h_3 is left deformation of displacement amplification structure corresponding to the load P. h_1 and

h_2 are right deformation and displacement of the indenter measured by laser displacement sensor and capacitance displacement sensor, respectively. P, h_1, h_2, h_3 and penetration depth h have relationships as follows

$$h_3 = \mu h_1 \tag{6}$$

$$P = \lambda h_1 \tag{7}$$

$$h = h_2 - h_3 = h_2 - \mu h_1 \tag{8}$$

where μ and λ are calibration coefficients.

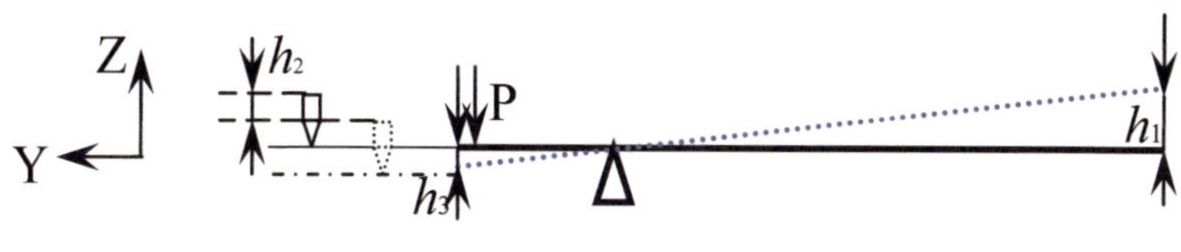

Fig. 4. Calculation model and the related parameters.

According to equations (6)—(8), penetration load P and depth h can be obtained, and material parameters can be calculated by the Oliver-Pharr method.

4. Design and analysis of precise driving units

Nanoindentation and scratch test technology mainly involves precise driving and precise measuring technology. For precise driving, there are many choices, for example, electromagnetic driver, shape memory alloy driver, micro-film, micro-beam and so on. Up to now, most of indentation devices are large because of using of electromagnetic and electrostatic drivers which also need complex control. Due to size limitation of SEM and TEM, these large indentation devices can not be used to realize in situ indentation test. So it is necessary to find more suitable driving mechanism to ensure miniaturization of indentation device. In this section, based on early research foundation on the piezoelectric-driven and flexure hinge, kinds of precise driving units realized by piezoelectric actuator and flexure hinge were designed.

4.1 Piezoelectric actuator

Principle of piezoelectric actuator is shown in Fig.5 which is based on the inverse piezo effect. The piezoelectric actuator will deform when an electric voltage signal is applied to it.

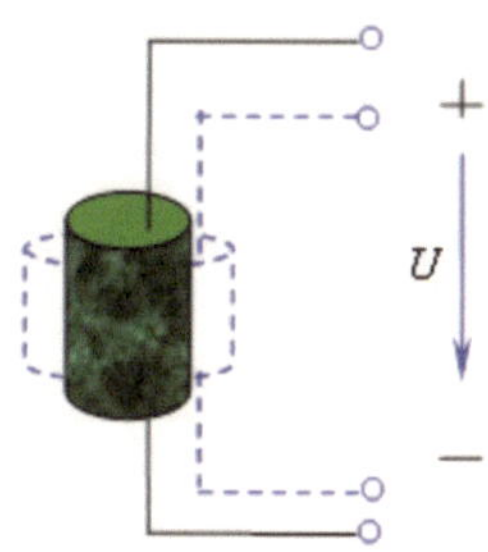

Fig. 5. Principle of piezoelectric actuator

The amount of movement is a function of the polarity of the voltage applied and the direction of the polarization vector.

Piezoelectric actuator takes many advantages of small size, unlimited resolution, large force generation, fast response, low power consumption and no wear. So it is widely used in fields of actuators, micro- and nano-positioning, laser tuning, active vibration damping, micropumps, and so on. In this chapter, two kinds of piezoelectric actuator are selected, PT200/10*10/40 piezoelectric stack used for z axis and AE0505D16F for x-y axis.

Hysteresis is an inherent property of piezoelectric ceramic. Hysteresis of the two kinds of piezoelectric stacks was measured and shown in Fig.6 and Fig.7.

In Fig.6 and Fig.7, there are two curves, respectively. One is the output displacement when voltage increases and the other is the output displacement when the voltage decreases. It is obvious that the displacement is different at the same voltage and the two curves are not symmetrical. The hysteresis H can be expressed

$$H = D_{DMAX} / D_S \times 100\% \tag{9}$$

where D_{DMAX} is the maximum difference of displacement at the same voltage; D_S is the total output displacement. In Fig.6, the maximum difference of displacement is about 1.15μm at voltage 45V and the total output displacement is 12.91μm. According to equation (9), the hysteresis is about 8.91% for AE0505D16F. In Fig.7, the maximum difference of displacement is about 5.49μm at voltage 45V and the total output displacement is 35.92μm. According to equation (9), the hysteresis is about 15.28% for PT200/10*10/40 piezoelectric stack. So hysteresis is different for different kinds of piezoelectric stacks and some measurements for example close-loop control should be taken to decrease the hysteresis for special application.

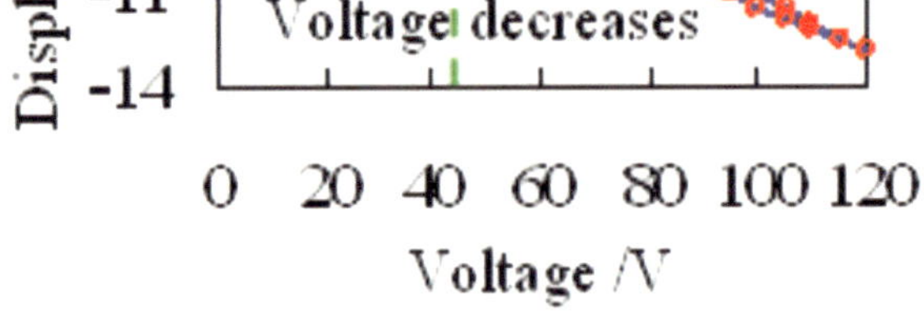

Fig. 6. Hysteresis curve of AE0505D16F

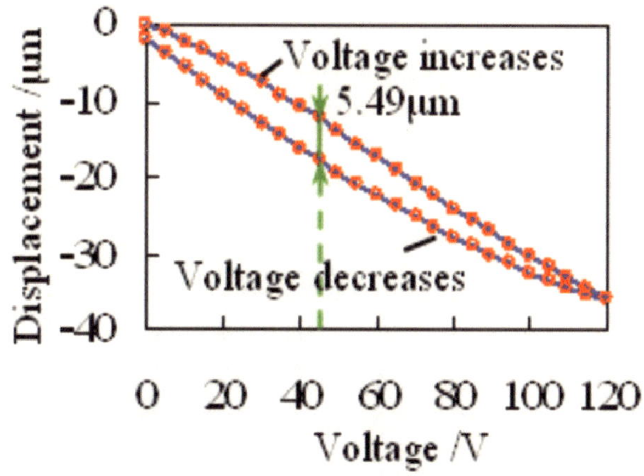

Fig. 7. Hysteresis curve of PT200/10*10/40

4.2 Flexure hinges

Materials and structures will deform under the external load and the deformation is usually very small and linear. Those are the working principle of flexure hinges. Compared to conventional mechanisms with sliding and rolling bearings, the flexure hinge takes many advantages of simple and compact structure, no lubrication and high positioning accuracy. For these reasons, flexure hinges have been widely used in fields of micro-positioning, micromanipulation, micro-gripper and so on.

Stiffness and output displacement of flexure hinges are contradictory to each other. Larger elastic deformation is hoped to ensure output displacement. On the other hand, enough stiffness is also very important to ensure the device having good dynamic characteristic and the anti-interference ability. Also internal stress of materials should not exceed permissible stress. Currently, there are four kinds of materials—beryllium bronze, aluminium, steel and titanium alloy to be used to fabricate flexure hinges. For these four kinds of materials, titanium alloy has the highest inherent frequency and best anti- interference ability, while the displacement is too small. In contrast, beryllium bronze has larger elastic deformation, but cost of these two kinds of materials is too high, and they are not suitable to make flexure hinges. Here, 65Mn was chosen to process them, which had numerous advantages of cheap price, high sensitivity, low elastic lag, high fatigue resistance, etc.

4.2.1 Z-axis flexure hinge

Z-axis precise driving unit consists of z-axis flexure hinge and z-axis piezoelectric actuator, and it is used to realize the precise loading and unloading of the indenter. Z-axis piezoelectric actuator was PT200/10*10/40 piezoelectric stack. Large output displacement was given to z-axis flexure hinge realized by the level-type enlarging structure as shown in Fig.8, which was convenient to estimate the initial contact point.

Static and modal analysis was carried out to evaluate strength and dynamic performance of z-axis flexure hinge by finite element method. Displacement load of 10μm was applied to the area where piezoelectric actuator was located. And analysis results were shown in Fig.9 and Fig.10. As shown in Fig.9, displacement of 42μm was obtained at the output end which indicated that magnification of the flexure hinge was about 4, while the maximum stress was 33MPa, which was less than permissible stress of 65Mn being 432MPa. The first three natural frequencies of the flexure hinge were about 1133.7Hz、1366.7Hz、4243.5Hz which indicated that z-axis flexure hinge had good stability in the indentation device working at low frequency condition.

Fig. 8. Model of z-axis flexure hinge

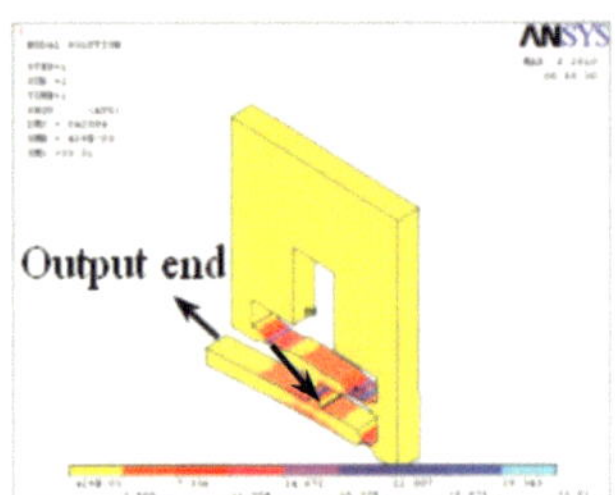

Fig. 9. Stress of z-axis flexure hinge

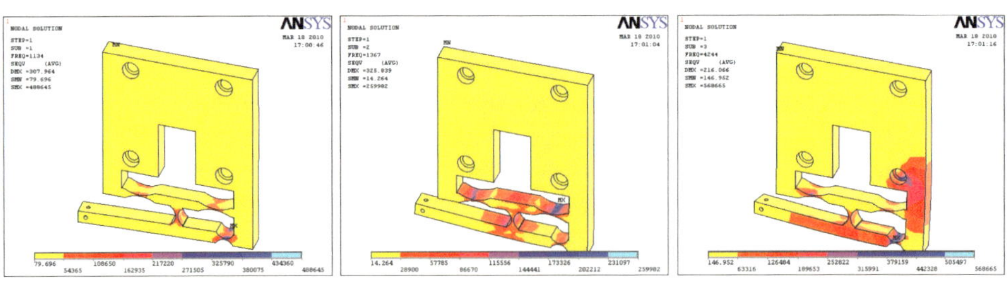

a) b) c)

Fig. 10. Mode shapes of z-axis flexure hinge (a) First mode shape (1133.7Hz); (b) Second mode shape (1366.7Hz); (c) Third mode shape (4243.5Hz)

4.2.2 x-y precise positioning hinge

x-y precise positioning platform including y-axis macro-adjusting mechanism, x-y precise positioning hinge and x-y piezoelectric actuators, is used to realize precise positioning of sample during indentation test and to realize precise motion of the sample during the scratch test. y-axis macro-adjusting mechanism as well as another two macro-adjusting mechanisms was bought directly and the models were GCM-1253001BM. x-y piezoelectric actuators were AE0505D16F piezoelectric stacks. The designed x-y precise positioning hinge was shown in Fig.11. Static and modal analysis results were shown in Fig.12 and Fig.13. The maximum stress was 158.6MPa, which was less than permissible stress of 65Mn being 432MPa. The first three natural frequencies of the flexure hinge were about 2669.5 Hz, 4831.0 Hz, 6281.8 Hz and the hinge had good dynamic performance.

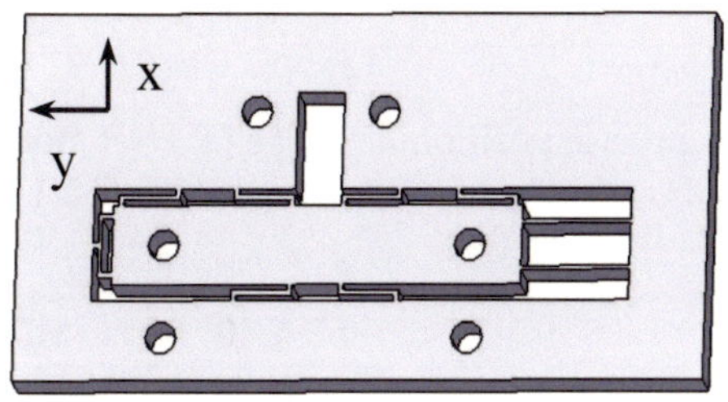

Fig. 11. Model of x-y precise positioning hinge

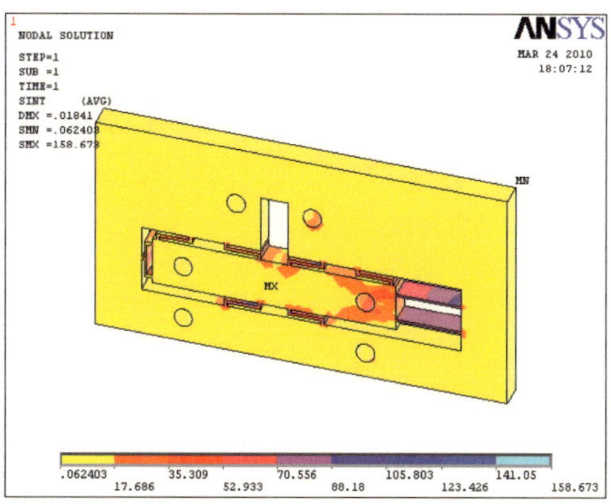

Fig. 12. Stress of x-y precise positioning hinge

5. Precise measuring unit

Parameters of materials are calculated by the penetration load and depth data. Because it is on very small scales, very high accuracy and resolution is required for sensors. As mentioned in section 3, penetration load and depth is obtained by indirect measurement method.

The displacement amplification structure and two displacement sensors are used to realize measurement. The laser displacement sensor LK-G10 which has resolution of 10nm is used to measure the output end (the right) of the displacement amplification structure, and the capacitance displacement sensor MDSL-0500M6-1 which has resolution of 10nm is used to measure the displacement of the indenter. Then the measurement data is collected by the A/D card and sent to the computer. The main parameters of the two sensors are shown in table 1.

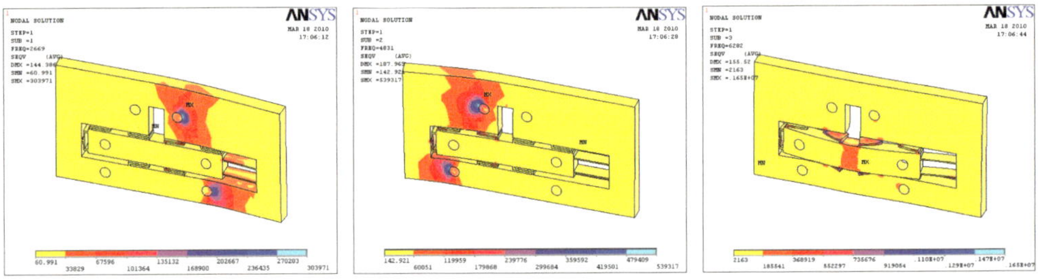

a) b) c)

Fig. 13. Mode shapes of x-y precise positioning hinge (a) First mode shape (2669.5Hz); (b) Second mode shape (4831.0Hz); (c) Third mode shape (6281.8Hz)

	LK-G10	MDSL-0500M6-1
Measurement range	±1mm	±0.5mm
Resolution	10nm	10nm
Accuracy	±0.02% F.S.	±0.02% F.S.
Reference distance	10mm	0mm
Linearity	±0.03% F.S.	±0.025% F.S.

Table 1. Main parameters of the two sensors

In this section, we will focus on design and analysis of the displacement amplification structure which plays an important role in measuring unit as well as entire indentation device. The designed displacement amplification structure with a lever amplification mechanism is shown in Fig.14. The sample is located on point A during the indentation test. Work principle is shown in Fig.15. Assumptions are as follows:

1. The upper thin plate rotates around the point O and the rotation angle is so small that the plate can be thought to be horizontal;
2. There is no bend deformation for the upper thin plate during the rotation.

Fig. 14. Model of amplification structure

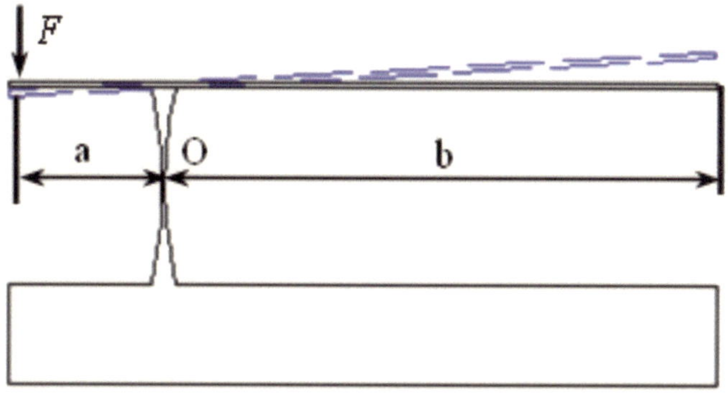

Fig. 15. Work principle of amplification structure

As shown in Fig.15, the displacement amplification structure not only works as a sample stage but also has the function of amplifying displacement signal. According to Fig.4 and Fig.15, the magnification factor is given by

$$k = \frac{|h_1|}{|h_3|} = \frac{b}{a} \tag{10}$$

where a is the horizontal distance between point A and the rotation point O; b is the horizontal distance between point B and the rotation point O.

In this chapter, the magnification factor k was designed to be 4. Static and modal analysis was carried out to evaluate the strength, output displacement and dynamic performance of displacement amplification structure. Displacement load of 10μm was applied to point A. Output displacement of point B was 38.2μm shown in Fig.16, and the maximum stress was 6.04MPa which was less than permissible stress of 65Mn being 432MPa. Fig.17 was the first three mode shapes and the first three natural frequencies were 170.53Hz, 407.42Hz, and 909.51Hz. The displacement amplification structure would bend or rotate at the structure' first three natural frequencies which were a little low. So the work frequency of the indentation device should be away from natural frequencies to avoid sympathetic vibration and also it is better to take measures to alleviate and isolate the vibration existing in the surroundings.

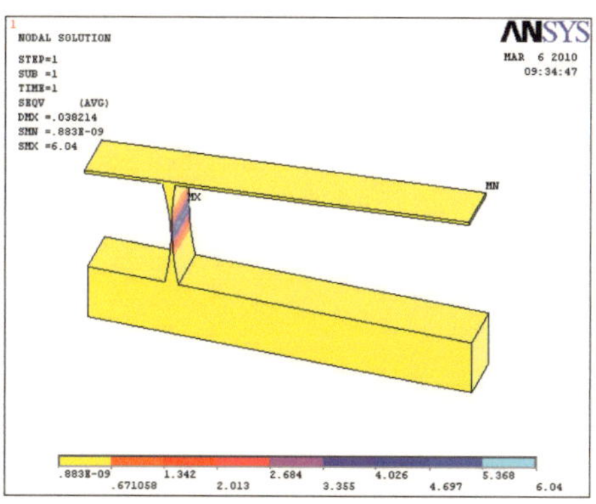

Fig. 16. Stress of amplification structure

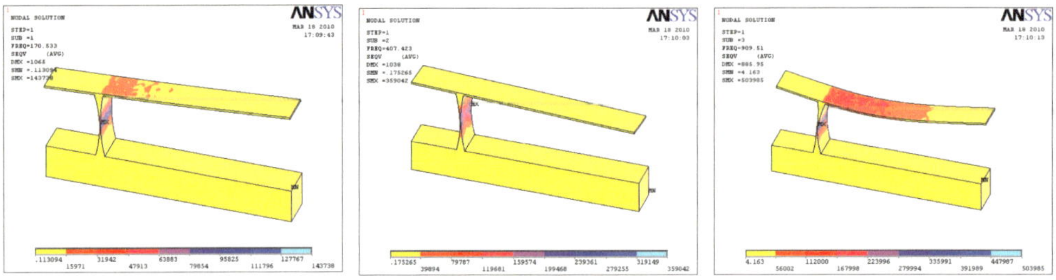

a) b) c)

Fig. 17. Mode shapes of displacement amplification structure (a) First mode shape (170.53Hz); (b) Second mode shape (407.42Hz); (c) Third mode shape (909.51Hz)

Output performances of the amplification structure under small load were analyzed by finite element method when the load *F* was 0.1mN and 1mN, respectively. Analysis results were shown in Fig.18 and Fig.19 respectively. In these two figures, the amplification structure had 43.3nm and 433nm output displacements corresponding to the loads 0.1mN and 1mN. The magnifications of input loads and output displacements were coincident which indicated that the structure had good linear output performance. Output displacement of 43.3nm can be detected easily by laser displacement sensor with the resolution of 10nm. That was to say the load resolution of the displacement amplification structure was higher than 0.1mN.

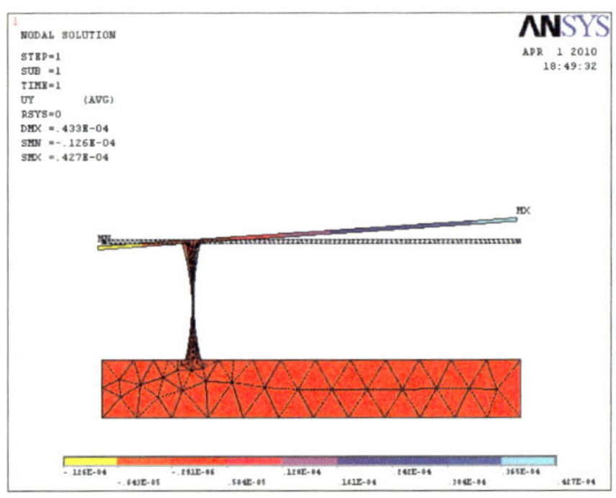

Fig. 18. Deformation of amplification structure when *F*=0.1mN

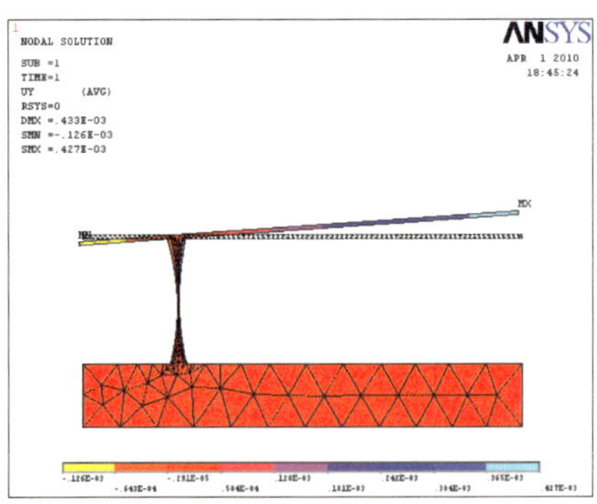

Fig. 19. Deformation of amplification structure when *F*=1mN

6. Prototype design

According to the analysis in the previous sections, the catia model of designed indentation device was shown in Fig.20. Parts were fabricated and the prototype was assembled as shown in Fig.21. The brief work processes are as follows:

1. Clear the sample surface;
2. Install the sample on the displacement amplification structure;
3. Install the indenter and lock it with the lock screw;
4. Adjust the macro-adjusting mechanism to make the laser displacement sensor in the suitable measuring range(the indicator light will be green);
5. Apply voltage to electronic components and wait for a moment to make the components stabilization;
6. Adjust the z-axis macro-adjusting mechanism to make the indenter close to the sample surface. When it is very close to the surface, stop macro-adjusting mechanism and apply voltage to the z-axis piezoelectric stack. Use the change of the read of the laser displacement sensor to judge the contact between the indenter and the sample surface;
7. Choose suitable voltage step to load and unload the indenter. During the process, use software to record the data sent by the A/D card. And then, process the data and obtain parameters of the sample.

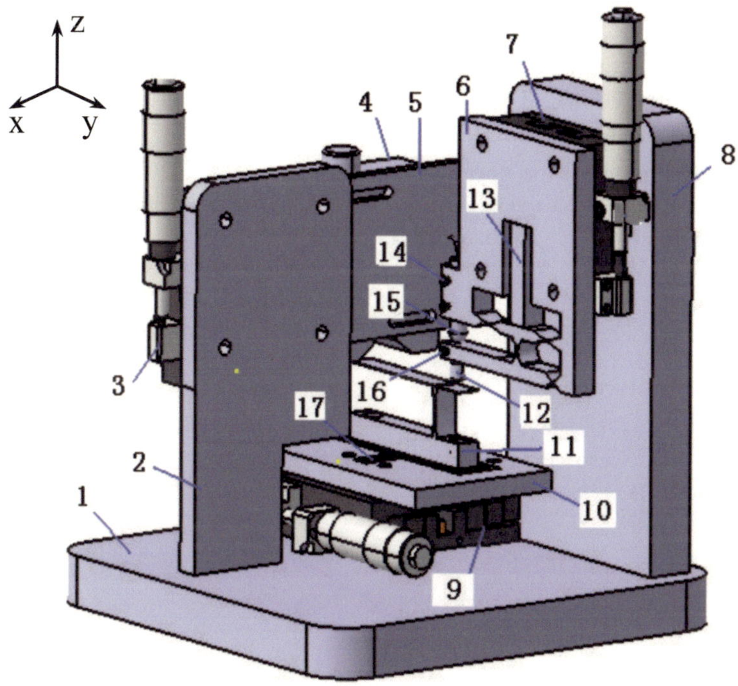

Fig. 20. Catia model of designed indentation device 1 Base; 2(8) Supporting plates; 3(7,9) Macro-adjusting mechanism; 4 Laser displacement sensor; 5 Connector; 6 z-axis flexure hinge; 10 x-y precise positioning hinge; 11 Displacement amplification structure; 12 Indenter; 13 z-axis piezoelectric stack; 14 Lock screws of the sensor ; 15 Capacitance displacement sensor; 16 Lock screw of the indenter; 17 x-axis piezoelectric stack.

Fig. 21. Prototype of designed indentation device

7. Experiments

In this section, experiments of the designed indentation device were carried out to evaluate its performances. These experiments mainly include calibration of laser and capacitance displacement sensors as well as the displacement amplification structure, output performance test of the designed x-y precise positioning hinge and z-axis precise driving hinge and indentation test of optical glass.

7.1 Calibration experiments of the sensors

Use z-axis precise driving unit to generate precise displacement signal. Use the laser and capacitance displacement sensors to measure the signal, respectively. And then, record the reading and the output voltage, respectively. The experiment data was processed with the criteria of least squares. Curves and equations of linear fitting were obtained, which were shown in Fig.22 and Fig.23. From these two figures, relation between measured

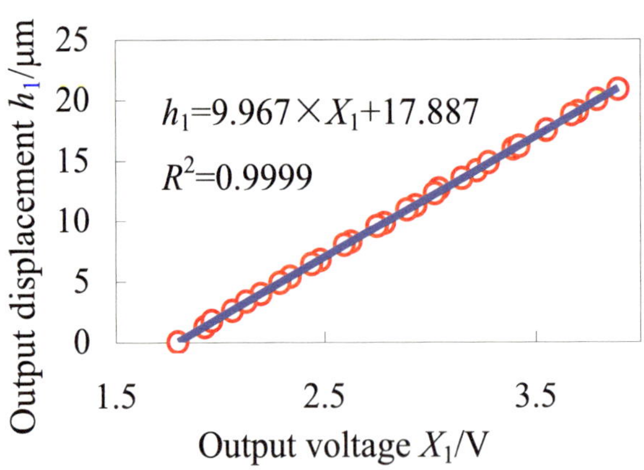

Fig. 22. Calibration curve of the laser displacement sensor

displacement h_1/μm and output voltage X_1/V of the laser displacement sensor was $h_1=9.967\times X_1-17.887$ and relation between measured displacement h_2/μm and output voltage X_2/V of the capacitance displacement sensor was $h_2=49.538\times X_2-194.27$. Their linear correlation coefficients R^2 were both close to 1, which showed the two sensors had high linearity. So the equations of linear fitting can be used in the experiment without correction.

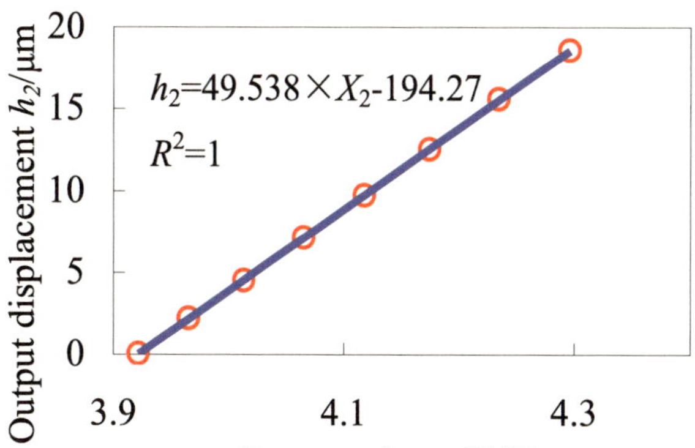

Fig. 23. Calibration curve of the capacitance displacement sensor

7.2 Calibration experiments of the displacement amplification structure

According to section 3, the displacement amplification structure plays an important role in the measuring unit as well as the entire indentation device. Calibration experiments were carried out to obtain the relation of load P and output displacement h_1 of point B as well as the relation of deformation h_3 of point A and output displacement h_1 of point B, and the results were shown in Fig.24 and Fig.25.

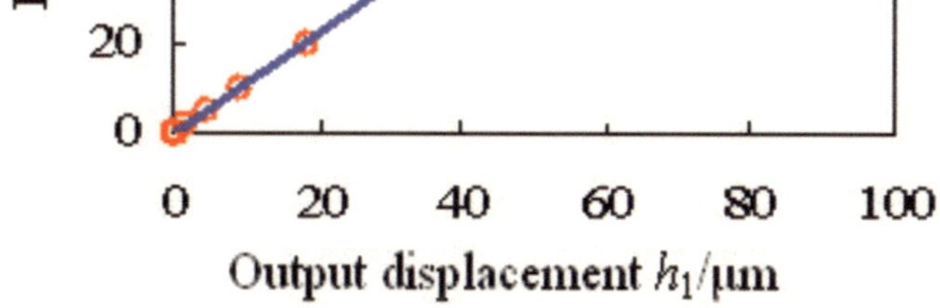

Fig. 24. Relation curve of load P and displacement h_1

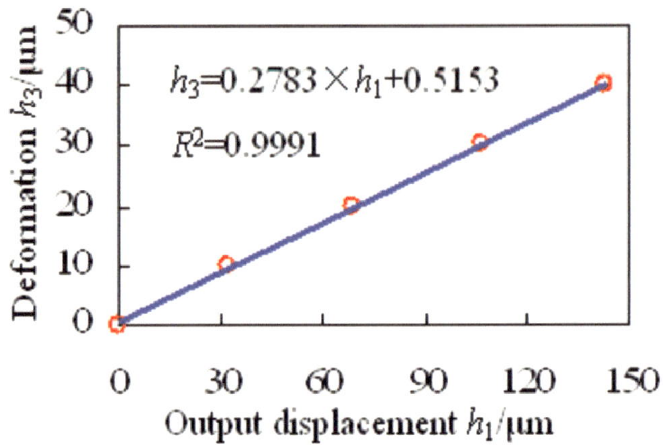

Fig. 25. Relation curve of displacement h_3 and h_1

From these two figures, relation between the load P/mN and output displacement h_1/μm of point B of the displacement amplification structure was P=1.1227Xh_1-0.426, and relation between deformation h_3/μm of point A and output displacement h_1/μm of point B is h_3=0.2783×h_1+0.5153. Their linear correlation coefficients R^2 were 0.9998 and 0.9991, which indicated that output of the structure was linear. Also equations of linear fitting can be used in the experiment without correction.

7.3 Output performance of x-y precise positioning hinge and z-axis precise driving hinge

Output performances of x-y precise positioning hinge and z-axis precise driving hinge were tested by laser displacement sensor. The range of applied voltage was form 0V to 120V for x and y piezoelectric stacks with step of 5V while the range was from 0V to 90V for z axis piezoelectric stack with step of 5V or various steps (5V to 1V). The testing results were shown in Fig.26 - Fig.29.

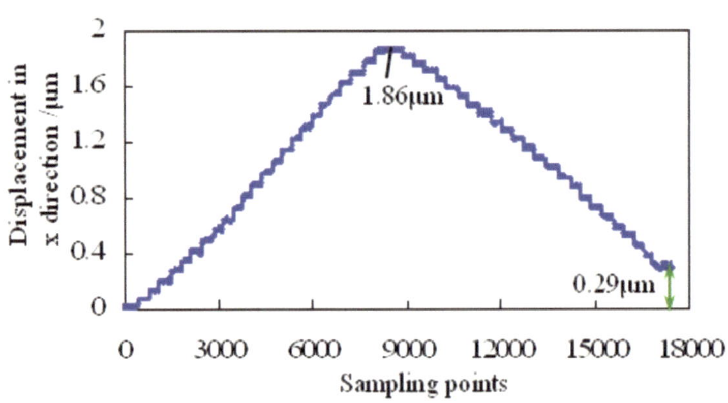

Fig. 26. Output displacement in x direction

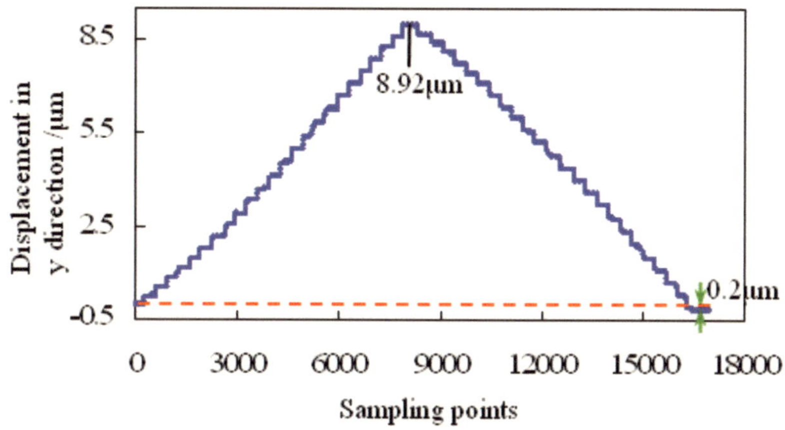

Fig. 27. Output displacement in y direction

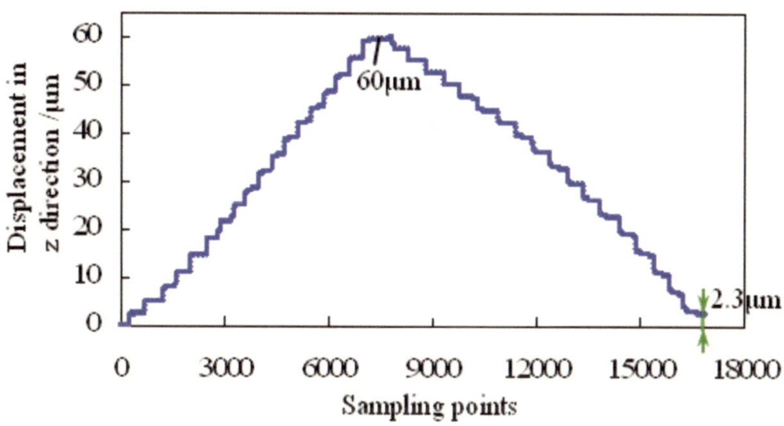

Fig. 28. Output displacement in z direction

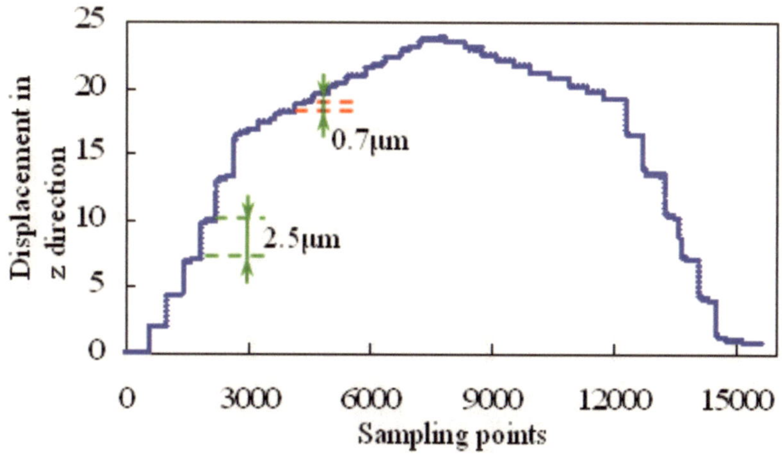

Fig. 29. Output displacement in z direction with various steps

As shown in Fig.26 and Fig.27, the output performances were large different for x and y axis. The maximum output displacement was 1.86μm in x direction while it was 8.92μm in y direction. For 24 steps, the total output displacement was 77.5nm and 371.7nm in x and y directions respectively. So it had different output displacement resolution in the two directions. These differences were caused by different stiffness of flexure hinges in x and y directions.
Fig.28 was the output curve of z-axis precise driving hinge when applied voltage was from 0V to 90V with step of 5V. The maximum output displacement was about 60μm at voltage of 90V, which was large than the maximum output displacement of z piezoelectric stack about 40μm. So the z-axis precise driving hinge had the function of displacement magnification. Fig.29 was the output curve of z-axis precise driving hinge with various steps from 5V to 1V. Due to the hysteresis of piezoelectric stack, the curve was asymmetrical. Through the manner of various steps, it was convenient to realize the judgement of contact between the indenter and the sample with large step and to realize loading and unloading process with small step. Also it can be used to research the mechanical performance of materials under different steps.
In order to evaluate performance of the unit, parameters were defined as follows. D_s was the maximum output displacement. D_r was the residual displacement. β was the absolute error. α was the average error of each step. And then, β and a can be expressed

$$\beta = D_r/D_s \times 100\% \tag{11}$$

$$a = D_r/n \tag{12}$$

where n was the total steps in a test circle.
According to equations (11) and (12), parameters were obtained and listed in Table 4. When the voltage step decreases, higher resolution will be obtained and unlimited resolution will be possible under ideal conditions.

	x axis	y axis	z axis
D_s(μm)	1.86	8.92	60
D_r(μm)	0.29	-0.2	2.3
β	15.6%	2.24%	3.83%
a(nm)	6.04	4.17	63.8
Resolution (μm /5v)	0.0775	0.3717	3.33

Table 4. Parameters of three directions

7.4 Indentation test of optical glass

Indentation experiments of optical glass were carried out and the *P-h* curves were shown in Fig.30 and Fig.31.

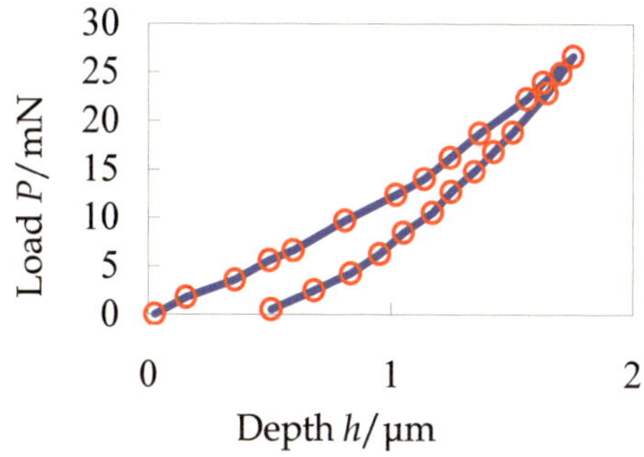

Fig. 30. *P-h* curve with maximum load 26mN

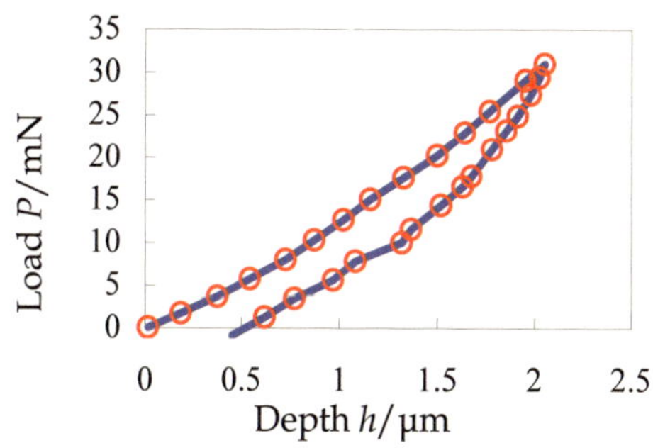

Fig. 31. *P-h* curve with maximum load 30mN

The three polynomial fitting was used to fit the curve of partial unloading data. Fig.32 was fitted curves and equations corresponding to Fig.30, and Fig.33 was fitted curves and equations corresponding to Fig.31. As shown in Fig.32 and Fig.33, the correlation coefficients were close to 1 which indicated that the selected order was suitable. So the relation between load *P* and the depth of partial unloading for the two experiments were given：

Test one with maximum load 26mN: $P= 22.197 \times h^3 - 85.055 \times h^2 + 131.78 \times h - 62.456$;

Test two with maximum load 30mN: $P= 107.74 \times h^3 - 570.35 \times h^2 + 1033.6 \times h - 619.31$.

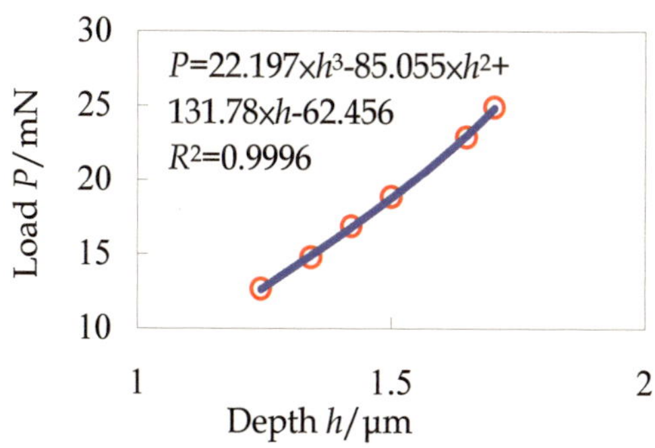

Fig. 32. Fitted curves and equations of partial unloading data of test one

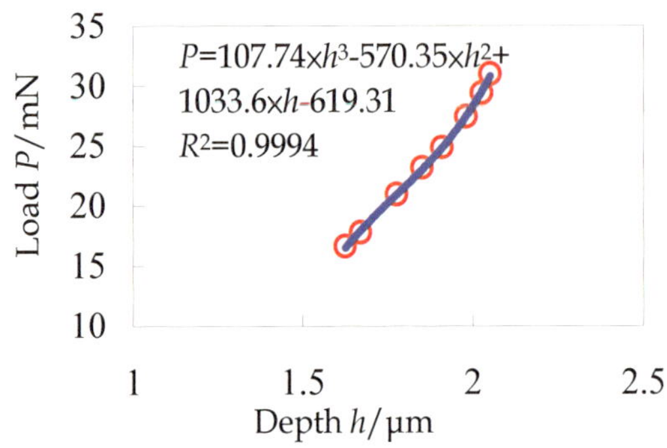

Fig. 33. Fitted curves and equations of partial unloading data of test two

According to equations mentioned in section 2, contact stiffness between indenter and optical glass of the two experiments was 35.0801mN/µm for test one and 53.49705mN/µm for test two. The relation between hardness *H* and load *P* at the unloading portion was obtained shown in Fig.34 (test one) and Fig.35 (test two) from which we can see that material's hardness would change with penetration depth but it would stabilise when the depth was larger than a certain value.

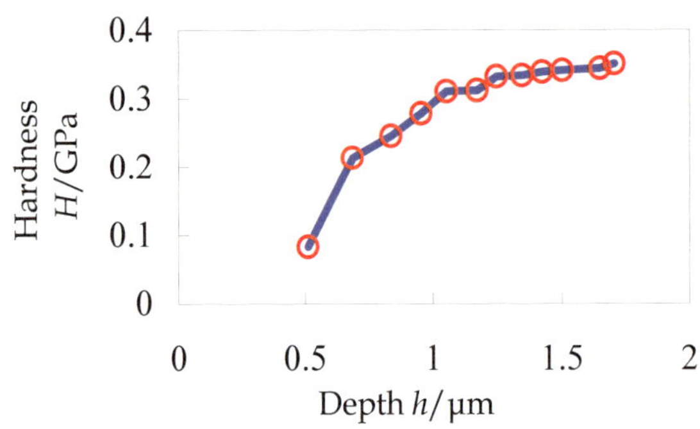

Fig. 34. Relation curve between hardness *H* and depth *h* of test one

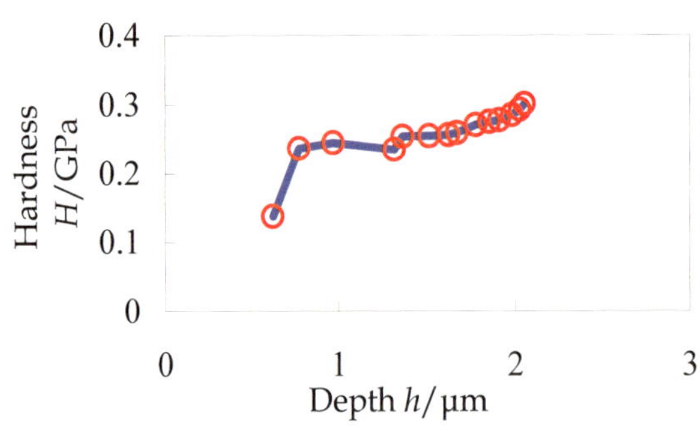

Fig. 35. Relation curve between hardness *H* and depth *h* of test two

Compared with the result by commercial indentation device shown in Fig.36 (Zhao et al., 2009), the measured penetration depth was larger at a same load which was caused by structure compliance, indenter installation as well as the sample surface process, and it would be reduced and eliminated in our future work.

The indentation morphology was obtained through high resolution optical microscope shown in Fig.37. The material generated some cracks which had significant value to analyze the damage mechanism especially that the full measuring process was monitored by SEM, TEM as well as other monitoring methods.

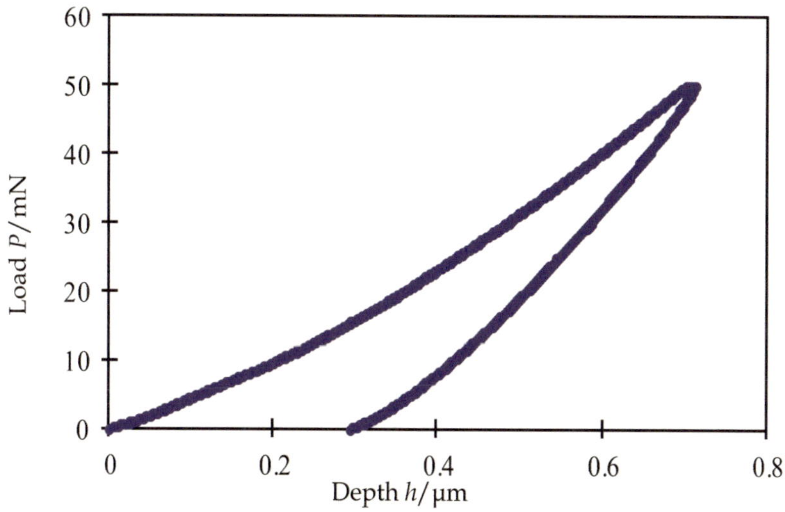

Fig. 36. *P-h* curve obtained from a commercial indentation device

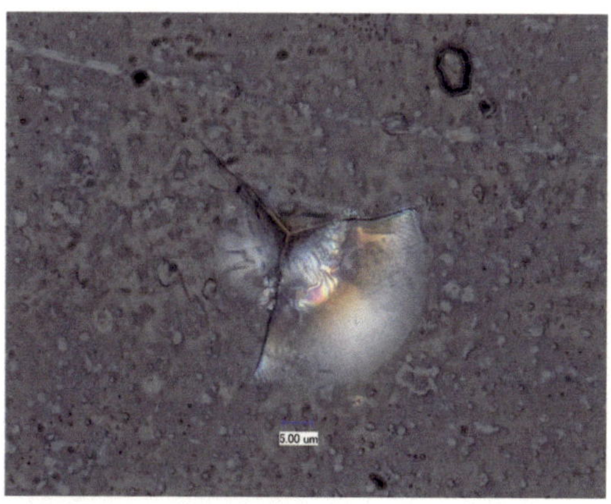

Fig. 37. Indentation morphology of optical glass

The designed indentation device integrating flexure hinges and piezoelectric actuators is smaller than commercial indentation devices but can realize high precise measurement which gives solid foundation for our future work. We will make the device more precise and design smaller indentation device that can be located on the platform of SEM to realize in situ measurement.

8. Conclusions

A new kind of indentation measurement method through two displacement sensors and a displacement amplification structure was proposed and established. Based on this method, a miniaturization indentation device was designed and analyzed. Hysteresis of two kinds of piezoelectric stacks were measured and analyzed. The hysteresis was different for different piezoelectric stacks.

Two key components of the device, the precise driving and precise measuring units, were designed and analyzed. Results from static and modal analysis indicated that designed hinges and amplification structure had enough strength. The dynamic performances of flexure hinges were good but the natural frequencies of the amplification structure were low. So the work frequency of the indentation device should be away from the natural frequencies of the amplification structure to avoid sympathetic vibration and also it is better to take measures to alleviate and isolate the vibration existing in the surroundings.

Calibration experimental results of the two displacement sensors and the amplification structure showed that they all had high linearity. The amplification structure had functions of stage and load-sensing, and can be used to realize precise load measurement.

Output performances of flexure hinges were measured and obtained. They all had stable output displacement and the output resolution was related with the voltage step applied to the piezoelectric stack.

Indentation experiments of optical glass were carried out. Relation curves between load and penetration depth were obtained. Contact stiffness and hardness were discussed. Curves obtained from the designed device and commercial device were compared and analyzed. The designed indentation device had smaller volume and can realize indentation test but the accuracy should be improved more.

The future work will focus on designing smaller indentation device and realizing in situ experiments.

9. Acknowledgements

This research is funded by the National Natural Science Foundation of China (Grant No. 50905073), Research Fund for the Doctoral Program of Higher Education of China (Grant No.200801831024), Research Fund of the Ministry of Science and Technology of China (Grant No.2009GJB10029), Young Scientist Fund of Jilin Province of China (Grant No.20090101), The International Scientific and Technological Cooperation Project (Grant No.2010DFA72000) and Graduate Innovation Fund of Jilin University (Grant No.20111058).

10. Reference

Abdel-Aal, H.A.; Patten, J.A. & Dong, L. (2005). On the thermal aspects of ductile regime micro- scratching of single crystal silicon for NEMS/MEMS applications. *Wear*, Vol.259, No.7-12, (May 2005), pp. 1343-1351, ISSN 0043-1648

Bhushan, B. (2007). Nanotribology and nanomechanics of MEMS/NEMS and BioMEMS/BioNEMS materials and devices. *Microelectron Eng.*, Vol. 84, No.3, (March 2007), pp. 387-412, ISSN 0167-9317

Bradby, J.E.; Williams, J.S. & Swain, M. V. (2003). In situ electrical characterization of phase transformations in Si during indentation. *Phys. Rev. B*, Vol.67, No.8, (February 2003), pp. 085205.(1-9), ISSN 1098-0121

Bruet, B. J. F.; Song, J.H.; Boyce, M.C. & Ortiz, C. (2008). Materials design principles of ancient fish armour. *Nat. Mater.*, Vol. 7, No.9, (July 2008), pp.748-756, ISSN 1476-1122

CSIRO.UMIS. Cited 2011; Available from: http://www.csiro.au/hannover/2000/catalog/projects/umis. Html

CSM instruments. Cited 2011; Available from: www.csm-instruments.com

De Hosson, J.T.M.; Soer, W.A.; Minor, A.M.; Shan, Z.W.; Stach, E.A.; Syed Asif, S.A. & Warren, O.L.(2006). In situ TEM nanoindentation and dislocation-grain boundary interactions: a tribute to David Brandon. *J. Mater Sci.*, Vol.41, No.23, (Novermber 2006), pp.7704-7719, ISSN 0022-2461

Diez-Perez, A.; Güerri, R.; Nogues, X.; Cáceres, E.; Peña, M.J.; Mellibovsky, L.; Randall, C.; Bridges, D.; Weaver, J.C.; Proctor, A.; Brimer, D.; Koester, K.J.; Ritchie, R.O. & Hansma, P.K. (2010). Microindentation for In Vivo Measurement of Bone Tissue Mechanical Properties in Humans. *J. Bone Miner. Res.*, Vol.25, No.8, (August 2010), pp. 1877-1885;

Doerner, M.F. & Nix, W.D. (1986). A method for interpreting the data from depth-sensing indentation instruments. *J. Mater. Res.*, Vol.1, No.4, pp.601-609, (August 1986), ISSN 0884-2914

Han, X.; Zhang, Z. & Wang, Z. (2007). Experimental nanomechanics of one-dimensional nanomaterials by in situ microscopy. *NANO.*, Vol.2, No.5, (October 2007), pp. 249-271, ISSN 1793-2920

Hansma, P.K.; Turner, P. J. & Fantner, G.E. (2006). Bone diagnostic instrument. *Rev. Sci. Instrum.*, Vol.77, No.7, (July 2006), pp. 075105(1-6), ISSN 0034-6748

Hemker, K.J. & Nix, W.D.(2008). Nanoscale deformation: Seeing is believing. *Nat. Mater.*, Vol.7, No.2 (February 2008), pp. 97-98, ISSN 1476-1122

Huja, S.S.; Hay, J.L.; Rumme, A.M. & Beck, F.M. (2010). Quasi-static and harmonic indentation of osteonal bone. *J. Dent. Biomech.*, Vol. 2010, (February 2010), pp. 736830(1-7) , ISSN 1758-7360

Hysitron Incorporated. Cited 2011; Available from: http://www.hysitron.com/

Jardret, V. & Morel, P. (2003). Viscoelastic effects on the scratch resistance of polymers: relationship between mechanical properties and scratch properties at various temperatures. *Prog. Org. Coat.*, Vol.48, No.2-4, (December 2003), pp. 322-331, ISSN 0300-9440

Koff, M.F.; Chong, L.R.; Virtue, P.; Chen, D.; Wang, X.; Wright, T. & Potter, H.G.(2010). Validation of cartilage thickness calculations using indentation analysis. *J. Biomech. Eng.*, Vol.132, No.4, (April 2010), pp. 041007(1-6), ISSN 0148-0731

Lim, C.T.; Zhou, E.H. & Quek, S.T. (2006). Mechanical models for living cells—a review. *J. Biomech.*, Vol.39, No.2, (January 2006), pp. 195-216, ISSN 0021-9290

Lucas, B.N. & Oliver, W.C. (1999). Indentation power-law creep of high-purity indium. *Metall. Mater. Trans. A*, Vol.30, No.3, pp. 601-610, (March 1999), ISSN 1073-5623

Michler, J.; Rabe, R.; Bucaille, J-L.; Moser, B.; Schwaller, P. & Breguet, J-M.(2005). Investigation of wear mechanisms through in situ observation during microscratching inside the scanning electron microscope. *Wear*, Vol.259, (May 2005), pp. 18-26, ISSN 0043-1648

Micro Materials Ltd. Cited 2011; Available from: www.micromatreials.co.uk

MTS NanoIndenter. Cited 2011; Available from: http://www.charfac.umn.edu/InstDesc/nanoindenterdesc.html

Nowak, R.; Chrobak, D.; Nagao, S.; Vodnick, D.; Berg, M.; Tukiainen, A. & Pessa, M. (2009). An electric current spike linked to nanoscale plasticity. *Nat. Nanotechnol.*, Vol. 4, No.5, (March 2009), pp. 287-291, ISSN 1748-3387

Oliver, W.C. & Pharr, G.M. (1992). An improved technique for determining hardness and elastic modulus using load and displacement sensing indentation measurements. *J. Mater. Res.*, Vol.7, No.6, pp.1564-1583, (June 1992), ISSN 0884-2914

Rabe, R. (2006). Compact test platform for in-situ indentation and scratching inside a scanning electron microscope (SEM). PhD thesis, EPFL, 3593 (2006)

Schuh, C.A.; Mason, J.K. & Lund, A.C. (2005). Quantitative insight intodislocation nucleation fromhigh-temperature nanoindentation experiments. *Nat. Mater.*, Vol.4, No.8, (August 2005), pp. 617-621, ISSN 1476-1122

Sneddon I.N. (1965). The relation between load and penetration in the axisymmetric boussinesq problem for a punch of arbitrary profile. Int. J. Eng. Sci., Vol.3, No.1, (May 1965), pp. 47-57, ISSN 0020-7225

Suresh, S.; Spatz, J.; Mills, J.P.; Micoulet, A.; Dao, M.; Lim, C.T.; Beil, M. & Seufferlein, T. (2005). Connections between single-cell biomechanics and human disease states: gastrointestinal cancer and malaria. *Acta. mater.*, Vol. 1, No.1, (January 2005), pp. 15-30, ISSN 1359-6454

Suresh, S. (2007). Nanomedicine: elastic clues in cancer detection. *Nat. Nanotechnol.*, Vol.2, No.12, (December 2007), pp. 748 - 749, ISSN 1748-3387

Tai, K.; Dao, M.; Suresh, S.; Palazoglu, A. & Ortiz, C. (2007). Nanoscale heterogeneity promotes energy dissipation in bone. *Nat. Mater.*, Vol.6, No.6, (May 2007), pp. 454 - 462, ISSN 1476-1122

Tao, P.J.; Yang, Y.Z. & Bai, X.J. (2010). Vickers indentation tests in a Zr62.55Cu17.55Ni9.9Al10 bulk amorphous alloy. *Mater. Lett.*, Vol.64, No.9, (May 2010), pp. 1102-1104, ISSN 0167-577X

Thurner, P.J. (2009). Atomic force microscopy and indentation force measurement of bone. *WIREs Nanomed. Nanobiotechnol.* Vol.1, No.6, (October 2009), pp. 624–649, ISSN 1939-0041

Yang, F.Q.; Geng, K.B.; Liaw, P.K.; Fan, G.J. & Choo, H. (2007). Deformation in a Zr57Ti5Cu20Ni8Al10 bulk metallic glass during nanoindentation. *Acta. mater.*, Vol.55, No.1, pp. 321-327, (January 2007), ISSN 1359-6454

Zhang, J.Z.; Michalenko, M.M.; Kuhl, E. & Ovaert, T.C. (2010). Characterization of indentation response and stiffness reduction of bone using a continuum damage model. *J. Mech. Behav. Biomed. Mater.*, Vol.3, No. 2, (February 2010), pp. 189-202, ISSN 1751-6161

Zhao, H.W.; Yang, B.H.; Zhao, H.J. & Huang, H. (2009). Test of nanomechanical properties of single crystal silicon. *Optics and Precision Engineering*, Vol.17, No.7, (July 2007), pp.1602-1608, ISSN 1004-924X

Zheng, Y.P. & Mak, A.F.T.(1996). An ultrasound indentation system for biomechanical properties assessment of soft tissues in-vivo. *IEEE T. Biomed. Eng.*, Vol. 43, No.9, (April 1996), pp. 912-918, ISSN 0018-9294

Zhou, J. & Komvopoulos, K.(2006). Nanoscale plastic deformation and fracture of polymers studied by in situ nanoindentation in a transmission electron microscope. *Appl. Phys. Lett.*, Vol.88, No.18, (May 2006). pp. 181908 (1-3), ISSN 0003-6951

Zhu, Y. & Espinosa, H.D. (2005). An electromechanical material testing system for in situ electron microscopy and applications. *Proc. Natl. Acad. Sci.*, Vol.102, No.41, (August 2005), pp. 14503– 14508, ISSN 1091-6490

Part 4

Vascular Biomechanics

9

Elements of Vascular Mechanics

Gyorgy L Nadasy
Clinical Experimental Research Department and Department of Human Physiology, Semmelweis University Budapest

1. Introduction

Between half and two thirds of human mortality in developed countries can be attributed to vascular diseases. Financial losses, human sufferings are increasing with aging of the population. Vascular diseases develop when some or many vessels in the body are unable to fulfill their functions. The main function of blood vessels is essentially a mechanical one: to conduct blood. Vessels are functioning in a unique in the body mechanical environment: they are continuously subjected to hemodynamic forces: to shear stress of flowing blood and to distending forces of pressure of the blood in the lumen. Vessels are so much adapted to these hemodynamic forces that it is impossible to understand their physiology, pharmacology and pathology without taking into consideration the unavoidable biomechanical steps in the complicated pathways of cellular and systemic physiological vascular feed-back control loops, to understand vascular drug action and pathomechanism of vascular disease (Lee 2000).

Biomechanics is thus at the very core of all vascular sciences. That is reflected in the high number of papers published in the area. 35 000 papers listed in the Ovid Medline between 1948 and 2010 included knowledge on vascular mechanics in its narrower sense (excluding papers dealing only with physiological and pharmacological means of vascular smooth muscle control). Deteriorating Windkessel function of the aged, of the chronic hypertensive, even after effective treatment of mean arterial pressure, geometric, biomechanical consequences of atheroscerotic focal remodeling of large arteries, contractile and elastic remodeling of resistance *arteries* with aging, with hypertension and with diabetes, remodeling of venous networks and the venous wall in chronic venous disease, inevitably draws the attention of clinicians and of pathologists to biomechanical questions. Recent developments in vascular mechanics, backed with many methodical improvements in the field (Berczi 2005, Cox 1974, Duling 1981, Huotari 2010, Mersich 2005, Nadasy 2001, Shimazu 1986, See Fig. 1.), integration of these results into the context of reliable older knowledge makes now a systemic overview of the most important aspects of vascular mechanics possible. We will see that an almost axiomatic approach to a phenomenological description of vascular biomechanics is now in sight. Methodical advancement in the field of cellular physiology, histochemistry and biochemistry (Discher 2009) identified many if *not all extra- and intracellular fiber types and molecules* contributing to the biomechanics of the vascular wall. Mechanical factors in intra- and extracellular fiber protein expression control are just being identified. The emerging debate whether mechanics or biochemistry controls vascular

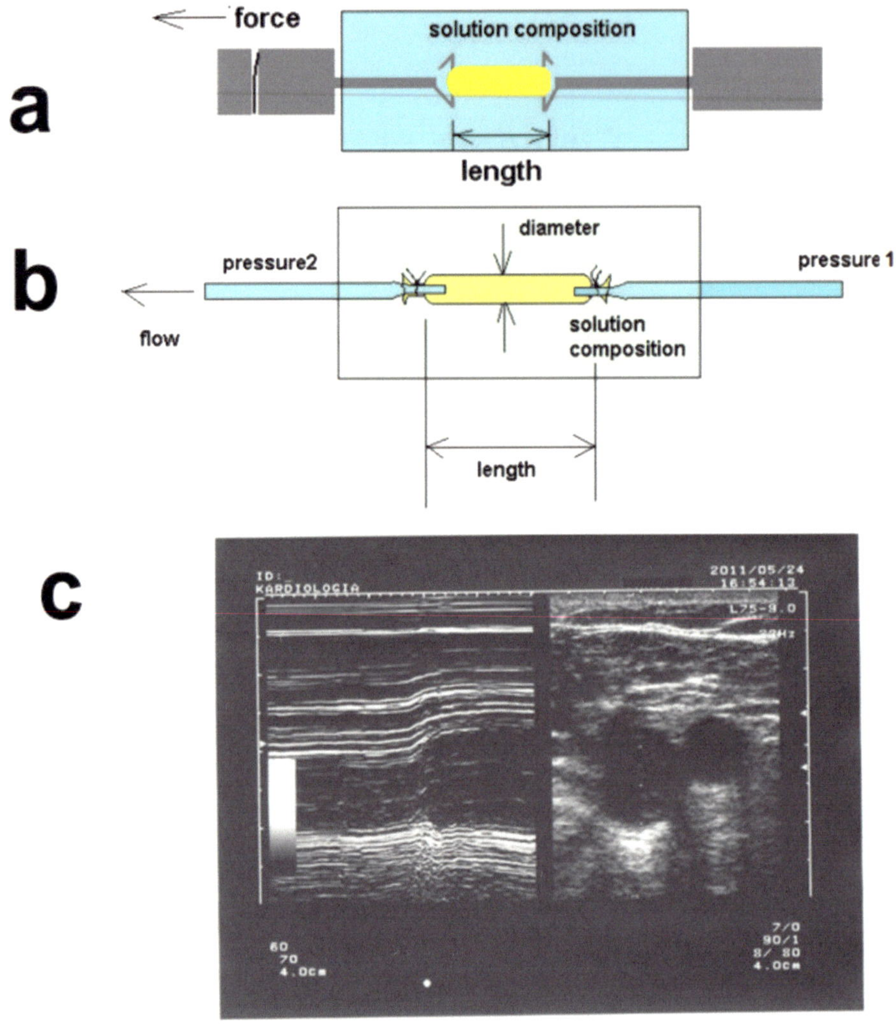

Fig. 1. Some methodics of vascular mechanics. a. In vitro wire myography. Circumferential vascular rings and strips are mostly studied. Frequently applied for isometric measurements of active forces in response to different vasoactive substances. Elasticity and tensile strength can also be studied. Geometric measurements (strip width and thickness) are needed to compute stress and to compare the situation with in vivo pressure loads. b. In vitro pressure arteriography. Cylindrical segments are mounted on cannulas in a glass-bottomed tissue bath. Devices have been developed both for macroscopic and microscopic vessels. Intraluminal pressure and flow can be altered to mimic in vivo situation, outer and inner diameters are measured optically. Mostly pressure-diameter plots are taken at different levels of smooth muscle tone, or diameter alterations are recorded at continuous pressures in response to vasoactive substances. c. In vivo ultrasonographic measurement of vascular lumen changes for biomechanical computations. Right, B-mode record of common femoral artery and vein. Left, elastic dilation of common femoral vein diameter in response to a controlled Valsalva attempt. M-mode record (as a function of time, courtesy of dr AÁ Molnar and of prof V Bérczi).

protein expression, we believe, is meaningless if the question is approached from the point of view of system physiology. Mechanical forces from hemodynamics can induce transmitter release, which, in turn might close the physiological control loop by acting back on hemodynamics. Or, biologically active substances inducing alterations in local tissue function, might, at the same time induce vascular changes supporting or just speeding up other existing vascular control loops adapting tissue circulation to altered tissue function. Such "feed-forward" loops are very common in physiology. We believe, that in many cases derailment of such optimized control processes in a situation that could not be phylogenetically expected, will be the reason for the observed "pathological effect" visibly acting against biomechanical control (Safar 2005). What is still missing now, is the mechanics at the molecular level. *The low-energy level steric deformations of force-bearing molecules*, determining the phenomenologically descriptionable mechanical behavior are not known, maybe, with the single exception of actomyosin crossbridges of vascular smooth muscle cells.

2. Biological background of vascular mechanics

2.1 Separation of the vascular space

The *closed vascular system of vertebrates* ensures fast nourishment of large neural and muscle masses, and fast exchange of materials in gills, lungs, kidneys, liver and intestines (Schmidt-Nielsen 1979). The term "closed" means, that blood vessels are lined internally with a fairly continuous endothelium (some exceptions do exist). Blood cells in vertebrate tissues are not forced to uncertain, zigzagging routes in extracellular space among neighboring cells (as e.g. in many worms), they will move through tissues not leaving the lumen of preformed vascular tubes (again, situations with exceptions do exist). Lesser friction makes faster blood flows possible with the same energy expense.

2.2 Distribution of blood flow in space: the network geometry

While diffusion routes of substances from blood to cells are by confinement of blood into vessels somewhat increased, owing to the rich network of minute capillary vessels few cells in the body will be farther than about a hundred micrometers from a neighboring small vessel. Such small exchange vessels, the capillaries should be very narrow and large in number to ensure optimal diffusion, and this increases friction of blood in them. This is minimized by the very specific molecular structure of both the luminal surface of the endothelial cell lining and of blood cells, ensuring easy sliding along each other. Friction is also limited by the fact that larger distances are traveled by the blood in larger vessels. Getting closer to their target tissues such larger vessels (arteries) will divide into smaller and smaller branches, finally forming the capillaries. Capillaries will be collected again by repeated confluences into larger vessels, the veins. That is the basic principle how *vascular networks* are built. We can also easily recognize that such a geometry ensures that blood flow to each piece of the body can be separately controlled by adjusting the diameter of the minute vascular tubes leading to it (Abramson 1962, Cliff 1976, Schwartz 1980).

2.3 Distribution of blood flow in time: periodic pump and elastic pressure reservoir

Convection of blood in tubes with real friction can be maintained by continuous investment of mechanical energy. In many lower animals, a peristalsis-like movement of the blood vessel wall propagates blood in the vascular system, but in all vertebrates, motoric force is centralized at a discrete site of the circulation, the heart. Vessels leading away from the heart toward the tissues will be the *arteries*, and vessels leading and emptying the blood back into the heart will be the *veins*. Motoric force of the heart is produced by the heart muscles. Rotational pumps might be the solution for modern left ventricular assist devices, heart chambers with muscular walls could produce pumping force only in two phases, filling and ejection, which means that pressures and flows produced are inherently periodic. Periodic flow in tubes is highly uneconomic. This problem is circumvented by the *elasticity* of the vessels, especially of those close to the heart. These are filled with blood during the ejection period of the heart, and they press the blood forward by their elastic contraction while the pump is idle during its filling phase (see *Windkessel* function). Higher blood flow means the possibility of a higher tissue metabolism, higher speeds of muscle contraction, higher rates of neural, renal, splanchnic and skin functions, all advantageous for the individual. To press viscous blood through a system of microvessels needs a pressure difference. The less is the tissue's hydrodynamic resistance and the higher is the difference between inlet and outlet pressures, the higher the tissue flow will be. Diffusion will be optimal from a set of very narrow vessels. That determines a certain resistance for the capillary segment of the circulation. Such adaptation took place in the pulmonary circulation of mammals where vascular resistance outside the pulmonary capillaries is negligible. An other possibility to elevate tissue blood flow is to elevate the pressure head. In the systemic circulation of vertebrates outlet pressure, that is venous pressure, cannot be further decreased, as blood returning to the heart has close to atmospheric pressure. Arterial pressure, however, seems to be increasing in more developed forms of vertebrates, mammals having higher arterial pressures than reptilians, amphibians and fishes (Altman 1974, Schmidt-Nielsen 1979, Schwartz 1980).

2.4 Economic and independent control of blood flow in space and in time: resistance arteries and further elevation in blood pressure

But surprisingly, not all the energy provided by high arterial pressures will be used up to keep tissue flows at high levels. Substantial part of this energy will be lost, seemingly useless, in a short segment of the arterial circulation, in the *resistance arteries*. In healthy humans the mean arterial pressure of approximately 95 mmHg of larger arteries (inner diameters over 1 mm) will be halved in the small arteries and arterioles (inner diameters from 600 μm down to about 20 μm), pressures in the arterial side of the capillaries being around 40 mmHg. What might be the advantages of such a situation? For economic reasons, tissue blood flow should be adjusted to metabolic or other physiological needs. E.g. working muscle requires 30-50 times larger flow per unit mass than in the resting state. Large difference between maximum and minimum blood flows will be characteristic also for the splanchnic, renal and skin circulations. The solution is that in resting tissue small arteries will have smaller lumina due to continuous *smooth muscle contraction*, which can be dilated quickly as tissue needs increase, increasing local flow. Dilatation of a larger population of such resistance arteries should induce the collapse of pressure in the arteries, with collapse of blood flow to many parallelly connected tissues and organs. To ensure their blood flow they should also dilate to a certain level, further decreasing arterial pressure, again, with further needs for adjustments in all vessels of the body. A relative high, controlled *mean arterial pressure*, however, provides a pressure reservoir, from which all capillaries are supplied through a control segment of the resistance arteries (Abramson 1962, Cliff 1976, Milnor 1982, Nadasy 2007a). The mean arterial pressure in the reservoir is then controlled by feed-back mechanisms, adjusting heart pumping function and actual levels of overall peripheral hemodynamic resistance. And now we can reach the conclusion that by this mechanism, very high blood flows can be provided for functioning tissues, with a certain independence from affecting the circulation of other organs and tissues.

2.5 The price: unceasing pulsatile stress on the arterial wall

We needed that flow of reasoning to touch on a central problem of mammalian biology, which is a biomechanical one: *The wall of the arteries will be subjected to continuous and periodically changing forces arising from the pulsatile arterial pressure throughout the life* of the individual. This is a very specific problem in animal biology (Toth 1998, Nadasy 2007a, 2007b). Hearts should beat continuously, but the periods between two contractions (diastole) guarantee some time for biochemical, metabolic and circulatory recovery. The same can be told for periodic contractions of skeletal muscle and subsequent tendon loads and for the compression forces in bone and cartilage. But the artery wall can never get rid of the effect of the hard distending pressure and its periodic systolic elevations. All components of the wall had to accommodate to the omnipresence of distending forces. One possibility to reduce force per square millimeter section of the wall, on individual vascular constituents is to increase the thickness of the wall. The aortic wall, with about six times higher pressures is much thicker than that of the pulmonary trunk. Thicker wall means larger diffusion distances to nourish the artery wall itself. Diffusion in case of large arteries will not be sufficient, the supplying vessels (vasa vasorum) should enter the wall. Still the innermost layers of the large arteries will be avascular, as the pressures in the wall would compress any vasa vasorum in it. Avascular tissues are but a few in the mammalian body, comprising geriatrically hectic areas (tooth enamel, eye cornea, lens, article hyaline cartilage).

2.6 Force-bearing histological elements of the wall

A substantial part of the periodic stress due to the pulsatile component of the blood pressure will be met by the *elastic membranes* (Apter 1966). Their amount is high in arteries close to the heart, decreasing toward the periphery and diminishing in the smallest arteries with inner diameters below about 120 μm (true arterioles). The other connective tissue component, *collagen* lends rigidity and high tensile strength to the wall. Still there is some mystery about the omnipresence of *smooth muscle* in the aorta and in the large arteries. Contraction (reduced circumference) in these vessels is not extensive, and if any, it will hardly affect blood flows in such large vessels. It is widely accepted, that their tone sets optimal *elasticity* of the artery wall. Contracting, they strengthen the cytoskeletal elements (*intermediate, actin and myosin filaments*) in the wall. These cells thus are among the parallelly and serially connected force-bearing elements of the vascular wall. The *dense bodies* are forming a lattice network with *intermediate filaments connecting* them. Parallel bunches of thin (actin) filments attach also to the dense bodies and to the *hemidesmosomes* of the smooth muscle membrane. *Thick (myosin) filaments,* interconnect opposing actin filament bunches, and with the ATP-fueled *actomyosin crossbridges* can pull them closer to each other. Active slide of actin and myosin filaments upon each other ensures thus smooth muscle contraction. Vascular smooth muscle, can characteristically form very slow cycling of cross-bridges even at actively shortened length, yielding the typical *latch contraction* (Rhee 2003, Somlyo 1968). And it can be proven that at least part of vessel wall *viscosity* has to be attributed to passive slide in their contractile apparatus. All smooth muscle cell is surrounded by a basement membrane. In addition, several proteins of the mechanical transmission between intra- and extracellular fibers and filaments forming the *mechanical anchoring structures* have been identified (Clyman 1990, Gabella 1984).

In resistance arteries, however, contraction of smooth muscle will massively affect blood flow to the affected territory. The relative thick wall of these vessels will result that a relative slight contraction of a circumferentially positioned smooth muscle cell at the outer surface will induce a much more effective reduction in the inner radius.

3. Mechanics of solid materials and fluids – their applicability for vascular mechanics

Blood vessels are subjected to general laws of physics and mechanics, several of the parameters applied to study non-living material and several of the general mechanical laws find a broad application in the field of vascular mechanics (Bergel 1961, 1964, Fung YC 1984, Gow 1972, Monos 1986). We must not forget, however, that *vascular (living) tissue is one of the most complicated semi-solid materials ever studied by specialists.* There are some specific characteristics rarely found in non-living material. Such is the build-up of the whole structure under conditions of periodic and continuous distending and shear forces. The geometry of the specimens, the amount, quality and direction of force-bearing fibers, their mechanical interconnections with each other specially adapt to the in vivo occurring mechanical forces. The *force bearing elements in the vascular wall are mostly fibers, arranged in direction of the forces, able to bear pulling forces only.* Pushing forces are rare in the wall, maybe they can be produced from compression of closed, deformable fluid compartments and after pathologic calcification of the tissue. That complicates the understanding of cyclic viscoelastic events. How then, elongation of viscous units can be restored? The ability, never seen in non-living material, to *produce active stress at the expense of chemical energy is the solution.* And in all mechanical studies, it is an ever present complicating factor. Smooth muscle tone will massively affect not only existing geometrical appearance (lumen size and wall thickness), but will modify elastic properties, affect tissue homogeneity and, as we will see yield a substantial part of tissue viscosity. For this reason, biomechanical measurements should be made either in vivo or under in vitro conditions that mimic the in vivo situation in composition of the tissue bath in which the vascular tissue is tested. The vascular smooth muscle tone should be set to supposed in vivo values, or, the measurements should be made at different levels of smooth muscle tone. Unfortunately, the smooth muscle tone itself does change in response to distending forces (myogenic response) or to endothelial shear of flow (endothelial dilation). For many non-living material the stress-strain characteristics will be conveniently linear at least in a certain segment of the curve. That allows the definition of a single elastic modulus to characterize elasticity. Rigidity of vascular walls, however, always heavily depends on actual values of wall stress, the higher is the stress, the steeper will be the stress-strain characteristic curve, providing higher values of their locally computed ratio (tangent), the incremental elastic modulus. Attempts to find a simple description how the elastic modulus of the vessel wall changes with stress failed until now. Hopes that the elastic modulus linearly changes with stress (an exponential shape for the stress-strain relationship) did not bear the critics of more accurate measurements. According to our experience, a double-exponential approach yields almost satisfactory results (Orosz 1999a, 1999b).

4. Network and branching geometry

We must not forget that hemodynamics will be determined at least as much by network properties of the whole networks than by properties of individual vascular segments. However, networks lend themselves to study and analysis with much more difficulty, both methodical and computational, than do individual segments. For this reason network properties are much less analyzed in the literature. For want of space we will refrain from a more detailed analysis of the effect of mechanical factors on the development of the network properties. Network developments seem to follow the law of minimum energy requirement (Rossitti 1963). That can be altered in aging networks and at chronically elevated pressure (Nadasy 2000, Lorant 2003). A well analyzed territory is the retinal arteriolar network. Rarefaction, that is, the decreased number of parallelly connected resistance arteries seems to be an important contributor to morphologically elevated vascular resistance in chronic hypertension (Harper 1978). The "chaos theory" seems to be one fruitful approach to describe general laws of geometric vascular network development (Herman 2001).

5. Segmental geometry

5.1 Optimal cylindrical symmetry

Most vessels, especially arteries are smooth lined, long cylindrical tubes, positioned in-between larger branchings (Schwartz 1980). This shape is optimal to ensure minimum loss of hydrodynamic energy provided by heart contractions and homogenous distribution of force around the circumference and along the axis, produced by intraluminal pressure. In real situations, however, especially in pathologic ones, deviations from this optimum do occur, in the axial, circumferential and radial directions.

5.2 Disturbances of axial symmetry

To reach their anatomical targets vessels should bend, but that *axial bending* is usually kept to a minimum by adjusting the axis to an arched curve with a large radius. Anatomical situation, however, can force the course of a vessel axis into a narrow bend. The typical anatomical pattern of the large artery system of mammals with the aortic arch itself forms a narrow bending for a very large mass of flowing blood. A sensitive area in human vascular anatomy is the base of the skull, here the inner carotid artery is forced into a narrow, S shaped bony channel, the carotid siphon. Arteries passing joints should follow the position of the joint. In mammalian embryology, a frequent situation is that vessels originally developing as branches deviating in an angle from mother vessel will enlarge their lumen and taking over the role of the distal main branch, which itself then regresses. The originally sharp angle of the axis in such cases will be later splayed to an arch as a rule. Somewhat similar situation can be observed in adult pathology, when developing collaterals bypass the site of slowly developing vascular stricture. Adjustment of the course of the axis is not as effective in such cases, and a broken course of an artery will be a frequent observation on X-ray angiography (coronary, leg). *Irregular course* is a frequent pathological feature in resistance arteries, too. It can be observed in retinal arteries in hypertension and in aging and is one of the main symptoms of the venous varicosity disease. One current explanation for pathomechanism of varicose notches is that as pressure-induced axial elongation will not be counteracted by sufficient axial prestretch and tether, the vessel axis bends first, then with increasing instability it irreversibly buckles into one direction.

Axial irregularities of lumen diameter and wall thickness are the very essence of vascular pathology. In fact other irregularities of lumen shape will frequently go on unnoticed until the events will develop toward local narrowing, disturbing flow or induce local distension, aneurysm, compressing neighboring tissues or endangering with imminent rupture and bleeding. However, there is a physiological disturbance of cylindrical symmetry at side branches of arteries. An endothelial cushion just over the orifice ensures that axial blood rich in red blood cells will be diverted into the side branch, preventing thus plasma skimming. *Focal pathologic processes* typical for arteriosclerosis will typically disturb cylindrical symmetries in all directions. On the other hand, such focal lesions in turn typically develop where bends, angles, side branches, strictures by impressions of surrounding tissues disturb cylindrical symmetry of vessel shape and laminar flow. Uneven lumen and wall thickness along the axis in many resistance arteries is almost the definition of the diabetic microangiopathy. This causes tissue flow disturbance and microaneurysms endangering with rupture.

5.3 Circumferential deviations from cylindrical symmetry

Slight circumferential deviations from cylindrical symmetry are inherent in case of vessels running on bony surfaces. Careful analysis shows that the thoracic and abdominal arteries are not fully circular, but of an ovoid shape with a somewhat wider base from which the intercostal and lumbar arteries emerge. Ellipticity of lumen cross section has been thought to be the very essence of venous mechanics. And really, certain veins, e. g. the lumen of human inner jugular vein forms but a narrow slit at low pressures, which is for this vessel, in the erect body position. Other veins, however, are surprisingly circular even at fairly low pressures. Not much deviation of the anteroposterior and mediolateral diameters of the human brachial and axial veins could be observed by in vivo ultrasonographic measurements in a wide pressure range (Berczi 2005). While increasing ellipticity is characteristic for cannulated venous segments in the low pressure range in vitro, in vivo, or even in situ, such collapse of one of the diameters is restricted by the radial tethering provided by surrounding fat and fascial tissue down to 0 mmHg transmural pressure (Nadasy unpublished). Disturbances of circumferential symmetry, however are occurring as a rule in case of focal atherosclerotic lesions and in any case of mural thrombosis. Present techniques at hand can analyze the differences of histologic composition around the vessel circumference (and in the wall along the radius), but the biomechanical consequences, uneven distribution of force on force bearing elements, are still poorly understood. We are convinced, however, that it is a key issue in the pathomechanism of the progressive development of the arteriosclerotic plaque. With destruction of the inner media in a sector of the wall, large pulsatile forces will be transmitted to the outer layers in this segment, with the consequence of accumulation of collagenous fibers and cessation of vasa vasorum flow. While some remodeling of the force-bearing elements of the wall can make revascularization possible, a necrotic nucleus, getting closer to the luminal surface and endangering with rupture into the vascular lumen, will be the most dangerous threat caused by the focal process. Some modern techniques raise the hope that distribution of force inside the vessel wall could be once directly studied. Greenwald has directly demonstrated the sequential strengthening of connective tissue elements (Greenwald 2007).

5.4 Radial asymmetry

Concerning the *radial asymmetry,* original views that a rigid adventitia could prevent further distention of the elastic media ("an elastic ball in a string bag" model), still vivid in the views of non-specialists has been opposed by direct elastic measurements on vessels from which the adventitia has been removed. Right now it seems that the adventitia, with its mostly loose connective tissue, is the site more for the axial tether, than for any contribution to circumferential force-bearing. Vasa vasorum, sympathetic nerves can run in it undisturbed by tissue pressure, the fibroblasts in it with their ability to differentiate into vascular smooth muscle cells can ensure an "appositional" medial thickening. There is an inherent, physiological radial inhomogeneity of the media itself in the wall of large arteries, circumferential elastic sheets (whith holes in them) and smooth muscle cells packed in angle with radius in a fish-bone pattern are forming alternative layers. Taking into consideration that intraluminal pressures at the inner surface should be decreased down to zero at the outer surface, it is really surprising to observe, still how similar are these layers and their elastic and smooth muscle components. This supports, unproven yet views that some equalization process in the media should exist, that distributes the large circumferential force to the similar wall constituents in a similar manner. Some radial inhomogeneity of force distribution, however, should exist in the wall. This can be proven by the elegant experiments of just cutting up vessel rings in the radial directions. The ring will be opened, the angle of which in such a state of zero stress can be measured and analyzed (Liu 1988). We must not forget, however, that the artery wall is never at zero at stress in vivo, fiber arrangement adapted to real pressurized wall tensions. In case of larger vessels the contribution of the endothelium to elastic properties of the wall is thought to be negligible. However, the basal membrane of the capillary vessels lends sufficient rigidity and tensile strength to these vessels. In addition, intimal thickenings of sclerotized vessels can take up a substantial part of wall stress, relieving thus the outer layer of the affected segments.

5.5 Vascular diameter

Vascular specialists with biomechanical backgrounds are rarely satisfied when "the" diameter of a vessel is mentioned. All vessels alter their diameter as a result of acute smooth muscle contraction, and the same measured intraluminal diameter could mean very different vessels at different levels of vascular smooth muscle tone (a larger vessel but with a larger tone), and different measured intraluminal diameters could mean a morphologically identical vessel segment but with a somewhat altered tone. The diapason between maximum and minimum contractions is routinely measured now in wire and pressure angiography, and such practice is more and more frequently applied in in vivo measurements and to some degree, even in clinical practice.

For the exact biomechanical analysis, we have to discriminate between the morphological diameter of the segment, best characterized by its fully relaxed state, its diameter in full contraction, which in healthy arteries below about 1 mm of inner diameter will be the fully closed segment, and the actual diameter measured in a discrete state at a given level of muscle tone. Things will be even more complicated when we realize that vascular lumen will also be dependent on transmural pressure, elasticity of the wall and also even on axial distension. Fortunately, relaxed vessels pressurized close to or somewhat over physiological pressures, turn to fairly rigid structures and do not further change much their lumina as a function of pressure. This stable diameter will well characterize the morphological lumen. In the scientific practice it is even more accurate to characterize morphological lumen with the whole course of the relaxed pressure-diameter curve.

5.6 Physiological control of the morphological lumen

The "passive" (morphological) lumen will be differently controlled and with more delay than the actual lumen is determined by the actual level of smooth muscle tone. The morphological control process needs a reorganization of the histological components of the wall. The terms *"remodeling"* (segmental remodeling, geometrical remodeling, wall remodeling) or *"long term control"* are used to describe it (Fig.2.and 3.).

It had been known for ages that vessels with larger flows have larger lumina. The problem can be reduced to the question, that branching of a larger mother vessel how will effect the lumens of the smaller and smaller daughter branches? Early analysis of pressure and flow in vascular networks have shown that while mean linear velocities are decreasing toward smaller branches (30 cm/sec in the aorta, a few hundred micrometers in the capillaries) there is an elevation of mean pressure drop per unit length. There is hardly any drop in mean arterial pressure in large arteries, substantial pressure drop occurs along a few cm length of small arteries with a few hundred μm of diameter, finally, a sharp drop of pressure happens in arterioles, a few mm of lengths, but with diameters between 30-150 μm. (Abramson 1962, Cliff 1976, Milnor 1982, Schwartz 1980). In the simplest case of symmetric branching, to maintain the mean linear velocity in daughter branches would need a ratio of radii of daughter (r_d) to mother (r_m) branches of $\Pi r_d^2 + \Pi r_d^2 = \Pi r_m^2$; from whence $2\, r_d^2 = r_m^2$ and $r_d / r_m = 1/\sqrt{2} = 0.707$. To maintain the unit pressure drop per unit length (with unaltered viscosity, following the Hagen-Poisseuille law) would need daughter to mother radius ratios of $Q=\Pi/8^* r_m^4/\eta^*\Delta p/\Delta l = 2^*\Pi/8^* r_d^4/\eta^*\Delta p/\Delta l$; from whence $2\, r_d^4 = r_m^4$ and $r_d / r_m = 1/\sqrt[4]{2} = 0.841$. Hemodynamic analysis of existing arterial networks thus leads us to the conclusion, that in case of symmetrical branching daughter to mother branch ratios should be in-between these two values, m $0.707 < r_d / r_m < 0.841$. Measuring many arterial diameters Murray has suggested, that in case of any types of branchings, the equation of $r_m^3 = r_{d1}^3 + r_{d2}^3 + r_{d3}^3 + r_{d4}^3 + \ldots$. will be valid. This seems a fairly good approach even in our days. How the vessel wall should "know" how much is the flow in its lumen? The answer was given in two classic works by the great American cardiologist, Rodbard, who supposed that endothelial shear is somehow sensed by the endothelial cells and is kept constant by chronic morphogenetic processes adjusting vascular lumen to flow. The value of endothelial shear rate (dv/dr) computed based on the Hagen-Poiseuille law is $dv/dr = 4Q/\Pi r^3$ (where Q is the volume flow and r is the inner radius) being in accordance with Murrays law. For our symmetric bifurcation, $dv_d/dr_d = dv_m/dr_m$ and $4(Q/2)/\Pi r_d^3 = 4Q/\Pi r_m^3$ from whence $2\, r_d^3 = r_m^3$ (the form corresponding to Murray's law) and finally, $r_d / r_m = 1/\sqrt[3]{2} = 0.794$. This latter number is just in-between 0.707 and 0.841 as required by common hemodynamic experience. Validity of such computations has been proven in analysis of several types of vascular branchings (Lorant 2003, Nadasy 1981, Pries 2005, Rodbard 1970, 1975, Zamir 1977).

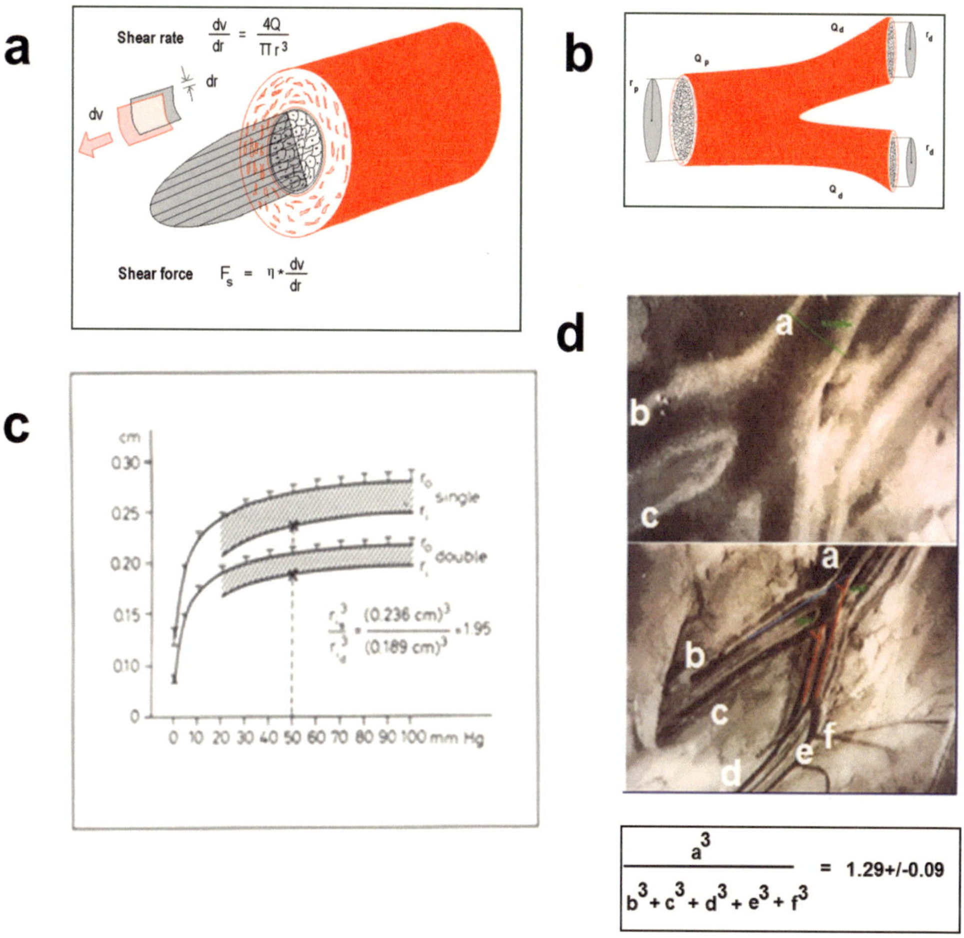

Fig. 2. Morphological (long-term) control of vascular lumen. a. Shear rate and shear stress in a cylindrical vessel with continuous flow. b. Geometry of a symmetric branching. c. Relevance of the Murray-Rodbard law: Pressure diameter plots of normal double and single (morphologically malformed) human umbilical artery segments. The ratio of the cubes of inner radii at physiological pressures is around, 2 which can be expected in case of doubled flow in single arteries. (From Nadasy 1981, with permission of Karger) c. Relevance of the Murray-Rodbard law: In vivo microprepared popliteal confluence of the rat saphenous vein. Video-microscopic records at two magnifications with normal pressure and flow in the lumen of anesthetized animals. The ratio of the cube of diameter of the mother branch to the sum of cubes of diameters of daughter branches is close to the expected 1 (from Lorant 2003, with permission of Physiol Res, Czech Academy of Sciences).

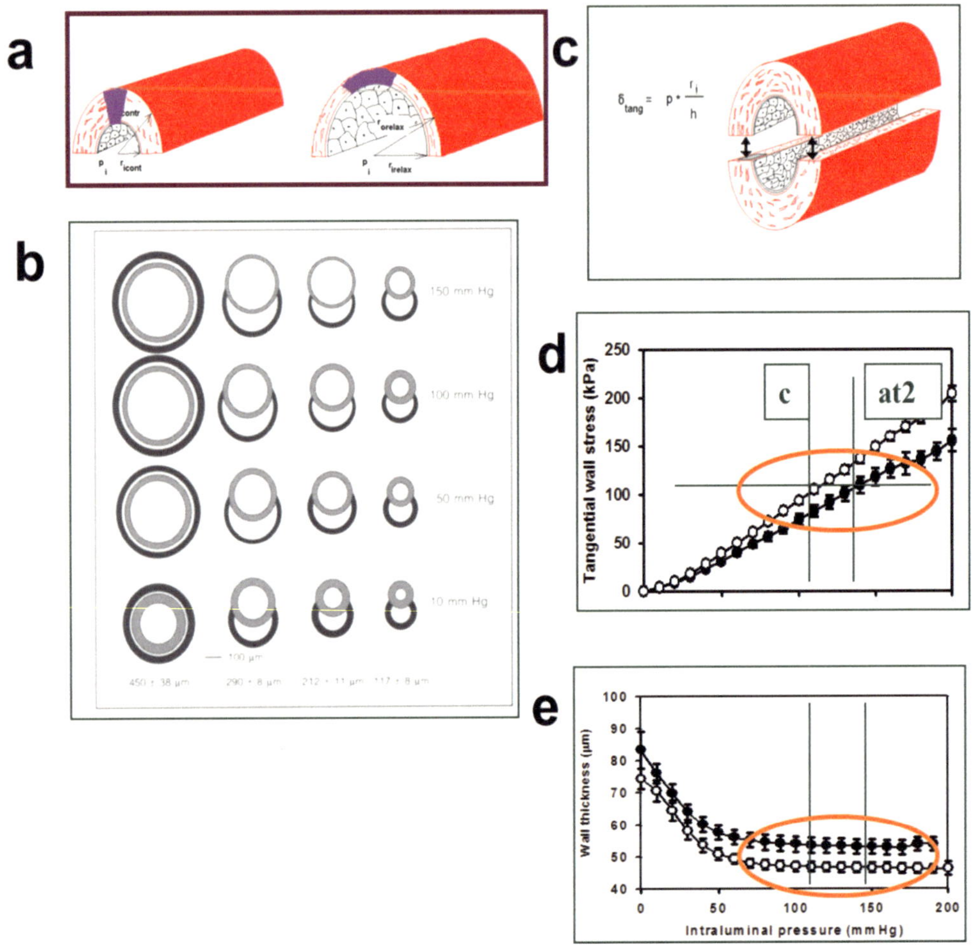

Fig. 3. Long-term control of vascular wall thickness. a. Vessels with thicker walls can more effectively control their inner diameter, a feature characteristic for resistance arteries. b. Series of intramural coronary resistance arteries from the rat. Contours of cross sections are shown. From left to right, morphologically different segments with diameters around 4-500, 300, 200 and 100 micrometers. From up to down, at different transmural pressures. Passive and contracted segments are shown. Note typical increasing thickness-to-diameter ratio toward smaller arteries, increasing effectivity of lumen control. (From Szekeres 1998, with permission of Karger) c. Parameters determining tangential wall stress. d and e. Wall stress and wall thickness of a resistance sized small artery from normotensive and hypertensive animals. Note that elevation of wall thickness (and reduction of lumen) just ensured unaltered wall stresses at elevated in vivo pressures as could be expected by the Folkow-Rodbard-Mulvany law (from Nadasy 2010, with permission of Karger).

At a more scrutinizing analysis, however, the situation will be more complicated than that. In fact, not shear rates but shear forces will be sensed. This latter will be the function of blood viscosity, a fairly elusive factor supposing its dependence on shear stress (the blood being a non-Newtonian fluid) and on the level of accumulation of red blood cells in the axial flow (according to the Fahreus-Lindquist law). At a fairly good flow, and especially in smaller vessels, almost clear plasma, devoid of cellular elements will slide along the luminal surfaces of endothelial cells, yielding a fairly continuous viscosity of 2-3 cP. The other problem is the changing levels of flow (depending on pressure and on more distal resistance) and changing levels of arterial tone. Based on experience on skeletal muscle circulation, we have to suppose that relative short periods with high flows will be sufficient stimuli to increase the morphological lumen. On the contrary, when maximum flows do not come for a while, shortened circumferences could get morphologically stabilized reducing the range of luminal vascular smooth muscle control (Nadasy 2010a). An other problem will be prominent when we compare arteries and veins. Mean velocity of blood flow is about half as fast as in corresponding arteries. We have to suppose that either other factors than blood flow contribute substantially to morphological lumen control in veins or, the venous endothelium is differently tuned to flow sensation than is the arterial. To solve these questions would be essential to reveal the pathomechanism of chronic venous diseases. An other complication is rising from wall thickness adjustment to elevated pressure, which, can reduce lumen below levels required by flow inducing thus resistance elevation and having a stabilizing effect on the high blood pressure. Which is the cause and which is the effect? Fine network analyses in different states of circulation and in different stages of the hypertension disease will be needed fully to describe the intermingling feedback networks of flow, resistance and pressure. We believe that a disturbed endothelium will not be sufficient to restore morphological lumen to rare flow maxima, or to counteract the lumen-narrowing effects of adaptive wall thickening. Extensive studies revealed a set of cytophysiologic and even molecular mechanisms how shear is sensed at endothelial luminal surfaces.

Summing up this chapter, we have to conclude that despite many remaining questions, the law, describing the long term morphological control of vascular lumen formation, originally found by Murray and by Rodbard (1970, 1975) has proven its validity for a substantial period of time (Kamiya 1980, Lorant 2003) and can be accepted as one basal law of normal vascular functioning. It can now stated with a high level of certainty that long term control processes in the vascular wall do exist that adjust morphological lumen to flow so that they tend to stabilize endothelial shear.

5.7 Physiological control of vessel wall thickness

Vessels with higher pressures have thicker walls. The pulmonary arteries have thicker walls than the caval veins and the aorta than the pulmonary trunk, despite similar flows. Higher pressure means higher tangential force per unit length (F) on vessel circumference, $F = p^*r_i$, where p is the transmural pressure and r_i is the inner radius (Fig.3.). That can be distributed on a vessel wall thickness of h, $\sigma = p^*r_i / h$, where σ is the tangential stress. Folkow observed that arterial wall thickness increases in a compensatory manner in hypertension (1971, 1990, 1995). An other assumption published by Rodbard (1975) stated that morphological thickness of the vascular wall is controlled to stabilize the value of tangential stress. But there are more problems with this observation. As there is hardly any drop in mean arterial pressure along Windkessel and distributing arteries (to about 600 μm of inner diameter) inner radius to wall thickness ratios should have remained unaltered along the whole arterial tree to make σ unaltered, too. In fact radius to wall thickness ratios decrease toward smaller arteries. What is even more contradicting, in more distal resistance arteries following a substantial pressure drop, radius to wall thickness ratios should increase to compensate for lower pressures to ensure stable values of tangential stress. Just the opposite is the case: small resistance arteries have relative thicker walls, and computed in vivo values for tangential stress decrease along the arterial tree reaching very low values in smallest resistance arteries. First we have to see what is the advantage of such a difference (Fig. 3a). Smooth muscle in large vessels will not contract to induce substantial reduction in diameter. That would be useless, as no substantial pressure drop could be reached taking into consideration real flow and viscosity values. But there is an opposite situation at the level of the resistance arteries: fast and effective acute changes in lumen are the very essence of their physiological functioning. Contraction of a helical smooth muscle cell ring at the outer surface of a resistance artery wall will be much more effective than the contraction at the inner site of the wall, the difference will be the higher the thicker the resistance artery wall is. A 20 % contraction at the inner circumference will induce 20% reduction of the lumen (2.4fold increase in segmental resistance). A similar 20% contraction at the outer circumference will induce 32.9, 47.0 and 100% reduction in lumen in vessels with 10:1, 8:1 and 6:1 radius to wall thickness ratios, respectively as they will push the inner vascular wall layers into the lumen. This will result respective segmental resistance elevations of 4.9 and 12.7 times in the first and second cases, while the lumen will be fully closed and flow will cease in the third situation. The question, however, can be raised, that as tangential stresses are so much less in smaller arteries, are the smooth muscle cells themselves in this vessels so much different? Several differences between smooth muscles of these vessels could be listed, but one outstanding difference is the presence and amount of elastic tissue which is diminishing toward the peripheral arteries practically synchronously with the reduction of the r_i/h ratio. A simple solution can be that while smooth muscle is similarly stressed in large and small arteries, the parallelly connected elastic membranes will bear a substantial part of the tangential strain. But later we will see that amount of elastic tissue will develop in response to pulsatile not steady stress. And we will also see that elastic tissue has also its impact in the lower part (below diastolic pressures) of the arterial pressure-diameter characteristics. And with that restriction the Folkow- Rodbard-Mulvany's law on long term control of the thickness of the vascular wall can now be valid (Rodbard 1970, 1975): *Vessel wall thickness develops to stabilize tangential wall stress – valid, when tissue composition is unaltered.*

Thickened wall of affected vessels is a main alteration in case of chronic arterial hypertension (Folkow 1971, 1990, 1995), chronic pulmonary hypertension and venous pressure elevation in chronic venous disease. In several clinical and experimental studies, hypertensive remodeling of arteries seemed just to stabilize in vivo tangential wall stress. Such control mechanisms should exist at least in a certain phase of the hypertension disease (Albinsson 2004, Dickhout 2000, Frisbee 1999, Hayashi 2009, Nadasy 2010a, Pries 2005).

6. Vascular elasticity

6.1 Significance of vascular elasticity

As we could see above, elasticity is a very important, inherent property of blood vessels, it ensures the fairly continuous pressure and flow in small vessels despite periodic functioning of the heart pump (Bergel 1961, 1964, Fung 1984, Gow 1972). Actual geometric properties of a vessel, determining their hydrodynamic resistance are in turn determined by their morphological geometry, the actual level of smooth muscle tone, transmural pressure, luminal flow inducing endothelial dilation, their axial stretch and the elasticity of the wall. The role of elasticity in forming the actual geometry will be more important in vessels with high distensibility, where small changes in transmural pressure will induce large changes in lumen volume in such vessels. As we will see later, most vessels are relative rigid at and especially above physiological pressures. That means that elasticity will be a central parameter in determining lumen size when pressures in the lumen decrease below physiological pressures. We have a good reason to think that elasticity of force-bearing fibers in the vascular wall helps to maintain a fairly even distribution of distending forces and stretches among parallelly and serially stressed wall components: elongation of a stressed fiber transfers part of the force onto the parallelly connected poorly stressed fibers. More elongation in the less stressed serially connected fibers will help an even distribution of stretch among the serially connected circumferential elements of the wall. One central problem in vascular pathology, we believe, is the elastic force-bearing capacity of the inner layers of larger vessels: by their force-bearing they relieve the outer layers. Decreasing hydrostatic pressures in the outer layers of the vessel wall make possible the proper functioning of the vasa vasorum and proper nourishment of the whole wall.

6.2 Parameters measuring elasticity

We can determine the elasticity of a vascular segment *in vivo* directly, by measuring geometrical parameters (inner, outer diameter, wall thickness) at different levels of pressure (Fung 1984). Modern *ultrasonography* (from the surface of the body and also intravascular) provides sufficient means for larger vessels (Berczi 2005, Molnar 2006, Mersich 2005, Shimazu 1986). A certain level of manipulation of pressure and smooth muscle tone is possible even in human volunteers (See Fig. 1c.). In vivo animal experiments of course give a wider and more accurate potential for that. Delicate measurements can be made on isolated, axially isometric, cylindrical vascular segments mounted in the tissue bath of a *pressure angiometer* (Cox 1974, Duling 1981, Nadasy 2001, see 1b.). The pressure-diameter characteristics can be recorded at different levels of smooth muscle contraction (with vasoconstrictors added) and in the so called passive state (fully relaxed smooth muscle e.g. with calcium-free incubation medium). We can cut rings, circular or helical strips and study them in a *wire myograph* (Fig. 1a.). To characterize circumferential (tangential) elasticity we can compute *compliance*, $C=\Delta V/\Delta p$, that is volume alteration in response to unit alteration in (transmural) pressure. The question, answered by the compliance value is, that how much the pressure will change if we press a certain amount of blood into the vessel. This parameter is frequently used to characterize venous elasticity and even for large arteries. One problem with it is, that a rigid, large vessel will have larger compliance than a small elastic one. To circumvent this problem the term *distensibility* has been applied, simply normalizing volume change to the initial volume (V_o), $D=\Delta V/ V_o *(1/\Delta p)$. Distensibility almost fully describes the elastic properties of the wall, the way it is taking part in hydrodynamic processes, and is used frequently in hydrodynamic models for this reason. However, it will not properly characterize the elasticity of the wall material as more elastic, but thicker walls can have the same distensibility. The Young's *elastic modulus* will be computed, which is the tangent of the stress-strain relationship. For cylindrical segments with inner and outer radii of r_i and r_o, respectively, with not negligible, but relative thin walls (valid for most vessels) the equation given by RH Cox (1974, 1975a, 1975b), the inventor of pressure angiometer (Fig. 1b.) is mostly accepted: $E= 2r_o r_i^2/ (r_o^2 - r_i^2)*(\Delta p/\Delta r_o)$, where Δp is the pressure change inducing an alteration of the outer radius, Δr_o. As we described it earlier, elasticity of the vascular wall will be heavily dependent on the conditions under which we have measured it. For this reason, we have to repeat the measurements at different levels of wall stress (intraluminal pressure) and at different levels of smooth muscle tone. Typically, pressure-diameter characteristics will be recorded at different levels of smooth muscle tone. If pressure alterations are slow enough, we can suppose that the wall is transiting a series of equilibrium states and each further infinitesimal elevation in pressure (stress) will induce an infinitesimal rise in circumference (strain) and the tangent of the normalized pressure-volume curve (incremental distensibility) and tangent of the stress-strain curve (incremental elastic modulus) can be computed and plotted as a function of pressure ("isobaric" parameters). Or, in case of the incremental elastic modulus, it will frequently be given as a function of computed wall stress. Such elastic modulus-tangential stress characteristics will characterize best the elasticity of the wall material itself. The vertical axis is usually logarithmic, but will not be linear even in this form. The question here can be raised how reproducible and how characteristic for the in vivo situation the elastic parameters measured this way will be? One problem is the different levels of smooth muscle tone. That has to be somehow stabilized, which is not so easy as it itself changes with changing wall stress. In case of a well developed myogenic response, as it is the case with many resistance arteries, elevated pressures will not produce elastic dilation but myogenic contraction (Kuo 1988, Osol 1985, Szekeres 2004). The term plasticity is used for a typical behavior pattern of vascular smooth muscle: the material of the wall somehow adapts to lasting pressure loads or pressure patterns. The mechanical past of the vessel (in the last few minutes) determines to some degree its present mechanical behavior.

For this reason, reproducibility should be ensured by preliminary incubation of the segment under controlled contractile and mechanical conditions (continuous or cyclic stress or pressure load). Stress or pressure changes during elastic measurements should be applied following a reproducible pattern: rises at continuous rates, stepwise rises, cyclic sinusoid or triangle patterns are widely used. Because of viscosity, characteristics taken with increasing and decreasing loads might differ. Elastic moduli in the range of 10^4 Pa can be considered as very low, found only at very low wall stresses, at a few mmHg intraluminal pressure in arteries, and close to 0 mmHg pressure in veins. Values of 10^6 Pa (10^5-10^7) are typical for many vessels in their physiological pressure range. Further elevating the pressure values in the lumen 10^7 Pa can be reached with any type of vessels (veins included!), but a damage to the wall (mostly reflected by reduced smooth muscle contractility) is in such cases imminent. The tensile strength of healthy vessels is surprisingly high, veins can be sutured as bypasses into the arterial system and arteries will endure for a shorter period 1 atmospheric transmural pressure. Tangential stresses in vivo range from the very low values of a few 10^4 Pa (e.g. 30 mmHg = 4 kPA pressure, r_i/h values around 3, for a thick walled, contracted arteriole, $\sigma= 1.2*10^4$ Pa) to highest measured values (e.g. 150 mmHg=20 kPa elevated systolic pressure for a thin walled large artery with r_i/h ratios around 10, $\sigma= 2*10^5$ Pa). Stresses in the range of 10^6 Pa will damage the wall as maximum forces produced by the smooth muscle are in the range of 3-5 atm (~3-5*10^5 Pa, 200-400 mmHg for a large vessel with a radius to thickness ratio of 10).

In in vivo experiments and in the clinical practice, too, vascular elasticity is frequently measured indirectly. The shape of the aortic pressure curve can be analyzed. The so called augmentation index gives an indirect information about the rigidity of the large arteries. Even more popular is the determination of the pulse wave velocity. These will be discussed in more detail in the chapter on the Windkessel arteries.

6.3 Effect of histological composition on vascular elasticity

After several decades of systemic investigations now we have a fairly good if not final picture how the histological composition of the vascular wall affects its elasticity (Apter 1966, Bergel 1961, 1964, Cox 1975a, 1978, Dobrin 1978, Fung 1984, 1995, Gow 1972, Greenwald 2007, Koens 2010, Oxlund 1986, Roach 1957, VanDijk 1984, Vidik 1982). The endothelium will not much contribute to the mechanical properties with the exception of capillaries and maybe, of the smallest other vessels. But we must not forget, that it is the endothelium that senses the shear at the inner surface of vessels and adjusts the diameter to it (see endothelial dilation, and vascular lumen control). Also, the endothelium, with the basement membrane underneath it, is a fairly mechanically stable structure forming the capillary walls. Capillaries in the renal glomeruli seem to be enforced by the leg processes of the podocytes there. Contrary to earlier expectations, the loose connective tissue of the adventitia will not restrict pressure-induced dilation of an elastic media ("elastic balloon in a string bag" model, Burton 1954). The collagenous fibers running in it will ensure some axial tether when, with decreasing pressures the axial extension of the arteries decreases. In case of veins at very low pressures, and in arteries at extremely large levels of medial smooth muscle contraction, we can suppose even some adventitial radial tether. Vascular smooth muscle if relaxed does not resist distending force until the intracellular fiber structure is not stretched. Vascular structures containing abundant smooth muscle (umbilical artery, resistance arteries) have very low elastic moduli at low stresses and high moduli at high stresses. At least part of the elasticity of the stretched vascular smooth muscle will be determined by elasticity in the actomyosin crossbridges ("series elasticity", Mulvany 1981, Siegman 1976), and should reflect the mechanical deformation of the myosin head or neck in response to distending forces. In fact, as each smooth muscle cell is surrounded by a socket of basal lamina, composed of collagenous fibers, and there is an intracellular scaffold in them formed of intermediate filaments and dense bodies, and to go further, bundles of actin filaments are more abundant in them than thick myosin filaments, contribution of these latter structures to "Series elasticity" cannot be excluded. We have found that elasticity of vascular tissue, rich in smooth muscle can be explained as originating from a "unit elasticity", possibly the elasticity of the actin filaments. Upon cyclic loading, the number of serially and parallely connected elastic units could adapt by breaking up and passive slide of overstressed latching actomyosin crossbridges as well as by spontaneous shortening in understressed ones. That ensured uniform stretch and load in affected filamentous units (Nadasy 1987, 2007b, Szekeres 1998).

Collagen lends rigidity and high tensile strength to the vascular wall, again, at high distending forces, while very moderate forces are sufficient to strengthen the coiled up collagenous bundles in the vessel wall. More collagen in the wall usually means higher rigidity (Apter 1966, Bergel 1961, 1964, Cox 1978, Dobrin 1978, Gow 1972, Greenwald 2007, Hegedus 1984, Oxlund 1986, Vidik 1982), but usually only at higher stresses. Collagen accumulation is typical in many types of diseased vessels ("sick vessel syndrome", Heistad 1995). Contrary to the other two wall constituents, *elastin* will resist distension even at very low stretches but will not be fully stretched even at high distending forces. The result is that its presence elevates elastic modulus (increases rigidity!) at low pressures, but decreases it at high tangential forces (Fig. 4b.). Frequent contradictions about connective tissue composition and measured elastic modulus can be prevented by exact analysis of the pressure and tangential stress levels where the measurements have been made. Plots of tangential elastic modulus against tangential stress usually are in good accordance with histological composition. Such plots are accepted as reflecting the elastic properties of the wall material itself. Still it is poorly understood how the elastic laminae are mechanically connected to smooth muscle. In rabbit aortic strips we have found that series and parallel elastic components of this elastic tissue changed parallelly upon passive stretch (Nadasy 2007b). But similar observations were made in the aneurysmic tissue fully devoid of muscle and elastic components (Toth 1998).

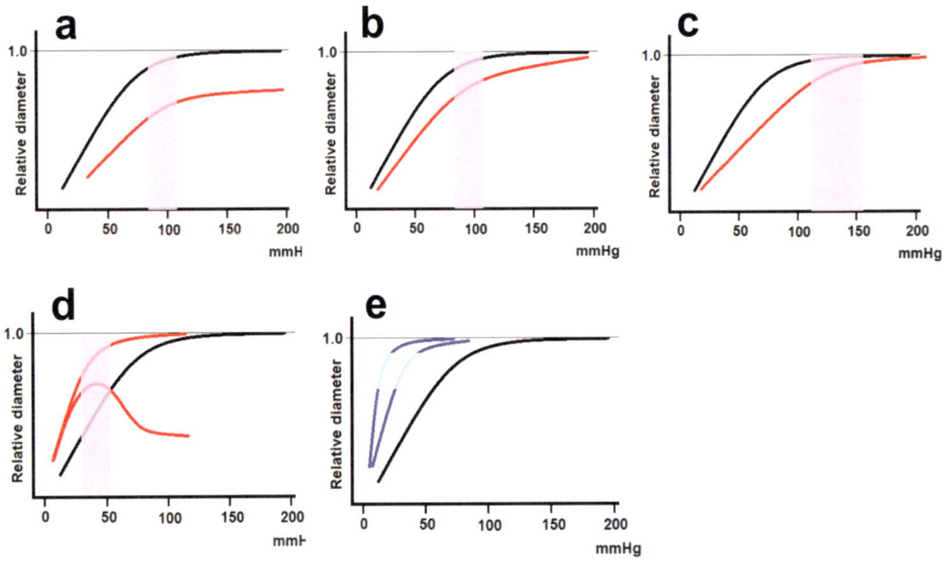

Fig. 4. Long and short term control of vascular elasticity. Comparisons of pressure-diameter characteristics of different vessels. Relative values of diameter are shown. Black lines represent the characteristics of a typical distributing artery in the relaxed state (same in each diagram). Shaded areas mark the range of in vivo pressures of vessels being compared (drawn in color) a. Red, the same artery in the contracted state. Smooth muscle contraction reduces elastic modulus as a function of pressure but increases it as a function of strain. (Cox-Dobrin law) b. Comparison with a more elastic (Windkessel) artery (red). Note decreased distensibility at low pressures and increased distensibility at high pressures c. Comparison with a hypertensively remodeled artery (red). Note upward dislocation (toward elevated in vivo pressures) of the transition between more and less distensible parts of the characteristic curve. d. Comparison with a resistance artery in the relaxed and contracted state (red). Note transition between more and less distensible parts of the characteristic curve toward lower pressures existing in vivo in these more distal resistance vessels. e. Comparison with elasticity of upper body (left, blue) and leg (right, blue) veins. Transition between more and less distensible parts of the pressure diameter characteristic curves also corresponds to in vivo occurring pressures in these vessels (Burton-Roach-Kadar law).

6.4 Typical shape of the vascular elasticity curve

As a result of the elastic properties of their components, *all vessels, without exception show, elastic characteristics with relative similar shapes* (Burton 1954, Dobrin 1978, Roach 1957, Wolinsky 1967, Fig. 4.). Their elastic modulus increases with stress in a such a manner that *they are distensible at lower than commonly occurring physiological pressures and turn rigid at higher pressures* (Burton-Roach-Kadar law). (Berczi 2005, Busse 1981, Cox 1975b, Gow 1972, Molnar A 2006. Molnar G 2010, Roach 1957, Stooker 2003, Szentivanyi 1998) The physiological working point of the vessel is somewhere at the turn of the pressure radius characteristic curve. The work-points, of course will be different for aortas, small resistance vessels, veins, embryonic vessels and hypertensive vessels (Fig. 4.). But that will hardly affect the validity of the above statement. Such an organization of vascular elasticity can be considered biologically logical: potential rises in pressure will not induce unlimited distension, while unexpected volume reductions will be "followed" by the elastic shrinkage of the wall, stabilizing to some degree against fast pressure drop. Larger smooth muscle tone can cause some complications (see next chapter).

6.5 Control of vascular elasticity

In addition to reduce inner radius (and to increase wall thickness) vascular smooth muscle contraction substantially alters the elastic properties of vessels (Apter 1966, Busse 1981,Cox 1975a, Dobrin 1969, 1978, Gow 1972, Greenwald 1982, Hudetz 1980, Monos 1979, Nadasy 1987, vanDijk 1984, Fig 4a.). In case of large arteries, where the lumen reducing effect practically will not affect flow-resistance, the setting of the elastic modulus can be considered one of the main functions of the smooth muscle in the wall. Effects of smooth muscle contraction on vascular elasticity have been described in some classical publications on vascular mechanics. We can set the rule that *smooth muscle contraction reduces the elastic modulus if plotted against pressure and increases it if plotted against radius* (Cox-Dobrin law) (Busse 1981, Cox 1975a, Dobrin 1969, 1978, Hudetz 1980, Monos 1979). One reason for the reduction of isobaric elastic modulus is the decreased stress (reduced inner radius-elevated wall thickness). The contracted pressure radius curves themselves can be less steep (lower pressures) and more steep (high pressures) than the relaxed ones. Right now we do not have a clear picture how the contractile apparatus of the smooth muscle cells is mechanically connected to the elastic lamina and to the more rigid collagenous components. Contracted pressure radius curves at high level of contraction can be very complicated resisting attempts to describe them as elasticity curves. Segments contracted at low pressures will give a very large hysteresis, that is the difference between the upward and downward routes of the pressure-radius curves will form a large loop. Under in vivo conditions, in arteries, periodic pulsatile pressure gives a "conditioning" effect, that reduces but does not diminish hysteresis. This effect can be mimicked in vitro by repeated pressure-radius cycles. In vessels with a pronounced myogenic effect (resistance arteries) elevating pressure will be responded by active myogenic contraction (Bayliss 1902, Jackson 1989, Kuo 1988, Szekeres 2004).

There is also a long-term control of vascular elasticity. Connective tissue components of the wall can be degraded and rebuilt, their amount can change with altering hemodynamic conditions and pathology (Arribas 1999, Briones 2003, Cox 1988, Greenwald 2007, Vidik 1982). Such alterations are very typical for segmental vascular remodeling processes, and of course, they will also affect the elastic properties of the vessels. As a rule we can state that such mechanically driven remodeling processes will ensure, with the exceptions of extreme pathologies, just that the shape of the pressure-radius curves will be adjusted to in vivo pressures as described above. The sole cellular component of the media is the vascular smooth muscle cell. In answer to different biomechanical and biochemical stimuli such cells will be transferred to the secretory state and secrete the components of the extracellular matrix. *Elastin production by them will be stimulated by pulsatile pressure changes* with consequent alterations in the pressure-radius characteristic curve (Kadar 1969). Several pathological processes will stimulate the superfluous production of collagen, increasing the vessel wall's rigidity. We have a good reason to think that this is the lesser harm: the high tensile strength of collagen helps prevent fatal rupture of the vascular wall (Toth 1998).

7. Vascular contractility

With the exception of the capillaries and of such extreme pathological states as late fibrosclerotic plaqes and advanced aneurysms, all vascular walls contain of smooth muscle cells in their media, able to contract. Smooth muscle contraction follows the biophysical laws characteristic for isometric and isotonic contraction of other smooth muscle (Herlihy 1973, Lundholm1966). Smooth muscle cells can contract to about 40% of their fully relaxed length. They are able to produce active shortening up to about 5 kp/cm^2 stress, a value similar to skeletal muscle, which latter is much richer in contractile proteins. The rate of vascular smooth muscle contraction is relative slow, full contractions are reached in a few tens of seconds. Typical are the tonic contractions, but certain vascular muscles also do produce periodic contractions (lymph vessels, portal vein, umbilical vessels?). A typical feature is the *spontaneous contraction*, which is more pronounced in microvessels than in large vessels. The smooth muscle slowly contracts in a medium to which no specific vasoconstrictor agent has been added. This spontaneous contraction is, at least partly responsible for the above mentioned "memory" effect. Vascular specimens should be subjected to a standardized equilibration process, lasting about 20-30 minutes to yield reproductive contractile and biomechanical responses. The explanation is the specific control of vascular smooth muscle contractility ensuring certain level of IC calcium without further stimulation. The other feature is the remaining tone. After contraction, in an in vitro contractility study, when the agonist has been washed off, the original resting level of strain or force will not return. This will be typical at low stresses. There is a temptation to apply larger than physiological stresses to prevent this to occur. The other frequent technical solution is the continuous readjustment of resting tone during the measurement. Because of the spontaneous and resting tones, it is a common requirement nowadays to test the passive length or force after incubation in calcium-free medium with fully to relaxed vascular smooth muscle. Vascular smooth muscles are prone to form "latch" contractions. That means that after some shortening their lengths against moderate stresses, the muscle will be able to resist very high passive stresses not distending at all, or distending only at a very slow rate. A slow cyclization of the actomyosin crossbridges is thought to be in the background of such a behavior (Somlyo 1968). Some new observations raised the possibility of a rearrangement of the cytoskeleton, ensuring such "plastic" behavior. We can also mention here that slow yielding to large stresses after latch-type contraction gives base for the stress-relaxation and creep. That is, latch contraction is responsible for at least part of viscotic behavior of the vascular wall (see also at viscosity). Modern cellular physiology has proven, that separate from contraction control molecular mechanisms will ensure the dephosphorylation of myosin light chains, terminating the actomyosin crossbridge cycle, which means that contraction and relaxation can be controlled somewhat separately in vascular smooth muscle (Schubert 2008). An other feature, we have to mention is the *myogenic contraction*. Passive stress on vascular muscle, especially from small arteries, will induce its active contraction. Such processes can be observed in vivo and form an important mechanism for tissue perfusion autoregulation.

While large arteries will not change their lumen to affect volume blood flow in a sensible manner, smaller arteries and veins can contract until their lumen fully disappears. The extent of contraction, the vascular "tone" is delicately set at different points of the circulation and in different times. Several ten types of cytoplasma membrane and some cytoplasmic receptors have been identified in vascular muscle affecting vascular contractility. Their amount and the extent of contraction or relaxation induced varies in different vascular territories. Also, thousands of drug molecules have been isolated or synthetized that affect vascular contractility, some of the most frequently used cardiovascular drugs are among them. While earlier it was thought that the amount of receptors is specific for the tissue, now we now that even receptor molecule expression is under physiological control, altered receptor expression and altered receptor sensitivity will form important part of vascular remodeling processes.

8. Viscosity of the vessel wall

For methodical reasons, because it is very difficult to study them under reproducible conditions, vessel wall viscosity is an unduly neglected area. Most authors agree that vessels are not only elastic, but viscoelastic (Apter 1966, Azuma 1971, T Bauer 1982, Bergel 1964, Craiem 2008, Fung 1984, Goto 1966, Greven 1976, Hasegawa 1983, Nadasy 1988, Orosz 1999a, 1999b, Steiger 1989, Toth 1998, Zatzman 1954). Vessels show all the three typical viscotic phenomena, the *creep* (viscotic elongation at continuous stress, Fig 5a.), the *stress relaxation* (decreasing stresses after unit-step elongation, Fig 5b.) and *hysteresis* loops (difference between upward and downward routes of the stress-strain curves, Fig. 5c.).

Viscosity might be essential in distributing the force among parallelly connected components of the wall, dampening sudden force elevations on them, preventing their rupture or overwear. There is an agreement that at least part of vascular viscosity will go on in the smooth muscle cells themselves. Our explanation was that *passive slide between actin and myosin filaments*, with breaking and reestablishment of latching cross-bridges could explain vascular viscosity. Viscous elongation this way could be restored by ATP dependent slow contraction and being reversible (Fig. 5a). In pathologic tissue, devoid of functionable smooth muscle cells, a *slow but inherent viscous dilation of extracellular connective tissue fibers* goes toward the fatal rupture of the wall (Fig 5b). Viscoelasticity of the wall can be modeled with Maxwell or Kelvin models, containing one viscous, one parallelly connected and one serially connected elastic units (Fung 1984, Orosz 1999a 1999b). In case of the simple acellular aneurysmal tissue we have identified a fairly continuous stoichiometric ratio between the three viscoelastic components which, first gives some insight into the molecular organizational principles of vascular viscoelasticity (Toth 1998).

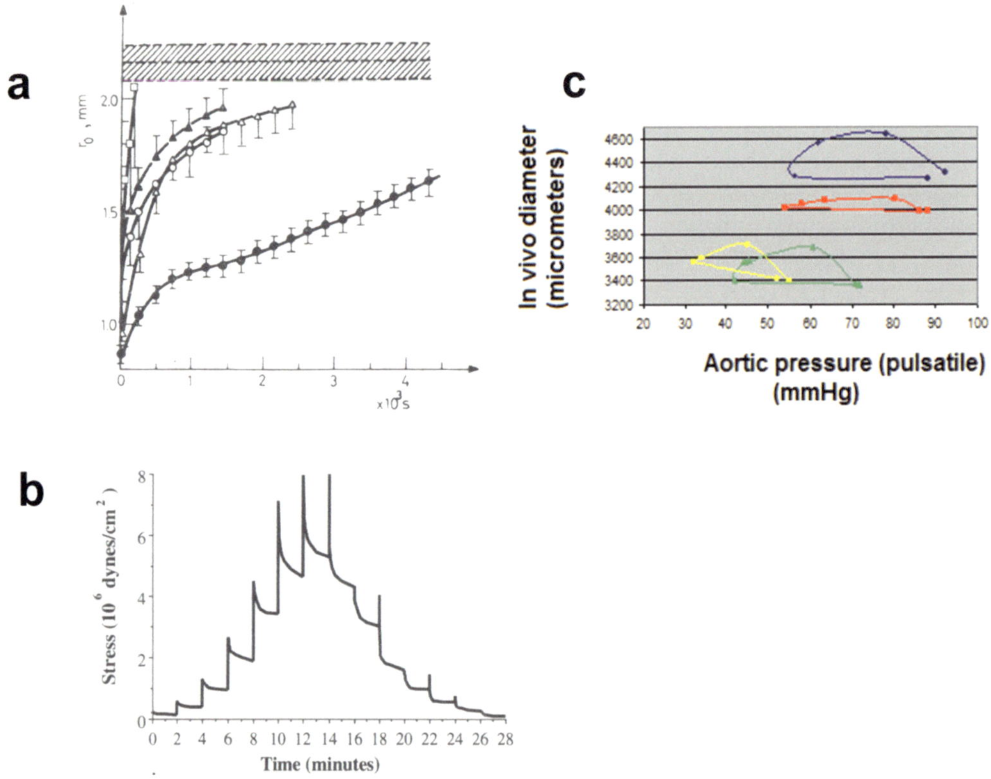

Fig. 5. Blood vessel wall viscosity. a. Viscous creep of contracted human umbilical arterial segments. Slow creep in oxygenized nKR (•), sped up by doubling distending pressure (□), or by applying smooth muscle relaxant, sodium nitrite (○) or with calcium-free solution (▲). Viscosity is also decreased by inhibiting the energy metabolism of smooth muscle cells by 2-deoxy-glucose (Δ). (From Nadasy 1988, with permission of Akadémiai Kiadó) b. Stress relaxation and tensile strength of human aneurysmic tissue. Strip from brain aneurysm sac. Stepwise elevation of length, force recorded as a function of time (with permission of Karger). c. In vivo pressure-diameter pulsatile hysteresis loops recorded in the rabbit thoracic aorta. Each loop corresponds to one cardiac cycle. Taken at different levels of bleeding hypotension. (Nadasy, Csaki, Porkolab and Monos, unpublished).

9. Biomechanics of different vascular segments

9.1 Windkessel artery and distributing artery biomechanics

Elasticity is the very essence of Windkessel artery function (Milnor 1982, Zieman 2005). With each ventricular contraction at rest about 70 ml of blood is pushed into the large arteries, close to the heart. These vessels are containing a fairly large number of concentric elastic sheets intertwined with layers of smooth muscle cells in a fishbone pattern, visibly connecting neighboring elastic sheets. (Clark 1985). At physiological stresses and above them these vessels are more elastic than more peripheral vessels with less elastic tissue (Stemper 2007, Fig. 4b.). With aging and hypertension, rigidity of these vessels increases with a concomitant increase of diameter (Farasat 2008, Giumelly 1999, Safar 2005). In vivo elasticity is frequently measured in form of *pulse wave velocity* (Huotari 2010, Westerhof 2007), aortic compliance (Long 2004, Mersich 2005), *input impedance* (Mazzaro 2005) or *augmentation index* (Safar 2005). Exercise training can stimulate elastin production and reduce high-stress stiffness (DeAndrade 2010). Elastin production is stimulated by periodic stress, that is, by pulse-pressure. The produced elastin will form parallelly connected sheets, that are fairly stretched even at physiological diastolic pressures and thus take part of the force from smooth muscle and collagen. Diameter to wall thickness ratios can thus be relatively large in elastic vessels. Too large periodicity in stretch, however, will speed up the disintegration of elastic lamellae, a typical feature in aged and chronically hypertensive large arteries (Greenwald 2007). An unsettled question is pulsatile viscosity. We have found a profound hysteresis of the pressure-diameter curves in vivo (Nadasy 2007 and unpublished, Fig. 5c.).

9.2 Resistance artery biomechanics

Resistance arteries have limited amount of elastic tissue, the real arterioles none at all. Their most important function is to offer a relative large but controllable resistance which makes controlled in space and time) flow distribution toward the tissues. They are characterized by relatively thick walls and a large diapason between most relaxed and most contracted diameters (Szekeres 1998, Fig. 3b.) and by massive myogenic response (Fig 4d.). Pulse pressure is dampened usually en route in large arteries, remaining undulations will support only a limited elastica production of medial cells. In hypertension, however pressure undulations can increase in resistance sized arteries with biomechanically and histologically observable elevation in elastin production. In later phases of the disease, however, these elastic lamellae will be disrupted. Similar alterations can happen with aging (Arribas 1999, Briones 2003, Gonzales 2005,2006, Intengan 1998, 2001, Laurant 1997, Nadasy 2010a, Takeuchi 2005). Even more important are the segmental geometry alterations. The great circulatory physiologist Folkow realized first that morphological wall thickening might reduce lumen and stabilize elevated resistance and hypertension. He supposed to happen it with an elevation of wall mass (hypertophic wall remodeling, Folkow 1971, 1990, 1995). Later, Mulvany has proven that morphological restriction of the lumen with increased wall thickness can happen without alteration in wall mass (eutrophic remodeling, Mulvany 1990, 1992). The idea emerged that what essentially happens first is a morphological stabilization of a contracted diameter (Mathiasen 2007, Nadasy 2010a). Now we have a picture that both in hypertension and aging there is a morphological lumen restriction of resistance vessels (Dickhout 2000, Frisbee 1999, James 2006, Jeppesen 2004, Kvist 2003, Matrai 2010, McGuffy 1996 Moreau 1998, Muller-Delp 2002, Mulvany 1996, Nadasy 2010a, 2010b, Najjar 2005, Orlandi 2006, Pose-Reino 2006, Riddle 2003, Rizzoni 2006, Rodriguez-Porce 2006, Stacy 1989, Varbiro 2000). We believe that the fact, that substances inducing immediate blood pressure rise have independent from biomechanical effects trophic action on the resistance artery walls is not contradictory to the biomechanical control theory. With their additional effects on vascular smooth muscle protein expression, in the real situation, they promote existing biomechanical control processes (Nadasy 2010a, Safar 1997, Simon 1994, Toyuz 2005). Even more important than changes in segmental geometry, can be the network alterations. Rarefaction and course deviations in hypertension also increase local resistance (Greene 1989, Harper 1978, Nadasy 2000, 2010b, Prasad 1995).

9.3 Biomechanics of veins

Veins are frequently referred to as being distensible. However, similarly to all vessels, veins also turn rigid when sufficiently stretched (Fig. 4e). Most in vitro and in vivo studies show that the transition between the distensible and rigid sections of the pressure-diameter characteristic curve - similarly to arteries and all other vessels - lies around typical physiological pressures (Berczi 2005, Molnar 2006, Molnar 2010, Monos 1983, 1995, 2003, Raffai 2008, Stooker 2003, Zamboni 1996,1998). That makes it possible to insert venous grafts into the arterial system (Monos 1983).

10. Conclusion

Geometry and viscoelasticity controlled both in the short and long runs. Viscoelastic units, the evidence of mechanically driven continuous vessel wall remodeling. The vascular mechanical failure: A biomechanical explanation for the thick vessel syndrome.
The possibility to produce mechanical work at the expense of chemical energy, the ability to restructure the active and passive force-bearing components, even degrade or synthesize them (vascular remodeling) makes the vascular wall an unusually complicated viscoelastic material.

Short term control of segmental geometry is most effective in resistance arteries. Contraction of the outer circumferential smooth muscle layer - because of the incompressibility of the wall - presses the inner layers into the lumen, inducing substantial decrease in lumen diameter and elevation in wall thickness. The hemodynamic effect will be much increased local vascular resistance. Short term control of elasticity will be an important physiological function of the smooth muscle of large arteries. When contracting, they stress upon the elastic membranes reducing high-stress isobaric elastic modulus of the wall. This improves adjustment of vascular impedance to altered ventricular function. Long term control of vascular lumen will be driven by endothelial shear (to keep it constant, Murray-Rodbard law). Normally, several mechanisms point toward such a balanced situation. Endothelial shear can alter several proteins' expression in the wall, the induced acute vasodilation can morphologically stabilize, agonists released in response to shear might contribute to alteration in the morphological lumen. Even substances with primary tissue effects might have additional direct or indirect vascular effects that help adjust vascular lumen to altered tissue function and blood flow needs (feed-forward control). In a phylogenetically unusual situation, however, such adaptation processes can "derail" and work against formation of an optimal morphological vascular lumen. Vessel wall thickness - on the long term - will be controlled to stabilize tangential stress - if there is no change in tissue composition (Folkow-Rodbard-Mulvany's law). In case of periodic stress, smooth muscle cells will be stimulated to produce elastin (Burton-Roach-Kadar's law), which reduces high-stress modulus. Elastic lamellae produced will bear part of the force, leaving less stress on parallelly connected smooth muscle and collagen, allowing thus lesser wall thicknesses. While the viscoelastic properties of the contributing molecules are poorly described, studies on blood vessels with extreme histological composition suggest that intracellular contractile fibers, elastic tissue and collagen are organized in viscoelastic units. The number of serially and parallely connected such units plastically adapts to lengths and forces applied. There seems to be a stoichiometrically determined connection between series and parallel elasticity and viscosity of such viscoelastic units. Viscosity - together with elasticity - helps even distribution of the forces among the parallelly connected elements of the vascular wall.

Restoration of elongated viscous units will be possible at the expense of ATP energy by smooth muscle contraction, if this viscous elongation happened by breaking up, passive sliding and reformation of "latching" actomyosin cross-bridges (intracellular viscosity). If viscous elongation happens between extracellular fibers, migration, adhesion and contraction of smooth muscle elements, with subsequent connective tissue production fixing the restored length might restore the original situation. Study of aneurysmic tissue, where no contractile elements are present to prevent slow but fatal viscous dilation, make it probable, that such restoring processes are continuously going on in healthy vascular tissues. Based on biomechanical experience, we can suppose that if common mechanisms to distribute the force to smooth muscle and elastic components fail, there is a possibility for the vascular wall to prevent fatal rupture to develop, by increasing the amount of collagen in the wall. By this, however, the adaptation to periodic stresses (large vessels), the ability to control resistance (small arteries) and the ability to reduce stress by contraction (veins) will be lost. With loss of smooth muscle, the "ropes" of collagenous tissue cannot be pulled and fixed together, new and new collagenous masses should be produced to prevent slow passive viscotic creep and fatal rupture. In case of large vessels that will alter the pressure distribution in the radial direction of the wall and will interfere with vasa vasorum blood supply of the vessel wall itself. The "blood vessel wall failure" will have a common course, independently of the original pathology that has induced it. That yields a simple biomechanical explanation for the "thick vessel syndrome" and for its amazing analogies with thc aging process.

11. Acknowledgement

This work and studies leading to this work have been supported by Hungarian National Grants OTKA TO 32019 and 42670, the Health Science Council of Hungary (ETT 128/2006) by the Hungarian Space Agency (BO 00080/03) as well as by the Hungarian Hypertension Society and the Hungarian Kidney Foundation.

12. References

Abramson DI Ed. Blood Vessels and Lymphatics Academic Press New York and London, 1962.

Albinsson S. Nordstrom I. Hellstrand P. Stretch of the vascular wall induces smooth muscle differentiation by promoting actin polymerization. J Biol Chem 279:34849-55,2004.

Altman PL, Dittmer DS eds. Biology Data Book 2nd edn. Vol III. Federation of American Societes for Experimental Biology, Bethesda, Maryland 1974.

Apter JT, Rabinowitz M, Cunnings DH. Correlation of viscoelastic properties of large arteries with microscopic structure. Circ Res 19:104-121,1966.

Arribas SM, Daly CJ, McGrath IC. Measurements of vascular remodeling by confocal microscopy. Methods Enzymol 307:246-273,1999.

Azuma T, Hasegawa M. A rheological approach to the architecture of arterial walls. Japn J Physiol 21:27-47,1971.

Bauer RD, Busse R, Schabert A. Mechanical properties of arteries. Biorheology 19:409-424,1982.

Bayliss WM. On the local reactions of the arterial wall to changes of internal pressure. J Physiol (London) 28:200-223,1902.

Bérczi V, Molnár A, Apor A, Kovács V, Ruzics Cs, Várallyay Cs, Hüttl K, Monos E, Nádasy GL. Non-invasive assessment of human large vein diameter, capacity, distensibility and ellipticity in situ: dependence on anatomical location, age, body position and pressure Eur J Appl Physiol 95:283-289,2005.

Bergel DH. The static elastic properties of the elastic wall. J Physiol 156:445-457, 1961.

Bergel DH. Arterial viscoelasticity. In: Pulsatile Blood Flow, Attinger ED ed. McGraw Hill, New York, 1964. pp. 275-292.

Briones AM, Gonzalez JM, Somoza B, Giraldo J, Daly CJ, Vila E, Gonzalez MC, McGrath JC, Arribas SM. Role of elastin in spontaneously hypertensive rat small mesenteric artery remodeling. J Physiol 552:185-195,2003.

Burton AC. Relation of structure to function of the tissue of the wall of blood vessels. Physiol Rev 34:619-642,1954.

Busse R Bauer RD, Sattler T, Schabert A. Dependence of elastic and viscous proerties on circumferential wall stress at two different muscle tones. Pflügers Arch 390:113-119,1981.

Clark JM, Glagov S Transmural organization of the arterial media. The lamellar unit revisited. Arteriosclerosis 5:19-34,1985.

Cliff WJ. Blood Vessels, Cambridge University Press, Cambridge, 1976.

Clyman RI, McDonald KA, Kramer RH. Integrin receptors on aortic smooth muscle cells mediate adhesion to fibronectin, laminin and collagen Circ Res 67:175-186,1990.

Cox RH. Three-dimensional mechanics of arterial segments in vitro: Methods. J Appl Physiol 36:381-384,1974.

Cox RH. Arterial wall mechanics and composition and the effects of smooth muscle activation. Am J Physiol 229:807-812,1975a.

Cox RH. Pressure dependence of the mechanical properties of arteries in vivo. Am J Physiol 229:1371-1375,1975b.

Cox RH. Passive mechanics and connective tissue composition of canine arteries. Am J Physiol 234:H533-H541,1978.

Cox RH, Bagshaw RJ. Effects of hypertension and its reversal on canine arterial wall properties. Hypertension 12:301-309,1988.

Craiem D, Rojo FJ, Atienza JM, Armentano RL, Guinea GV. Frctional-order viscoelasticity applied to describe uniaxial stress relaxation of human arteries. Physics Med Biol 53:4543-4554,2008.

DeAndrade Moraes-Teixeira J, Felix A, Fernandes-Santos C, Moura AS, Mandarim-de-Lacerda CA, deCarvalho JJ. Exercise training enhances elastin, fibrillin and nitric oxide in the aorta wall of spontaneously hypertensive rats. Exp Mol Pathol 89:351-357,2010.

Dickhout JG, Lee RM. Increased medial smooth muscle cell length is responsible for vascular hypertrophy in young hypertensive rats. Am J Physiol Heart Circ Physiol 279:H2085-H2094,2000.

Discher D, Dong C, Fredberg JJ, Guilak F, Ingber D, Janmey P, Kamm RD, Schmid-Schonbein GW, Weinbaum S. Biomechanics: Cell research and applications for the next decade. Ann Biomed Eng 37:847-859,2009.

Dobrin PB, Rovick AA. Influence of vascular smooth muscle on contractile mechanics and elasticity of arteries. Am J Physiol 217:1644-1651, 1969.

Dobrin PB. Mechanical properties of arteries Physiol Rev 58:397-460,1978.

Duling BR, Gore RW, Dacey RG Jr, Damon DR. Methods for isolation, cannulation and in vitro study of single microvessels. Am J Physiol 241:H108-H116, 1981.

Farasat SM, Morrell CH, Scuteri A, Ting CT Yin FCP, Spurgeon HA, Chen CH, Lakatta EG, Najjar SS Pulse pressure is inversely related to aortic root diameter. Implications for the pathogenesis of systolic hypertension Hypertension 51:196-202,2008.

Folkow B. The hemodynamic consequences of adaptive structural changes of the resistance vessels in hypertension. Clin. Sci 41:1-12,1971.

Folkow B. "Structural factor" in primary and secondary hypertension. Hypertension 16:89-101,1990.

Folkow B. Hypertensive structural changes in systemic precapillary resistance vessels: how important are they for in vivo haemodynamics? J Hypertens 13:1546-1559,1995.

Frisbee JC, Lombard JH. Development and reversibility of altered skeletal muscle arteriolar structure and reactivity with high salt diet and reduced renal mass hypertension. Microcirculation 6:215-22,1999.

Fung YC. Biodynamics. Circulation. Springer Verlag, New York, 1984.

Fung YC, Liu SQ. Determination of the mechanical properties of the different layers of blood vessels in vivo. Proc Natl Acad Sci US 92:2169-2173,1995.

Gabella G. Structural apparatus of force transmission in smooth muscles. Physiol Rev 64:455-477,1984.

Giummelly P, Lartaud-Idjouadiene I, Marque V, Niederhoffer N, Chillon JM, Capdeville-Atkinson C, Atkinson J. Effects of aging and antihypertensive treatment on aortic internal diameter in spontaneously hypertensive rats. Hypertension 34:207-211,1999.

Gonzalez JM, Briones AM, Starcher B, Conde MV, Somoza B, Daly C, Vila E, McGrath I, Gonzalez MC, Arribas SM. Influence of elastin on rat small artery mechanical properties. Exp Physiol 90:463-468,2005.

Gonzalez JM, Briones AM, Somoza B, Daly CJ, Vila E, Starcher B, McGrath JC, Gonzalez MC, Arribas SM. Postnatal alterations in elastic fiber organization precede resistance artery narrowing in SHR. Am J Physiol Heart Circ Physiol 291:H804-812,2006.

Goto M, Kimoto Y. Hysteresis and stress relaxation of the blood vessels studied by a universal tensile-testing instrument. Jap J Physiol 16:169-184,1966.

Gow BS. The influence of vascular smooth muscle on the viscoelastic properties of blood vessels. In: Cardiovascular Fluid Dynamics, Bergel DH ed. Academic, New York, 1972. pp. 66-110.

Greene AS, Tonellato PJ, Lombard LJ, Cowley AW Jr. Microvascular rarefaction and tissue vascular resistance in hypertension. Am J Physiol 256:H126-H131,1989.

Greenwald SE, Newman DL, Denyer HT. Effect of smooth muscle activity on the static and dynamic elastic properties of rabbit carotid artery. Cardiovasc Res 16:86-94,1982.

Greenwald SE. Ageing of the conduit arteries. J Pathol 211:157-172,2007.

Greven K. The time course of creep and stress relaxation in the relaxed and contracted smooth muscle. Bulbring E, Shuba MF eds, Raven Press, New York, 1976. pp. 223-228.

Harper RN, Moore MA, Marr MC, Watts LE, Hutchins PM Arteriolar rarefaction in the conjunctiva of human essential hypertensives Microvascular Research 16:369-372,1978.

Hasegawa M. Rheological properties and wall structures of large veins. Biorheology 20:531-545,1983.

Hayashi K, Naiki T. Adaptation and remodeling of vascular wall; biomechanical response to hypertension. J Mech Behav Biomed Materials 2:3-19,2009.

Hegedus K Some observations on reticular fibers in the media of the major cerebral arteries. A comparative study of patients without vascular disease and those with ruptured berry aneurysms. Surg Neurol 22:301-307,1984.

Heistad DD, Armstrong ML, Baumbach GL, Faraci FM. Sick vessel syndrome. Recovery of atherosclerotic and hypertensive vessels. Hypertension. 26:509-513,1995.

Herlihy JT, Murphy RA. Length-tension relationship of smooth muscle of the hog carotid artery. Circul Res 33:275-283,1973.

Herman P, Kocsis L, Eke A. Fractal branching pattern in the pial vasculature in the cat. J Cerebr Blood Flow Metabol 21:741-753,2001.

Hudetz AG, Mark G, Kovach AGB, Monos E. The effect of smooth muscle activation on the mechanical properties of pig carotid arteries. Acta Physiol Acad Sci Hung 56:263-273, 1980.

Huotari MJ, Maatta K, Nadasy GL, Kostamovaare J. A photoplethysmographic pulse wave analysis for arterial stiffness in extremities. Artery Research 4: 155,2010. (A)

Intengan HD, Schiffrin EL. Mechanical properties of mesenteric resistance arteries from Dahl salt-resistant and salt-sensitive rats: role of endothelin-1. J Hypertens ;16:1907-1912,1998.

Intengan HD, Schiffrin EL. Vascular remodeling in hypertension: roles of apoptosis, inflammation, and fibrosis. Hypertension 38:581-587,2001.

Jackson PA, Duling BR. Myogenic response and wall mechanics of arterioles. Am J Physiol 257:H1147-H1155,1989.

James MA, Tullett J, Hemsley AG, Shore AC. Effects of aging and hypertension on the microcirculation. Hypertension 47:968-974,2006.

Jeppesen P, Gregersen PA, Bek T. The age-dependent decrease in the myogenic response of retinal arterioles with the Retinal Vessel Analyzer. Grafes Arch Clin Exp Ophthalm 242:914-919,2004.

Kadar A, Veress B, Jellinek H. Relationship of elastic fibre production with smooth muscle cells and pulsation effect in large vessels. Acta Morphol Acad Sci Hung 17:187-200,1969.

Kamiya A Togawa T. Adaptive regulation of wall shear stress to flow change in the canine carotid artery. Am J Physiol 239:H14-H21,1980.

Koens MJW, Faraj KA, Wismans RG, van der Vliet JA, Krasznai AG, Cuijpers VMJI, Jansen JA, Daamen WF, van Kuppevelt TH Controlled fabrication of triple layered and molecularly defined collagen/elastin vascular grafts resembling the native blood vessel. Acta Biomaterialia 6:4666-4674,2010.

Kuo L, davis MJ, Chilian WM. Myogenic activity in isolated subepicardial and subendocardial coronary arterioles. Am J Physiol 255:H1558-H1562,1988.

Kvist S, Mulvany MJ. Contrasting regression of blood pressure and cardiovascular structure in declipped renovascular hypertensive rats. Hypertension 41:540-545,2003.

Laurant P, Touyz RM, Schiffrin EL. Effect of pressurization on mechanical properties of mesenteric small arteries from spontaneously hypertensive rats. J Vasc Res 34:117-125, 1997.

Lee RT, Huang H. Mechanotransduction and arterial smooth muscle cell:new insight into hypertension and atherosclerosis. Ann Med 32:233-235,2000.

Liu SQ, Fung YC. Zero-stress state of arteries. J Biomech Eng 110:82-84,1988.

Long A, Rouet L, Bissery A, Goeau-Brissoniere O, Sapoval M. Aortic compliance in healthy subjects: Evaluation of tissue Doppler imaging. Ultrasound Med Biol 30:753-759,2004.

Lorant M, Nadasy GL, Monos E: Changes in network characteristics of saphenous vein after long-term head-up tilt position of the rat. Physiol Res 52:525-531,2003

Lundholm L, Mohme-Lundholm E. Length at inactivated contractile elements, length-tension diagram, active state and tone of vascular smooth muscle. Acta Physiol Scand 68:347-359,1966.

Mathiasen ON, Buus N, Larsen ML, Mulvany JM. Small artery structure adapts to vasodilation rather than to blood pressure during antihypertensive treatment. J Hypert 25:1027-1034,2007.

Matrai M, Mericli M, Nadasy GL, Varbiro Sz, Szekeres M, Banhidy F, Acs N, Monos E, Szekacs B: Gender differences in biomechanical properties of intramural coronary resistance arteries of rats, an in vitro microarteriographic study J Biomech 40:1024-1030,2007.

Matrai M, Szekacs B, Mericli M, Nadasy GL, Szekeres M, Banhidy F, Bekesi G, Monos E, Sz Varbiro: Biomechanics and vasoreactivity of female intramural coronaries in angiotensin II induced hypertension. Acta Physiol Hung 97:31-40,2010.

Mazzaro L, Almasi SJ, Shandas R, Gates PE. Aortic imput impedance increases with age in healthy men and women. Hypertension 45:1101-1106,2005.

McGuffee LJ, Little SA. Tunica media remodeling in mesenteric arteries of hypertensive rats. Anat Rec 246:279-292,1996.

Mersich B, Rigo J Jr, Besenyei C, Lenard Z, Studinger P, Kollai M. Opposite changes in carotid versus aortic stiffness during healthy human pregnancy. Clin Sci 209:103-107,2005.

Milnor WM. Hemodynamics Wiulliams and Wilkins, Baltimore/London 1982.

Molnár AA, Apor A, Kristóf V, Nádasy GL, Preda I, Hüttl K, Acsády G, Monos E, Bérczi V: Generalized changes in venous distensibility in postthrombotic patients Thromb Res 117:639-45, 2006.

Molnar G, Nemes A, Kekesi V, Monos E, Nadasy GL. Maintained geometry, elasticity and contractility of human saphenous vein segments stored in a complex tissue culture medium Eur J Vasc Endovasc Surg 40:88-93,2010.

Monos E, Hudetz AG, Cox RH. Effect of smooth muscle activation on incremental elastic properties of major arteries. Acta Physiol Hung 53:31-39,1979.

Monos E. Csengôdy J Does haemodynamic adaptation take place in the vein grafted into an artery? Pfluegers Archiv 384:177-182,1983.

Monos E. Biomechanics of the Vascular Wall, Medicina, Budapest, 1986 (In Hungarian)

Monos E, Berczi V, Nadasy GL. Local control of veins: Biomechanical, metabolic, and humoral aspects Physiol. Rev. 75:611-666, 1995.

Monos E, Lóránt M, Dörnyei G, Bérczi V, Nádasy Gy: Long-term adaptátion mechanisms in extremity veins supporting orthostatic tolerance. (Review) News Physiol Sci 18:210-214,2003

Moreau P, d'Uscio LV, Luscher TF. Structure and reactivity of small arteries in aging. Cardiovasc Res 37:247-253,1998.

Muller-Delp J, Spier SA, Ramsey MW, Lesniewski LA, Papadopoulos A, Humphrey JD, Delp MD. Effects of aging on vasoconstrictor and mechanical properties of rat skeletal muscle arterioles. Am J Physiol Heart Circ Physiol 282:H1843-1854, 2002.

Mulvany HJ, Warshaw DM. The anatomical location of the series elastic component in rat vascular smooth muscle. J Physiol (London) 314:321-330,1981.

Mulvany MJ, Aalkjer C. Structure and function of small arteries. Physiol Rev 70:921-961,1990.

Mulvany MJ. The development and regression of vascular hypertrophy. J Cardiovasc Pharmacol 19 (Suppl 2):S22-S27,1992.

Mulvany MJ. Effects of angiotensin converting enzyme inhibition on vascular remodelling of resistance vessels in hypertensive patients. J Hypertens 14(Suppl.6.):S21-S24,1996

Nádasy G.L., E. Monos, E. Mohácsi, J. Csépli, A. G. B. Kovách: Effect of increased luminal blood flow on the development of the human arterial wall. Comparison of mechanical properties of double and single umbilical arteries in vitro. Blood Vessels 18:139–143, 1981.

Nádasy G.L., E. Mohácsi, E. Monos, J. L. Lear, A. G. B. Kovách: A simple model describing the elastic properties of human umbilical arterial smooth muscle. Acta Physiol Hung 70:75–85, 1987.

Nádasy G. L., Monos E., Mohácsi E., Kovách A.G.B.: The background of hysteretic properties of the human umbilical arterial wall. Smooth muscle contraction and hysteresis of the pressure–radius curves. Acta Physiol Hung 71:347–361, 1988.

Nadasy GL, Varbiro S, Acs N, Szekacs B, Lorant M, Jackel M, Kerenyi T, Monos E: Intramural coronary resistance artery network remodelling in chronically angiotensin II-infused female rats J Physiol (London) 526(Suppl. S):133P, 2000.

Nadasy GL, Szekeres M, Dezsi L, Varbiro Sz, Szekacs B, Monos E: Brief communication Preparation of intramural small coronary artery and arteriole segments and resistance artery networks from the rat heart for microarteriography and for in situ perfusion video mapping Microvasc Res 61:282-286,2001.

Nádasy G, Mericli M, Mátrai M, Várbíró Sz, Szekeres M, Ács N, Monos E, Székács B: Gender differences in biomechanics of intramural coronary resistance arteries of the rat. Acta Physiol Hung 92:287-288,2005.

Nádasy Gy: Arterial blood supply and tissue needs. In: Physiology and Maintenance., OOP Hanninen and M Atalay eds., Chapter 6.54.7.2. UNESCO Encyclopedia of Life Support Systems, EOLSS, www.eolss.net (2007a).

Nádasy GL, Monos E. Biomechanical principles of vascular wall design in health and in disease: some mathematics of angiogenesis. Acta Phsiol Hung 94:377-379,2007b.

Nadasy GL, Varbiro Sz, Szekeres M, Kocsis A, Szekacs B, Monos E, Kollai M: Biomechanics of resistance artery wall remodeling in angiotensin-II hypertension and after its recovery Kidney Blood Press Res 33:37-47,2010a.

Nadasy GL, Szekacs B, Varbiro Sz, Sekeres M, Wappler E, Szalai E, Simon A, Monos E. Analog biomechanical effects of aging and hypertension in resistance blood vessels Acta Physiol. Hung. 97: 462, 2010b.

Najjar SS, Scuter A, Lakatta EG. Arterial aging: is it an immutable cardiovascular risk factor? (Review) Hypertension 454-462,2005.

Orlandi A, Bochaton-Piallat ML, Gabbiani G, Spagnoli LG. Aging, smooth muscle cells and vascular pathobiology: Implications for atherosclerosis. Atherosclerosis 221-230, 2006.

Osol G, Halpern W. Myogenic properties of cerebral blood vessels from normotensive and hypertensive rats. Am J Physiol 249:H914-H921,1985.

Orosz M, Molnarka G, Toth M, Nadasy GL, Monos E: Viscoelastic behavior of vascular wall simulated by generalized Maxwell models - a comparative study Med Sci Mon 5:549-555, 1999.

Orosz M, Molnárka Gy, Nádasy Gy, Raffai G, Kozmann Gy, Monos E: Validity of viscoelastic models of blood vessel wall Acta Physiol Hung 86:265-271, 1999.

Oxlund H. Relationship between the biomechanical properties, composition and molecular structure of connective tissues. Connect Tissue Res 15:65-72,1986.

Pose-Reino A, Rodriguez-Fernandez M, Hayik B, Gomez-Ulla F, Carrera-Nouche MJ, Gude-Sampedro F, Estevez-Nunez JC, Mendez-Naya I. Regression of alterations in retinal microcirculation following treatment for arterial hypertension. J Clin Hypertens 8:590-595,2006.

Prasad A, Dunhill GS, Mortimer PS, MacGregor GA. Capillary rarefaction in the forearm skin in essential hypertension. J Hypertens 13:265–268,1995.

Pries AR, Reglin B, Secomb TW. Remodeling of blood vessels: responses of diameter and wall thickness to hemodynamic and metabolic stimuli. Hypertension 46:725-731,2005.

Raffai G, Lódi C, Illyés G, Nádasy G, Monos E: Increased diameter and enhanced myogenic response of saphenous vein induced by two-week experimental orthostatsis are reversible Physiol Res 57:175-183,2008.

Rhee AY, Brozovich FV. Force maintenance in smooth muscle: analysis using sinusoidal perturbations. Archives of Biochemistry & Biophysics. 410:25-38,2003.

Riddle DR, Sonntag WE, Lichtenwalner RJ. Microvascular plasticity in aging. (Review) Ageing Research Reviews 2:149-168,2003.

Rizzoni D, Agabiti-Rosei E. Small artery remodeling in hypertension and diabetes. Curr Hypert Rep 8:90-95,2006.

Roach MR , Burton AC. The reason for the shape of the distensibility curves of arteries. Can J Biochem Physiol 35:681-690,1957.

Rodbard S. Negative feedback mechanisms in the architecture and function of the connective tissue and cardiovascular tissues. Perspect Biol Med 13:507-527, 1970.

Rodbard S. Vascular caliber Cardiology 60:4-49,1975.

Rodriguez-Porcel M, Zhu XY, Chade AR, Amores-Arriaga B, Caplice NM, Ritman EL, Lerman A, Lerman LO. Functional and structural remodeling of the myocardial microvasculature in early experimental hypertension. Am J Physiol Heart Circ Physiol. 290:H978-H984,2006.

Rossitti S, Lofgren J. Vascular dimensions of the cerebral arteries follow the principle of minimum work. Surgery 54:347-350,1963.

Safar ME, van Bortel LM, Struijker-Boudier HA. Resistance and conduit arteries following converting enzyme inhibition in hypertension. J Vasc Res 34:67-81, 1997.

Safar ME. Systolic hypertension in the elderly: arterial wall mechanical properties and the renin-angiotensin-aldosterone system. J Hypertension 23:673-681,2005.

Schmidt-Nielsen K. Animal Physiology. Adaptation and Environment, Cambridge University Press, Cambridge 1979.

Schubert R, Lidington D, S-S Bolz. The emerging role of calcium sensitivity regulation in promoting myogenic vasoconstriction. Cardiovasc Res 77:8-18,2008.

Schwartz CJ, Werthessen NT, Wolf A. Eds. Structure and Function of the Circulation. Volumes 1-2-3. Plenum Press, New York and London, 1980.

Shimazu T, Hori M, Mishima M, KIbatake A, Kodama K, Nanto S, Inoue M. Clinical assessment of elastic properties of large coronary arteries: Pressure-diameterv relationship and dynamic incremental elastic modulus. Int J Cardiol 13:27-45,1986.

Siegman MJ, Butler TM, Mooers SV, Davies RE. Crossbridge attachment, resistance to stretch and viscoelasticity in resting mammalian smooth muscle. Science 191:383-385,1976.

Simon G, Abraham G, Altman S. Stimulation of vascular glycosaminoglycan synthesis by subpressor angiotensin II in rats. Hypertension 23(1 Suppl):I148-51,1994.

Somlyo AP, Somlyo AV. Vascular smooth muscle I. Normal structure, pathology, biochemistry and biophysics. Pharmacol Rev 20:197-272, 1968.

Stacy DL, Prewitt RL. Effects of chronic hypertension and its reversal on arteries and arterioles. Circul Res 65:869-879,1989.

Steiger HJ, Aaslid R, Reule HJ. Growth of aneurysms can be understood as passive yield to blood pressure. Acta Neurochir (Wien) 100:74-78,1989.

Stemper BD, Yoganandan N, Pintar FA. Mechanics of arterial subfailure with increasing loading rate J Biomech 40:1806-1812,2007.

Stooker W, Gok M, Sipkema P, Niessen HWM, BaidoshviliA, Westerhof N, Jansen AK, Wildevuur CRH, Eijsman L Pressure-Diameter relationship in the human greater saphenous vein. Ann Thorac Surg 76:1533-1538,2003.

Szekeres M, Nadasy GL, Dezsi L, Orosz M, Tôkés A, Monos E: Segmental differences in geometric, elastic and contractile characteristics of small intramural coronary arteries J Vasc Res 35:332-344,1998.

Szekeres M, Nádasy GY L, Kaley G, Koller A: Nitric oxide and prostaglandins modulate pressure-induced myogenic responses of intramural coronary arterioles J. Cardiovasc Pharmacol 43:242-249,2004.

Szentiványi M, Nádasy GL, Tóth M, Kopcsányi V, Jedrákovits A, Monos E: Biomechanics of the saphenous artery and vein in spontaneous hypertension Pathophysiology 4:295-302,1998.

Takeuchi K, Ideishi M, Tashiro T, Morishige N Yamada T, Saku K, Urata H. Higher small arterial elasticity in hypertensive patients treated with angiotensin II receptor blockers. Hypertens Res. 2005,28:639-644,2005.

Toth M, Nadasy GL, Nyary I, Kerenyi T, Orosz M, Molnarka G, Monos E: Sterically inhomogenous viscoelastic behavior of human saccular aneurysms J Vasc Res 35:345-355,1998.

Touyz RM. Intracellular mechanisms involved in vascular remodelling of resistance arteries in hypertension: role of angiotensin II. Exp Physiol. 90:449-455,2005.

VanDijk AM, Wieringa PA, Van der Meer , Laird DJ. Mechanics of resting isolated single vascular smooth muscle cells from bovine coronary artery Am J Physiol 246:C277-C287,1984.

Várbiro S, Nádasy GL, Monos E, Vajo Z, Ács N, Miklós Z, Tőkés A, Székács B Effect of ovariectomy and hormone replacement therapy on small artery biomechanics in angiotensin-induced hypertension in rats. J Hypertension 2000; 18:1587-1595,2000.

Vidik A, Danielsen CC, Oxlund H. On fundamental and phenomenological models, structure and mechanical properties of collagen, elastin and glycosaminoglycan complexes. Biorheology 19:437-451,1982.

Westerhof BE, Guelen I, Stok WJ, Wesseling KH, Spaan JAE, Bos JB, Stergiopulos N Arterial pressure transfer characteristics:effects of travel time. Am J Physiol 292:H800-H807,2007.

Wolinsky H, Glagov S. Structural basis for the static mechanical properties of the aortic media Circ Res 20:99-111,1967.

Zamboni, P. Marcellino, MG. Portaluppi, F. Manfredini, R. , Feo, CV. Quaglio, D. Liboni, A. The relationship between in vitro and in vivo venous compliance measurement. International Angiology 15, 149-152,1996.

Zamboni, P. Portaluppi, F. Marcellino, MG. Quaglio, D. Manfredini, R. Feo, CV. Stoney, RJ. In vitro versus in vivo assessment of vein wall properties. Annals of Vascular Surgery 12:324-329,1998.

Zamir M. Shear forces and blood vessel radii in the cardiovascular system. J Gen Physiol 69:449-461,1977.

Zatzman M, Stacy RW, Randall J, Eberstein A. Time course of stress relaxation in isolated arterial segments. Am J Physiol 177:299-307,1954.

Zieman SJ, Melenovsky V, Kass DA . Mechanisms, pathophysiology, and therapy of arterial stiffness Arterioscler Thromb Vasc Biol 25:932-943,2005.